Moder … niques

Compiled by C.B. Faust

School Teacher Fellow

The Royal Society of Chemistry

1991–92

Modern Chemical Techniques

Compiled by Ben Faust
Edited by John Johnston and Neville Reed
Designed by Imogen Bertin
Spectra digitised by Alan Steel

Published by the Education Division, The Royal Society of Chemistry
Printed by the Royal Society of Chemistry
Registered Charity Number 207890

Reprinted 1994, 1995, 1998, 2001

For further information on other educational activities undertaken by the Royal Society of Chemistry write to:

The Education Department
The Royal Society of Chemistry
Burlington House
Piccadilly
London W1V OBN

British Library Cataloguing in Publication Data

Faust. C.B.
Modern Chemical Techniques
I.Title
543

ISBN 1–870343–19–0

Unilever

THE ROYAL SOCIETY OF CHEMISTRY

Contents

THE ROYAL SOCIETY OF CHEMISTRY

Unilever

Foreword

Analytical techniques in chemistry are powerful tools in a chemist's armoury. Spectroscopic data and chemical information are used routinely in laboratories to elucidate a chemical structure or to follow a chemical reaction. However, the sophistication of the analytical techniques used changes rapidly, hence the routinely used method of today can all too readily be superseded by the new technology of tomorrow.

This book identifies some of the important chemical techniques in use today and their applications which I hope will illustrate how chemistry is using state-of-the-art technology to push back the frontiers of the subject.

Sir Rex Richards CChem FRSC FBA FRS
President, The Royal Society of Chemistry
July 1992

Unilever

THE ROYAL SOCIETY OF CHEMISTRY

Introduction

Ben Faust

In 1989 a series of Royal Society of Chemistry 'hands-on' symposia for teachers began with a workshop at Fisons Pharmaceuticals, Loughborough. This was the first of what has turned out to be a popular programme of events that Ian Poots and Philip Rodgers (of St Paul's School) and I have organised. Through the generous support of a number of companies and institutes of higher education teachers have been able to update chemical skills developed during their initial training, or learn new ones. They have been able to see applications of some of the newest methods and techniques available, and use state-of-the-art instrumentation.

At many of the workshops resources, handouts and spectra have been generated, and teachers have been able to take these back to their schools and colleges and use them in their own teaching. The quality of the resource material produced has been very high, and the number of places available on each symposium limited, so it was felt that every effort ought to be made to make at least some of the resource material available to all teachers.

During my appointment as School Teacher Fellow for 1991–92, one of my projects was to take the handouts and develop them so that they could be used by teachers who had not attended the symposia. This book is the result of my endeavours.

THE ROYAL SOCIETY OF CHEMISTRY

Unilever

How to use this book

Chemistry teachers come from a variety of backgrounds, and this book caters for teachers who are familiar with modern chemical techniques, as well as for those teachers that will find them unfamiliar. Consequently, in each chapter much of the basic theory has been included without emphasising the mathematics and physics involved. Where appropriate descriptions of the instrumentation and sample preparation have been included and these are followed by applications of the techniques. Many manufacturers and research organisations have volunteered information on current trends in the techniques, and attempts have been made to crystal ball gaze, based on their data. Finally, where possible some problems (with solutions !) have been included but it should be noted that the solutions offered are not necessarily the only way that the problem can be solved. For the interested reader advanced ideas have also been included in sections that are boxed, and consequently they should not disadvantage the reader if they are omitted.

It has been difficult to restrict the text to SI units because they are frequently not the ones used in practice; in most cases both the units used are quoted, and a SI equivalent.

Although not all of the techniques appear on examination syllabuses at present, it is hoped that readers will find the time to read through them all, particularly the chapter on chromatography. The applications in this chapter indicate the linking of techniques that is now possible.

It would be naive to think that any one method can always be used authoritatively, and to illustrate this a description of the spectra obtained from the synthesis of one pharmaceutical product, ibuprofen, has been included.

Unilever

THE ROYAL SOCIETY OF CHEMISTRY

Acknowledgements

The Royal Society of Chemistry would like to thank Unilever plc for supporting the printing and dissemination of this book to all secondary schools in the UK. The production of this book was only made possible because of the advice and assistance of a large number of people. To the following, and to everybody who has had any involvement with this project, both the author and the Royal Society of Chemistry would like to express their gratitude. Thanks are due to:

Mr M Ashton	Unilever Research, Port Sunlight;
Mr D Baker	Coalville Technical College, Coalville;
Dr D Ballard	Derby Tertiary College, Derby;
Mr M Barratt	Unilever Research, Colworth;
Dr R Bee	Unilever Research, Colworth;
Mr M Bloomfield	Hewlett Packard, Stockport;
Mr P Bradburn	Repton School, Repton;
Dr R Brown	Fisons Pharmaceuticals, Loughborough;
Mrs P Chalmers	Horseracing Forensic Laboratory, Newmarket;
Dr C Clemett	Unilever Research, Port Sunlight;
Mr D Cooper	Unilever Research, Port Sunlight;
Mr R Cybulski	Fisons Pharmaceuticals, Loughborough;
Mr C Dacombe	Unilever Research, Colworth;
Dr A Davis	Fisons Pharmaceuticals, Loughborough;
Dr M Dixon	Fisons Pharmaceuticals, Loughborough;
Dr B Doggett	Ratcliffe College, Leicester;
Dr P Duffin	Esso, Abingdon;
Dr J Dunlop	St Paul's Catholic School, Rickmansworth;
Mr N Dunnett	Horseracing Forensic Laboratory, Newmarket;
Mr N Entwistle	Fisons Pharmaceuticals, Loughborough;
Mr M Flanagan	Unilever Research, Port Sunlight;
Dr R Fletton	Glaxo Group Research, Greenford;
Mr T Gaskell	Walford High School, Northolt;
Mr A George	Unilever plc, London;
Mrs J Hammond	ICI Pharmaceuticals, Macclesfield;
Dr G Haran	Boots Pharmaceuticals, Nottingham;
Ms K Hasu	Watford Grammar School for Girls, Watford;
Dr N Hughes	ICI Specialities, Blackley;
Mr D Hunter	Fisons Pharmaceuticals, Loughborough;
Mrs A Jablonski	Ockbrook School, Derby;
Dr R Johnson	Unilever Research, Port Sunlight;
Dr K Jones	King's College London;
Dr R Jowitt	Aylesbury Grammar School, Aylesbury;
Dr S Kelley	Open University, Milton Keynes;
Dr P Kolker	ICI Pharmaceuticals, Macclesfield;
Dr T Lester	BP, Sunbury;
Mr D Linsdell	Hotpoint, Peterborough;
Mr B Long	Ecclesbourne School, Derby;
Mr A McArthur	Unilever Research, Colworth;
Mrs D McDonald	Fisons Pharmaceuticals, Loughborough;
Mr I Moss	Aldrich Chemical Company, Gillingham;
Mr K Nelson	Robert Pattinson School, Lincoln;
Dr K Nichol	Boots Pharmaceuticals, Nottingham;
Mr R Nixon	Ashby Grammar School, Ashby-De-La-Zouch;

THE ROYAL SOCIETY OF CHEMISTRY

Unilever

Mrs R Owen	Simon Balle School, Hertford;
Mr M Parry	Boots Pharmaceuticals, Nottingham;
Mr J Peet	St Crispin's School, Workingham;
Miss R Pendlington	Unilever Research, Colworth;
Dr I Poots	St Paul's School, London;
Ms A Reddecliffe	St Paul's RC Comprehensive School, Leicester;
Mr J Reid	Unilever Research, Port Sunlight;
Mr G Roberts	Hewlett Packard, Stockport;
Dr P Rodgers	St Paul's School, London;
Dr M Rose	British Mass Spectrometry Society;
Mr N Rowbotham	Loughborough Grammar School, Loughborough;
Mr S Shelley	Olympus Optical Company, London;
Dr K Sinclair	Napp Laboratories, Cambridge;
Dr R Smith	Aldrich Chemical Company, Gillingham;
Mr M Stubbs	Napp Laboratories, Cambridge;
Mr D Taylor	Fisons Pharmaceuticals, Loughborough;
Dr P Tickle	Ratcliffe College, Leicester;
Mr M Tigwell	VG Elemental, Winsford;
Dr J Turner	Boots Parmaceuticals, Nottingham;
Dr S Upton	Perkin Elmer, Beaconsfield;
Mrs S Walker	Napp Laboratories, Cambridge;
Mr S Watts	Fisons Pharmaceuticals, Loughborough;
Dr A Wheway	Oakham School, Oakham;
Mr K Whiting	Boots Pharmaceuticals, Nottingham;
Dr P Williams	Lion Laboratories, Barry;
Dr S Williams	Queen Mary and Westfield College, London;
Mr T Williams	Yarborough High School, Lincoln;
Mr R Windle	Chesham High School, Chesham.

One of the difficulties experienced by teachers is in obtaining spectra suitable for use in the classroom, therefore as many spectra and diagrams have been included in each chapter as possible. The Society is extremely grateful to the following for supplying spectra and diagrams:

Aldrich Chemicals for:
infrared spectra on pages 64, 69, 70, 73, 74, 75, 76, 77, 78, 79, 83, 85, 87, 88, 89, 90, 91
NMR spectra on pages 40, 42, 44, 45, 46, 48, 56, 57, 58, 59, 60

Boots Pharmaceuticals for all spectra in chapter 7

Fisons Pharmaceuticals for:
infrared spectra on pages 68 and 86
mass spectrum on page 12
NMR spectra on pages 43 and 49
ultraviolet/visible spectra on pages 96, 98, 99, 102, 103, 104, 107, 108, 109, 110, 114

Hewlett Packard for diagram on page 127

Horseracing Forensic Laboratory for HPLC plots on pages 156 and 157

Unilever

THE ROYAL SOCIETY OF CHEMISTRY

ICI for:
mass spectra on pages 14, 15, 16
ultraviolet/visible spectrum on page 92

Lion Laboratories for diagrams on pages 82 and 84

Queen Mary and Westfield College Magnetic Imaging Unit for images on pages 53, 54, 55

Unilever Research for:
electron microscope images on pages 166, 167, 168, 169, 170, 171
mass spectra on pages 5, 9, 10, 11, 19, 21, 25, 26, 27, 28
HPLC plots on pages 144 and 147
NMR spectra on pages 33, 37, 47, 50, 51
ultraviolet/visible spectrum on page 112.

The Society would like to thank Fisons Pharmaceuticals for providing office accommodation for this School Teacher Fellowship post, and the Head and Governors of Loughborough Grammar School for seconding Ben Faust to the Society's Education Department.

The author would like to thank Neville Reed and Jacqui Clee of the Royal Society of Chemistry for their guidance and support during this project.

THE ROYAL
SOCIETY OF
CHEMISTRY

Unilever

Bibliography

Readers will find more detailed explanations of the techniques covered in this book in the text books listed below:

R J Abraham, J Fisher and P Loftus *Introduction to NMR spectroscopy*, 1988.
C N Banwell *Fundamentals of molecular spectroscopy*, 3rd edn., 1983.
H E Duckworth, R C Barber and V S Venkatasubramanian *Mass spectroscopy*, 2nd edn., 1986.
D C Harris *Quantitative analysis*, 2nd edn., 1987.
J Tyson *Analysis – what analytical chemists do*, 1988.
D H Williams and I Fleming *Spectroscopic methods in organic chemistry*, 4th edn., 1989.

Unilever

THE ROYAL SOCIETY OF CHEMISTRY

1. Mass spectrometry

Mass spectrometry in some form is familiar to many people whose chemistry education has gone beyond compulsory education.

The theory

Diagrams of simple mass spectrometers, such as *Fig. 1*, are common. The principles of such spectrometers are quite straightforward:

1. a sample is introduced into the spectrometer and vaporised;
2. ions of charge z (where z is a multiple of the charge on an electron) are produced by bombarding the sample with electrons in the ionisation chamber;
3. the ions are accelerated by an electric field so that they have similar kinetic energy;
4. the ions of mass m are deviated by a magnetic field such that ions of low or high m/z value strike the sides of the spectrometer, but ions of one particular m/z value continue along the spectrometer body;
5. by varying the magnetic field strength all ions are sequentially focused into the detector; and
6. the ions are detected and the mass spectrum plotted.

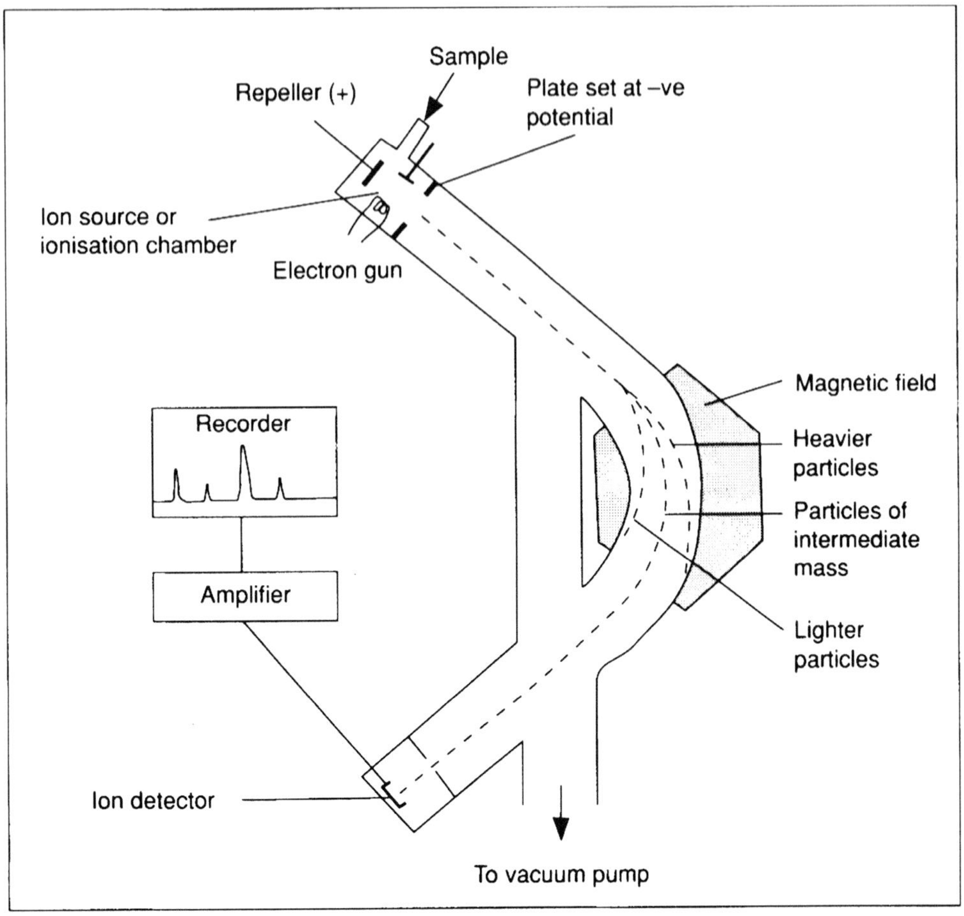

Figure 1 The mass spectrometer

Because the ions have to travel distances from 25 cm up to maybe 3 m inside the mass spectrometer it is important that they are not scattered or otherwise interfered with by other species – *ie* by reaction with neutral molecules. This is most likely to occur when air is present. This problem is overcome by subjecting the inside of the spectrometer to a high vacuum, typically 1.33×10^{-5} Nm^{-2} (10^{-7} mmHg) pressure. (1 $Nm^{-2} \equiv$ 1 Pa; 1 torr $\equiv$ 1 mmHg; 1 torr $\equiv 1.33 \times 10^{-2}$ Nm^{-2}). There are two ways of determining the *m/z* values of the ions produced: the accelerating electric field can be kept constant and the magnetic field scanned (in practice an electromagnet is used); or the magnetic field can be kept constant and the electric field scanned. It is usually the magnetic field that is scanned.

If masses are required to one atomic mass unit (one dalton) a single focusing instrument such as the one described will suffice. However, an accuracy of 1 in 10^6 – *ie* an accuracy that allows determination of precise atomic composition rather than nominal mass – can be achieved by using a double-focusing spectrometer. The principles for this type of spectrometer are identical to those for the single-focusing instrument, but include an electric sector before separation by a magnetic field (*Fig. 2*). Ions with a similar kinetic energy (ke) enter the electrostatic analyser and are focused into a narrower energy range. The magnetic field can then separate the ions, which have a much smaller energy range, giving a much greater resolving power.

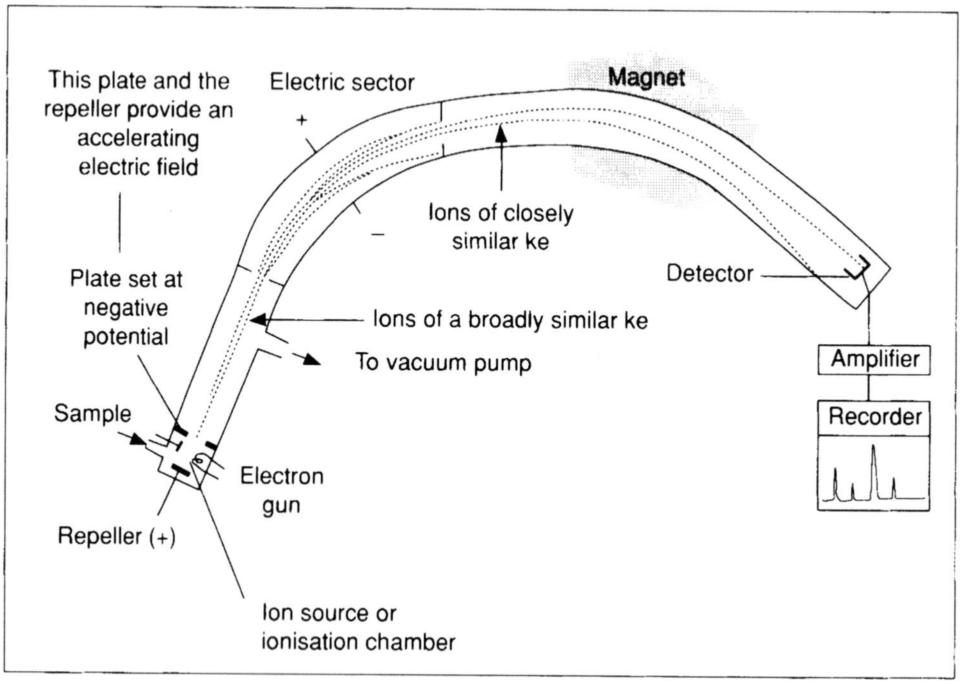

Figure 2 The double-focusing mass spectrometer

Practical considerations

The sample

Mass spectrometry is about 1000 times more sensitive than infrared (IR) or nuclear magnetic resonance (NMR) spectroscopy. Only a microgram or less is required to record a mass spectrum (many modern spectrometers require nanogram samples).

Different methods for introducing the sample in the vapour phase are used. Gaseous samples can be allowed to diffuse into the spectrometer – a 10^{-5} dm^3

(10 microlitre) sample would be ample for this. Volatile liquids can be injected in, but using rather smaller volumes, because the sample will vaporise under the low pressures present in the spectrometer.

Involatile liquids and solids are vaporised by placing the sample on a ceramic tip or in a glass capillary made of disposable soda-glass or reusable quartz (which has to be heated to very high temperatures to clean – *ie* to remove any residue) or in a metal crucible (typically 5 mm deep and 2 mm in diameter). These are inserted into the spectrometer (right up to, but not inside, the ion production area itself) and are subjected to temperatures of up to 300 °C (*Fig. 3*). The vapour is then allowed to diffuse into the ion production area. Solids dissolved in a solvent can be used by putting the solution into a crucible and allowing the solvent to evaporate.

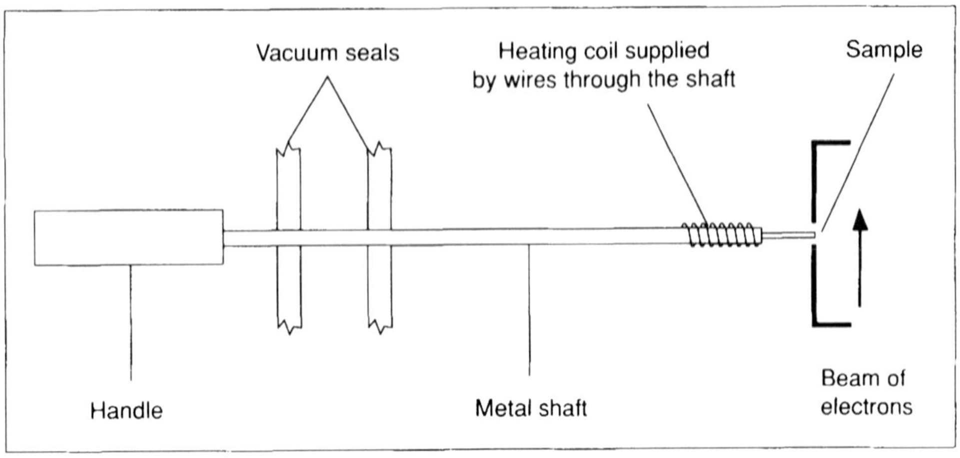

Figure 3 Vaporisation of solid samples

The ionisation chamber of the mass spectrometer is heated to 150–250 °C to ensure that the vaporised sample remains in the gas phase. Thermally stable non-polar organic molecules – *eg* perfluorokerosene (fully fluorinated kerosene) – with masses up to 1000 daltons can generally be vaporised at temperatures below 300 °C. If polar groups are present (*eg* OH, COOH and NH_2 groups) the volatility decreases, and molecules with moderate polarities will only vaporise readily if their masses are below 500 daltons.

The rate of production of vapour is important – too slow and insufficient ions will be produced to obtain an appreciable signal; too rapid and the dominant ions will saturate the detector and information regarding relative ion intensities will be lost. If there are too many ions in the ionisation chamber of the spectrometer ion molecule interactions might occur, forming ions with a mass greater than the mass of the sample molecules.

Ion production

The method of ionisation most commonly discussed is electron impact (EI). Electrons are 'boiled off' from a heated filament, which is made the cathode with respect to an anode set typically at +10 to +70V (*Fig. 4*). As the electrons accelerate towards the anode they can collide with the vaporised sample. Their energies are therefore up to 70 eV (1 eV ≡ 96.5 kJ mol^{-1}; 70 eV ≡ 6750 kJ mol^{-1}).

Collision between the high energy electron and the sample 'knocks out' an electron from an electron orbital, generally from the highest energy level:

$$X_{(g)} + e^- \longrightarrow X^+_{(g)} + e^- + e^-$$

$X_{(g)}$	e^-		$X^+_{(g)}$	e^-	e^-
sample	fast moving electron		ion	slower moving electron	electron from sample

(Multiple charged ions can be formed. Doubly charged ions are detected at half their mass value on the final spectrum, which has mass/charge as its horizontal axis.)

Some negative ions are formed (less than 0.1 per cent of the positive ions formed), and those that are formed are attracted away from the electric sector and magnet. However, it is possible to set the spectrometer to monitor negative ions by reversing the polarity of the repeller.

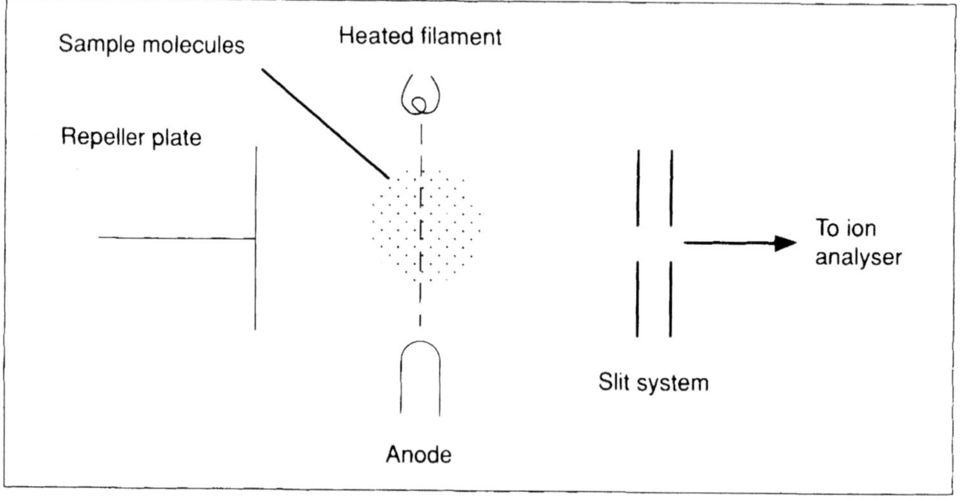

Figure 4 Electron impact mechanism

Ionisation of an element requires from 381 kJ mol^{-1} (francium) to 2370 kJ mol^{-1} (helium) (*ie* 4–25 eV), and organic molecules need about 600–1000 kJ mol^{-1} (7–10 eV). With organic molecules the energy of the ionising electrons is so great compared with the bond energies in the molecules that fragmentation is possible – *nb* the strongest common single bond (bond energy 485 kJ mol^{-1}, 5 eV) is C–F. Some of the residual energy (possibly up to 600 kJ mol^{-1}) of the fast moving electron might also be transferred to the ion as internal energy.

When an organic molecule forms a positive ion, it becomes a radical-cation because one electron has been removed from a pair in a filled orbital.

$$M \longrightarrow M^{\bullet +} + e^-$$

By convention, the radical electron is omitted from the representation of these ions in many texts.

If the radical-cation fragments, the molecular ion can lose either a radical or a neutral molecule *eg* the butyl ethanoate molecular ion (radical-cation) fragments as follows:

$$CH_3COOCH_2CH_2CH_2CH_3^+ \longrightarrow CH_3C^+{=}O + C_4H_9O \qquad 1$$

(molecular ion)

$$CH_3COOCH_2CH_2CH_2CH_3^+ \longrightarrow C_4H_8^+ + CH_3COOH \qquad 2$$

The remaining ions can fragment further to give the spectrum shown in *Fig. 5*. The molecular ion peak is not significant because the ion is unstable.

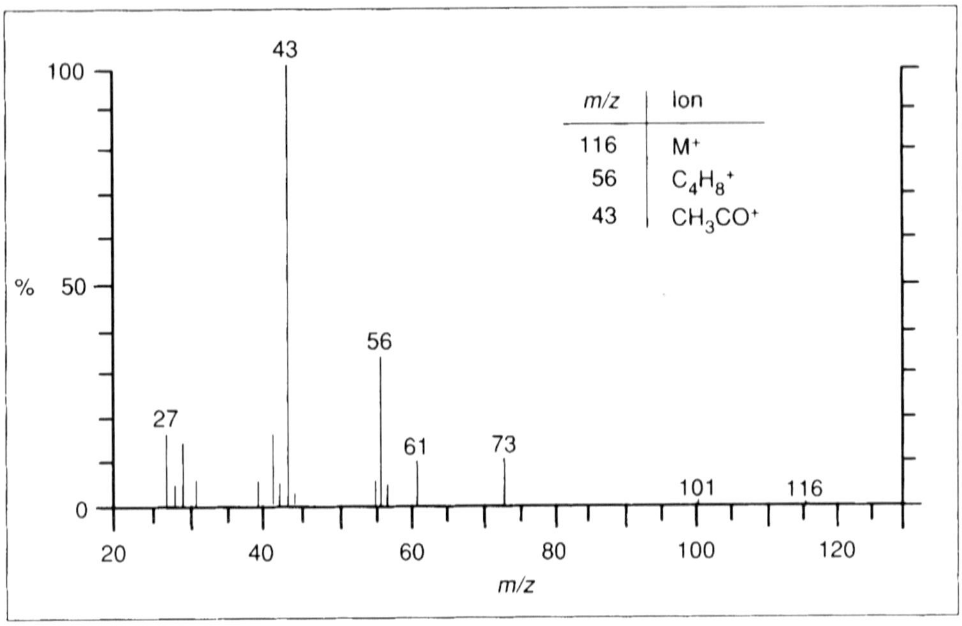

Figure 5 Mass spectrum of butyl ethanoate

The ionising electrons for this spectrum had energy 70 eV. These cause more fragmentation according to equation *1*. The peak at m/z = 43 is due to $CH_3C^+{=}O$, and the one at m/z = 56 due to $C_4H_8^+$. If ionising electrons of energy 15 eV had been used, the peak at m/z = 56 would have been the most abundant – *ie* fragmentation would predominantly have been according to equation *2*. However, in each case the molecular ion would not be significant.

So, although it is possible to obtain a unique spectrum which will act as a 'fingerprint' for an organic compound, all the operating conditions (including the energy of the ionising electrons) must be quoted. This is especially true if the spectrum obtained is to be compared with library spectra for identification.

Ionisation using lower potentials often gives more detailed information about the molecule because less energy is transferred to the sample and the molecular ion (M^+) can become more significant. A disadvantage of this method is that fewer ions are produced. One way of increasing the amount of molecular weight information obtained on a compound is to use a 'soft' ionisation technique such as chemical ionisation (CI).

The molecular ion and the arrangement and sizes of the peaks formed by the breaking apart (the fragmentation pattern) of an organic compound can often be used to identify it. However, there are cases when electron impact ionisation is inappropriate because it is impossible to ionise the molecule without it breaking up. The molecular ion can usually be observed by using other ionisation techniques, such as CI and FAB (see below). Other, more elaborate methods exist, such as laser desorption, secondary ion mass spectrometry (SIMS), electrospray and californium plasma desorption, which are compatible with large, polar molecules. (Descriptions of these techniques can be found in the bibliography.)

Deflection of the ions

The ions are accelerated through a series of plates set at increasingly negative potentials and emerge with a broadly similar energy, before passing into the electric sector where they are deviated so that the emergent ions all have a more sharply defined range of energies. Deflection in the electric field is independent of mass, but for a given charge it is proportional to the energy of the ions.

In the radial electrostatic field within the analyser the ions follow a circular path of radius r, broadly given by the equation:

$$r = \frac{2V}{E}$$

where V = electrostatic potential
E = energy of ion

The ions continue in a straight path outside the influence of the electric field, until they enter the magnetic field where they are deflected so that only the ions of a specific mass to charge ratio will continue on towards the detector, (Box).

Dependence of ion deflection on accelerating voltage and magnetic flux density

The kinetic energy of the ions after being accelerated is given by:

$$\tfrac{1}{2} mv^2 = zV \qquad 1$$

where m = mass of the ion
z = charge on the ion
v = velocity of the ion

The force on the ions in the magnetic field is described by:

$$\text{force} = Bzv \qquad 2$$

where B = magnetic flux density.

The force on a body as it accelerates towards its centre of curvature is:

$$\text{force} = \frac{mv^2}{r} \qquad 3$$

where r = radius of circular path.

Combining equations *2* and *3* we get

$$Bzv = \frac{mv^2}{r}, \text{ so } v = \frac{Bzr}{m} \qquad 4$$

Substituting equation *4* into equation *1*,

$$\tfrac{1}{2} m \frac{B^2z^2r^2}{m^2} = zV \quad \text{thus} \quad \frac{m}{z} = \frac{B^2r^2}{2V}$$

So, for a given m/z value, the radius of curvature of the deflected ion is dependent on the magnetic field strength and on the accelerating voltage.

The mass spectrum can be obtained by varying the accelerating voltage, *V*, or by varying the magnetic field (magnetic scanning). Voltage scanning can be done at high speed but this usually gives a distorted spectrum – the relative abundances of the fragments decreases as their mass increases. Magnetic scanning is the norm although it is a little slower, because it is restricted by the response time of the analyser magnet.

The double-focusing mass spectrometer is so precise that the relative mass of a compound can be determined to such an accuracy that its molecular formula can be assigned. The molecular ion peak of the unknown in *Fig. 8* is 122.036776, which can be assigned the molecular formula $C_7H_6O_2$ if it is assumed that only carbon, hydrogen, nitrogen and oxygen are present (Table 1). However without further information the precise isomer cannot be identified.

Table 1 Some formulae corresponding to nominal *m/z* = 122

Formulae	Actual mass
$C_4H_4N_5$	122.046668
$C_4H_{10}O_4$	122.057903
$C_6H_4NO_2$	122.024201
$C_6H_6N_2O$	122.048010
$C_6H_8N_3$	122.071819
$C_7H_6O_2$	122.036776
C_7H_8NO	122.060585
$C_7H_{10}N_2$	122.084394
$C_8H_{10}O$	122.073161
$C_8H_{12}N$	122.096970
C_9H_{14}	122.109545

These are based on the following relative atomic masses:

C = 12.0000000
H = 1.0078246
N = 14.0030738
O = 15.9949141

Ion detection

Many modern mass spectrometers use electron multipliers (*Fig. 6*). The ions strike a plate (the conversion dynode) made of a material (*eg* a copper/beryllium alloy) that emits electrons when struck by energetic particles . This is set at *ca* –1.4 kV. Secondary electrons are emitted from it, and they are accelerated and focused onto the second and subsequent dynodes, which are set at potentials progressively closer to earth. At each dynode there is an increase in the number of electrons emitted, such that at the end of the multiplier a gain of approximately 10^6 has been achieved. (A multiplier might contain 10 –14 dynodes.) Since ions are sharply focused onto one position the conversion dynode degrades after a period of time and must then be replaced.This occurs after perhaps two years on an instrument that is in daily use.

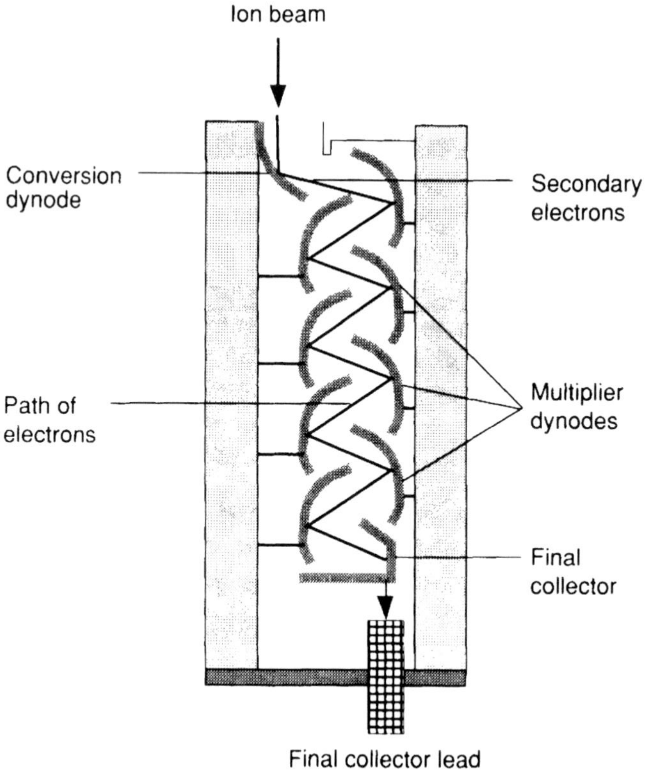

Figure 6 Schematic diagram of an electron multiplier

The amount of information obtained from a mass spectrometer is so vast that it is only realistic to capture the data using a microprocessor. The signal from the electron multiplier is fed to an analogue to digital converter, then stored in a computer. The computer stores the mass value at the centre of each ion distribution, and a maximum value which corresponds to the number of ions of that mass (*Fig. 7*). The computer can then plot mass spectra. If plotted graphically the computer will set *m/z* values on the x-axis and relative abundance on the y-axis, setting the most abundant ion as 100 per cent abundance (*Fig. 8*). The intensities of other ions are then plotted relative to this peak.

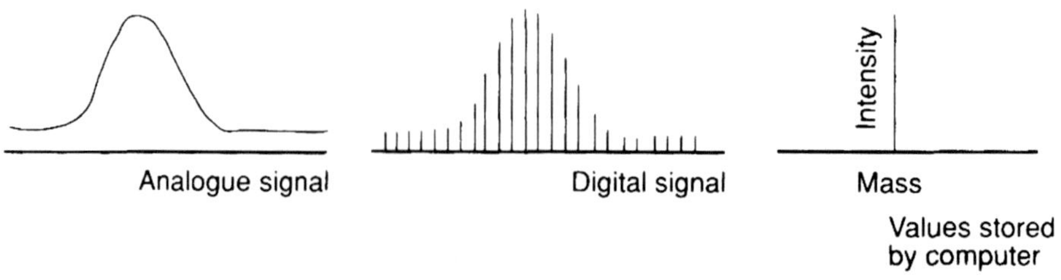

Figure 7 Recording of data from the mass spectrometer

Unilever

THE ROYAL SOCIETY OF CHEMISTRY

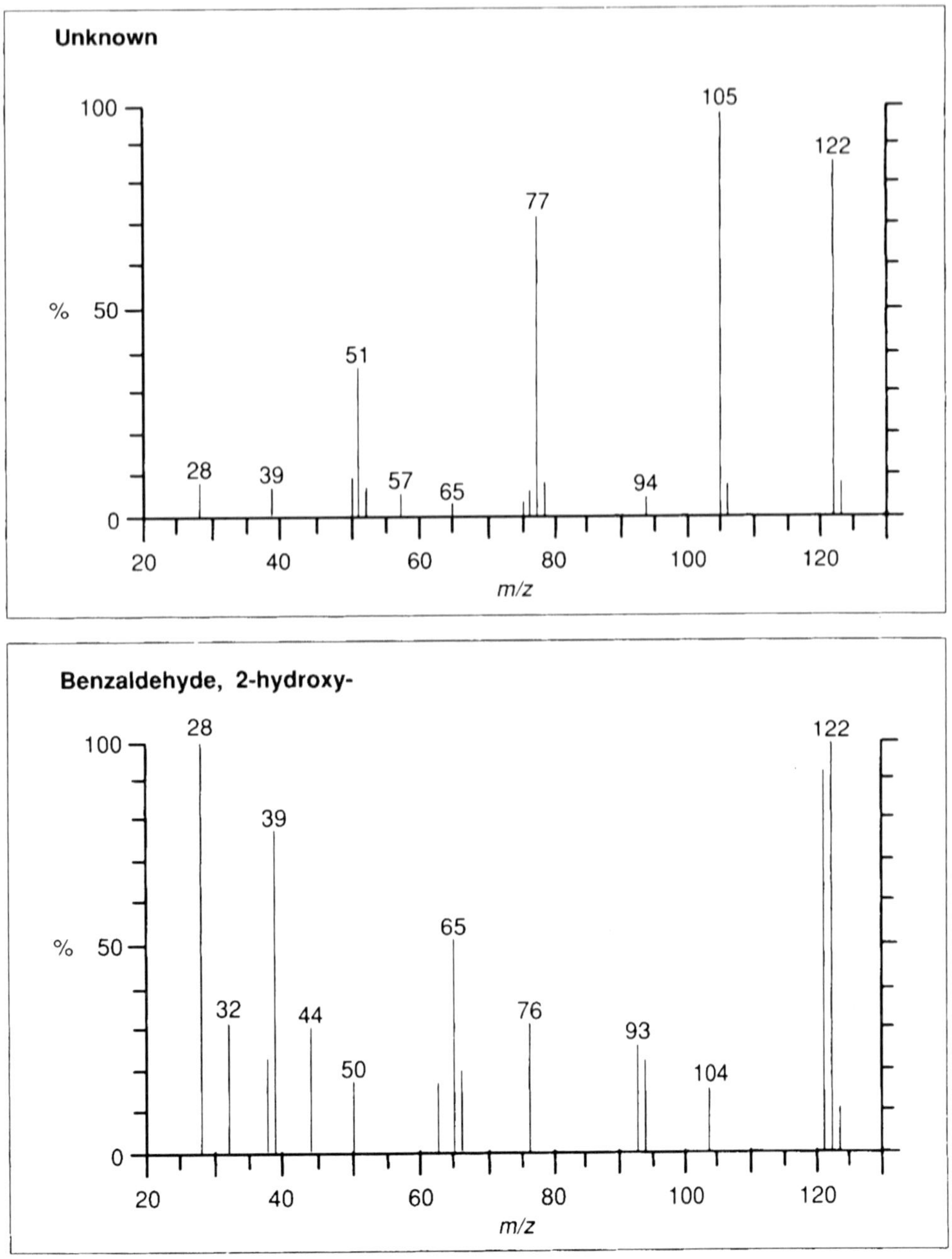

Figure 8 Mass spectra of an unknown and five isomers of the same molecular formula (continued overleaf)

The vacuum system

It is vital that the path of the ions in the spectrometer is not affected by residual gas molecules or by neutral species formed during the fragmentation or deflection of particles. The average distance travelled by ions between successive collisions is known as the mean free path, and is of the order of several hundred metres when the internal pressure of the spectrometer is 10^{-5} Nm^{-2}. In modern spectrometers two types of pumps are used to achieve this low pressure. A rotary, or backing, pump reduces the pressure to approximately 10^{-2} Nm^{-2}, in tandem with a diffusion pump (which traps gas molecules in the droplets as high speed jets of oil vapour condenses), or a turbomolecular pump (which has a powerful fan that turns very quickly).

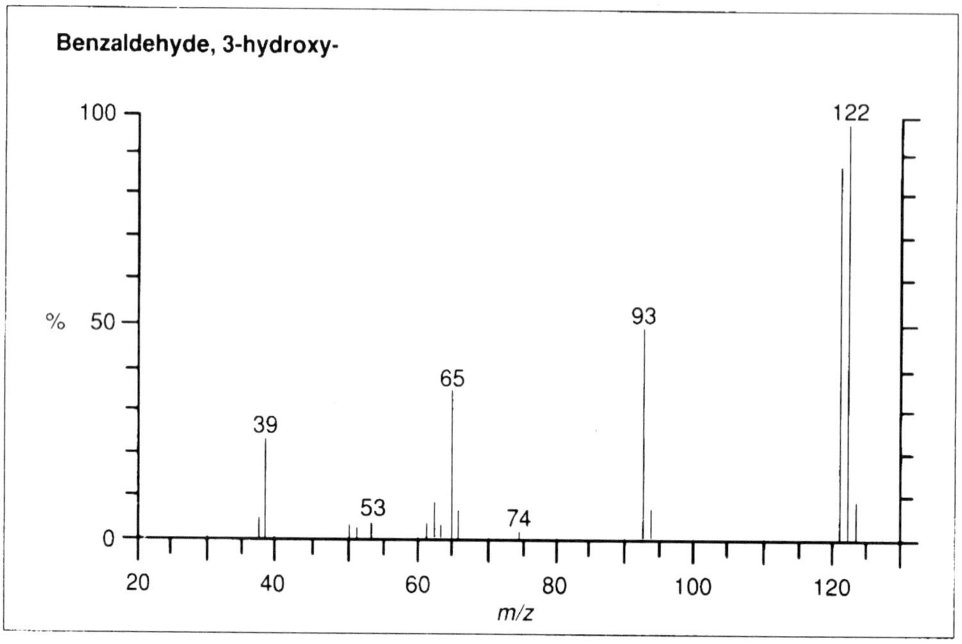

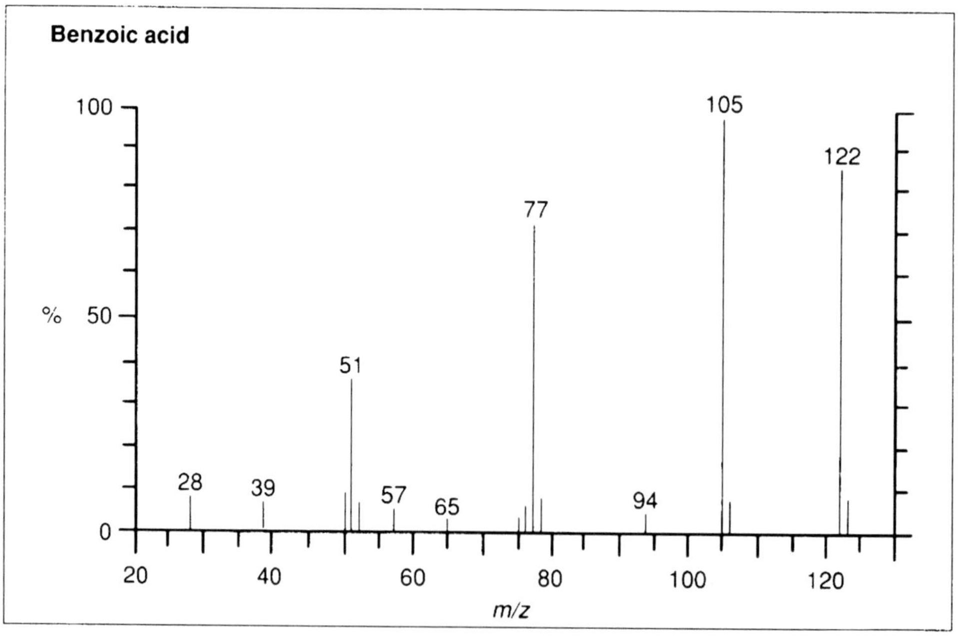

Figure 8 continued

Ions that do not reach the detector can pick up an electron from the walls of the spectrometer, and if of low relative mass they become gaseous and can be removed by the vacuum line, along with any particles that were not ionised in the first place. If the species are involatile, they condense on the inside walls of the spectrometer and are periodically removed by surrounding the instrument with electric band heaters and vaporising them off. Heavy or stubborn deposits have to be removed by mechanically stripping the instrument down and cleaning it. Fortunately, this does not happen very often.

The ionisation chamber of the spectrometer is particularly susceptible to contamination by condensed samples, therefore it is likely to need to be stripped down and mechanically cleaned every three months.

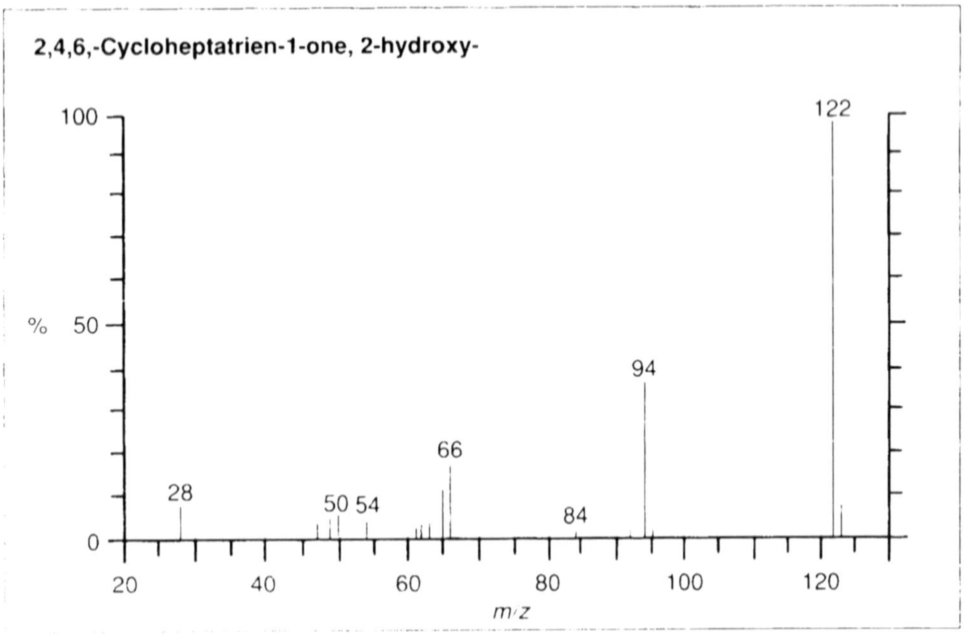

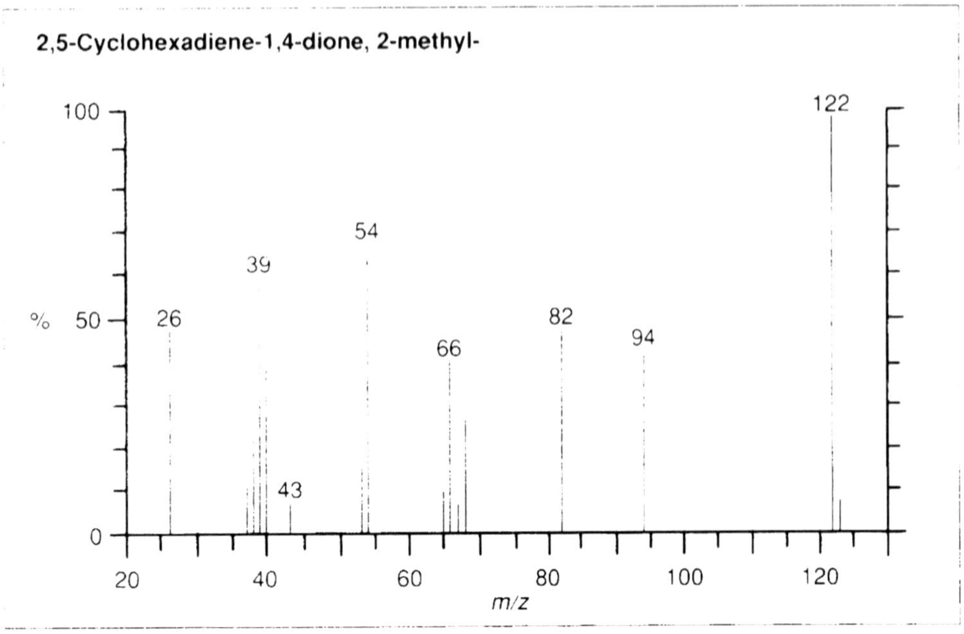

Figure 8 continued

Calibration

If low resolution data is sufficient, perfluorokerosene, for example, (containing fully fluorinated hydrocarbon molecules – *ie* no hydrogen atoms remain) can be used as a reference to calibrate the instrument. Many peaks are observed, separated by 12 or 50 mass units *(Fig. 9)*. Once calibrated, the spectrometer can function without recalibration for about one month.

More detailed information (higher resolution data) can be obtained by putting a reference material in with the sample so that the instrument can be calibrated at the same time as analytical data are obtained. The reference will depend on the mass of the compound, but is likely to be perfluorokerosene with a mass range appropriate to the approximate mass of the unknown. (Mass ranges of perfluorokerosene are available in the same way as boiling point ranges of petroleum ether.) The peaks due to the reference material are used for calibration, and are then subtracted from the final spectrum by a computer (the data system). It is unlikely that the reference peaks will coincide with the sample peaks because fluoro compounds tend to have masses lower than the nominal mass, whereas most other compounds have higher masses.

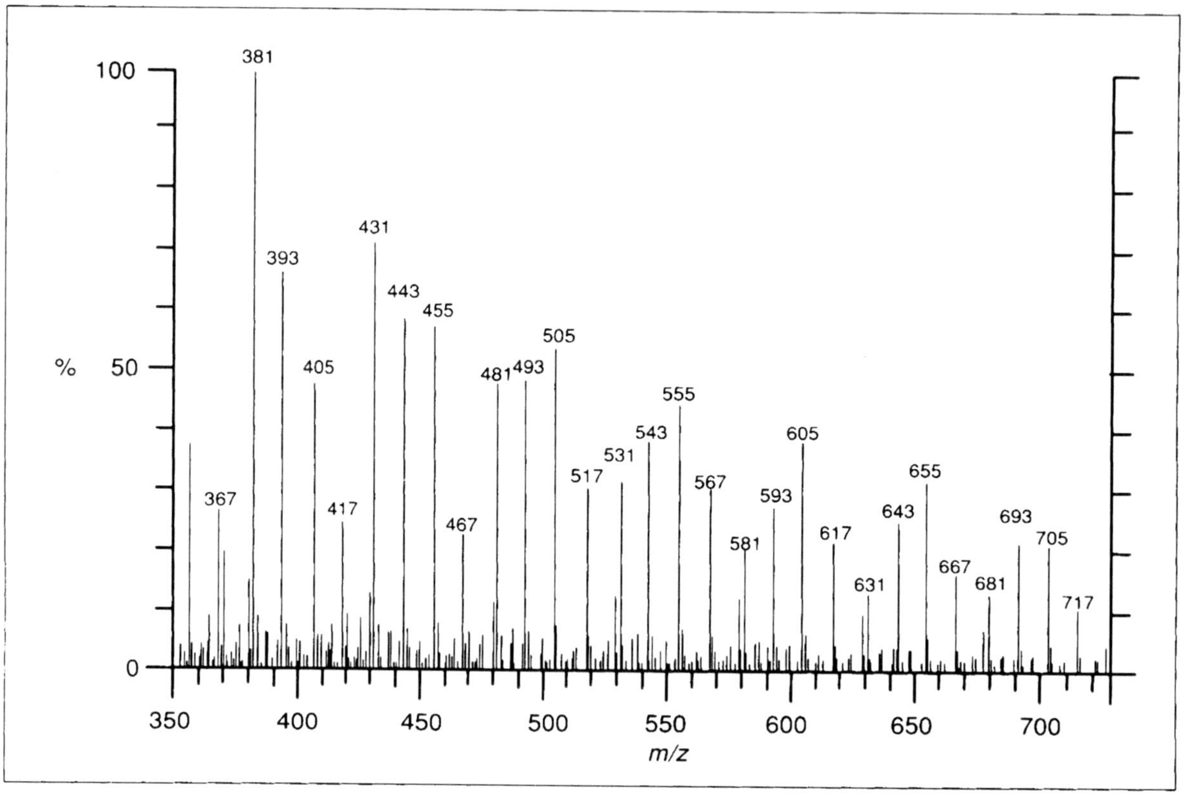

Figure 9 Mass spectrum of a perfluorokerosene, used for calibrating a mass spectrometer

Other ionisation techniques

Electron impact frequently fails to yield the information required from substances. A variety of alternative methods for ionising substances exist.

Chemical ionisation (CI)

In this technique a reagent gas such as methane, methylpropane or ammonia is ionised by electron bombardment and is then allowed to react with a neutral molecule to produce a molecular ion, *eg*

	CH_4	+	e^- fast moving electron	$\longrightarrow$	CH_4^+ molecular ion	+	e^- electron from reagent gas	+ e^- slower electron
then	CH_4^+	+	CH_4	$\longrightarrow$	CH_5^+ reactant ion	+	$CH_3\bullet$	

The reactant ion, CH_5^+ in this case, then reacts with the sample gas and protonates it:

$$M + CH_5^+ \longrightarrow MH^+ + CH_4$$

In this way positive ions are produced, but these are one mass unit higher than the parent molecule. The ions produced almost always have less internal energy than ions formed by electron impact, and thus fragment less as a consequence.

In practice the reagent gas is allowed into the ion chamber at a partial pressure of 10^2 Nm^{-2} (Pa). The amount of sample in the ion source can be very small – on the nanogram scale. The bombarding electrons come from a hot filament, and their energy is in the range 9650–48250 kJ mol^{-1} (100–500 eV) – higher than for electron impact. Statistically the reagent gas is more likely to be ionised than the sample.

The choice of reagent gas depends on the ease of fragmenting the sample. The internal energy of the species MH^+ decreases in the order $CH_5^+ > C_4H_9^+ > NH_4^+$. This results from the relative strength of the bonding between the reagent gas and the proton it transfers to the sample: ammonia bonds most strongly and methane least. Consequently, if the ammonium ion is the protonating agent it transfers very little energy to the sample. Conversely, the CH_5^+ ion readily protonates samples and the energy liberated is transferred as the internal energy of the sample – it is then likely to fragment more extensively. However, one disadvantage of using methane or methylpropane as the reagent gas is that the spectra obtained are complicated above the $(M+H)^+$ molecular ion because adducts such as $(M+CH_3)^+$ and $(M+C_2H_5)^+$ are also formed.

The difference between the spectra obtained by electron impact and chemical ionisation is immediately apparent in *Figs. 11 a* and *b*, where the spectra of Tenormin (*Fig. 10*) a drug produced by ICI, used for treating heart conditions are shown.

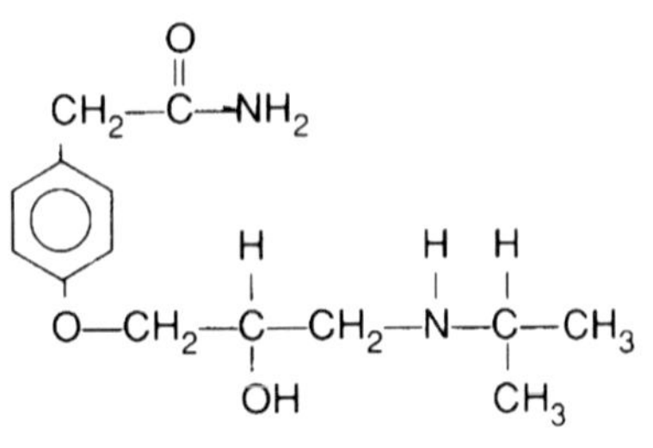

Figure 10 Structure of Tenormin

Under CI conditions equal abundances of positive and negative ions are usually produced. The processes involved in negative ion formation are complicated, but one way that they can be formed is:

$$M + \underset{\text{reactant ion}}{CH_3^-} \longrightarrow (M\text{-}H)^- + CH_4$$

The molecular ion peaks obtained by CI are predominantly $(M+H)^+$ and $(M-H)^-$. One major advantage of negative ion chemical ionisation is that some compounds can be detected at pico- (10^{-12}) or femto- (10^{-15}) gram levels.

Clearly the potentials inside the spectrometer have to be altered to enable negative ion spectra to be measured, but good, reproducible data can be recorded.

THE ROYAL SOCIETY OF CHEMISTRY

Unilever

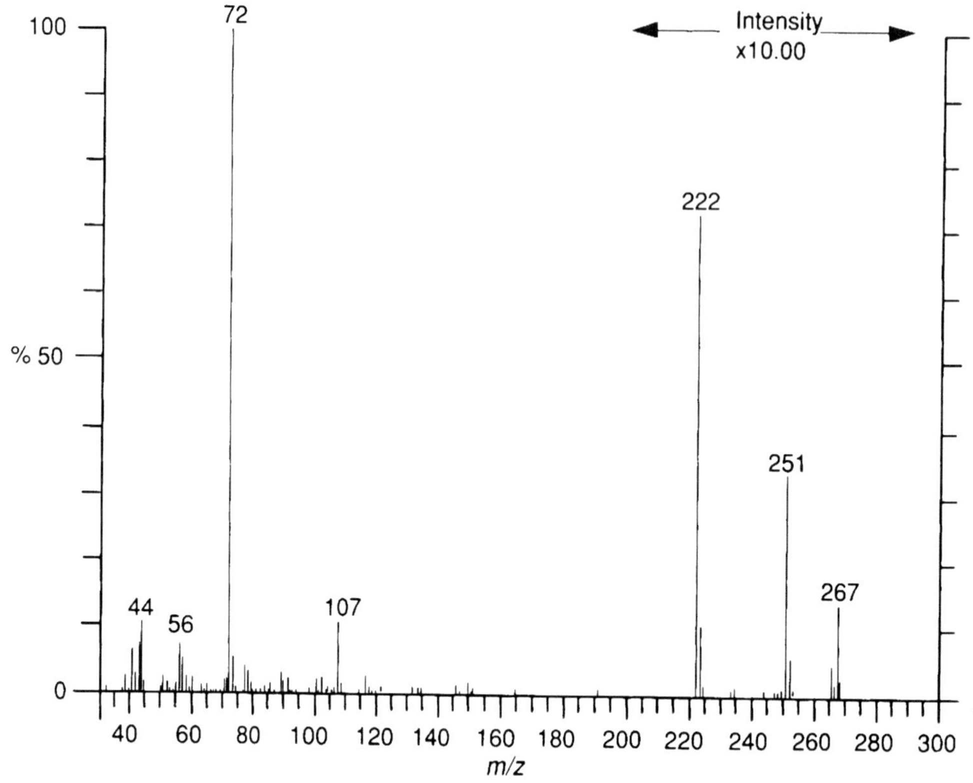

m/z	Ion
267	$(M+H)^+$ – sometimes seen in N containing compounds that fragment extensively
266	M^+
265	$(M-H)^+$
251	$(M-CH_3)^+$
107	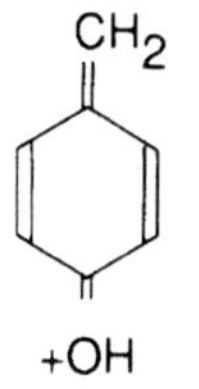
72	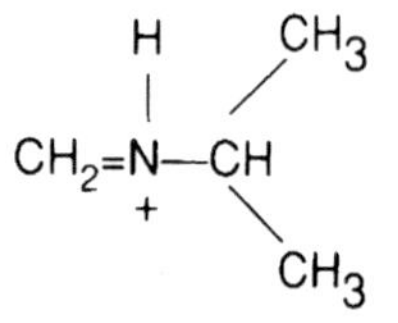
44	$(CH_2=CH-OH)^+$ and $O=C=\overset{+}{N}H_2$

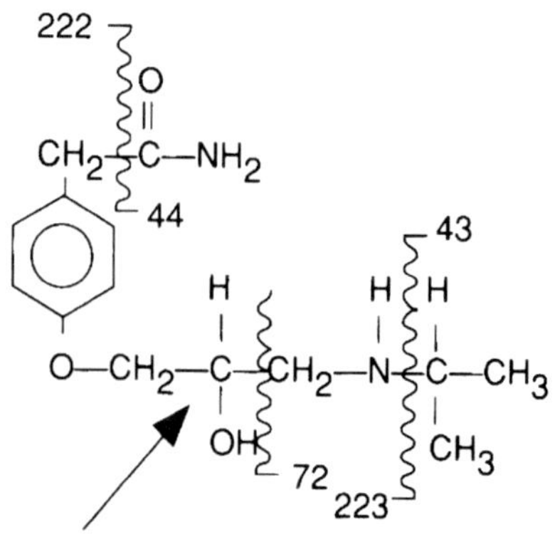

m/z = 222 also arises from loss of C_2H_4O by rearrangement involving loss of this carbon

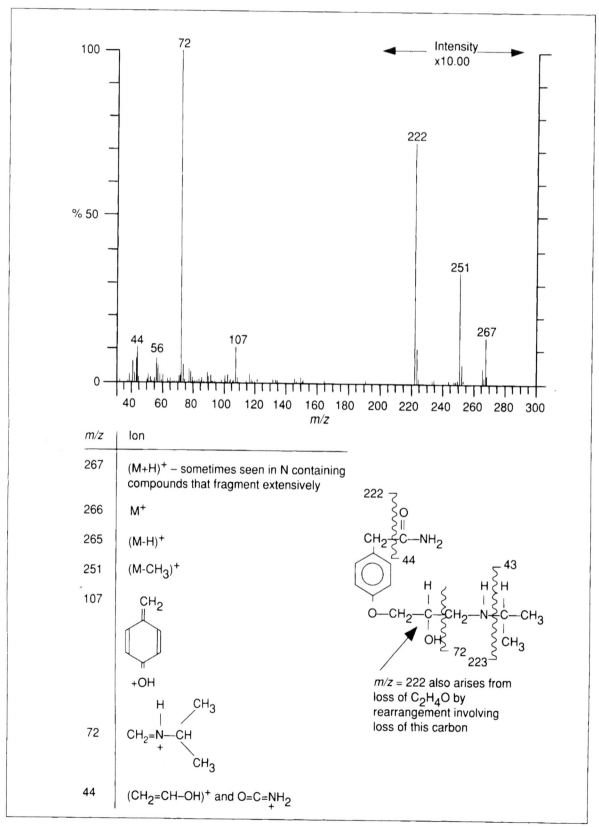

Figure 11a Mass spectrum of Tenormin using electron impact.

Unilever

THE ROYAL SOCIETY OF CHEMISTRY

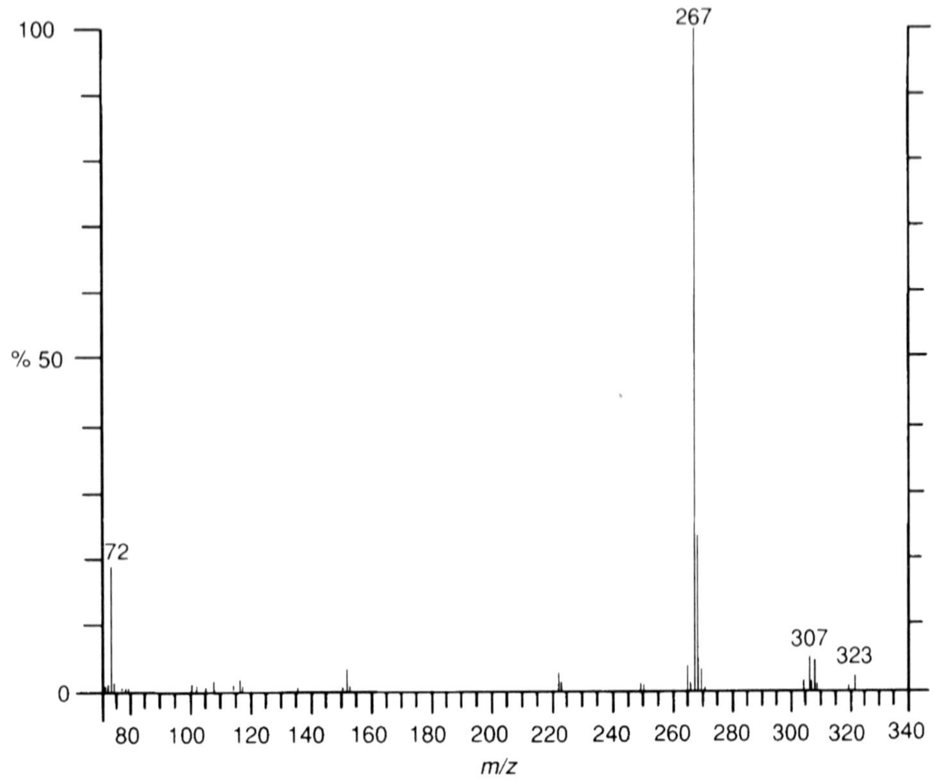

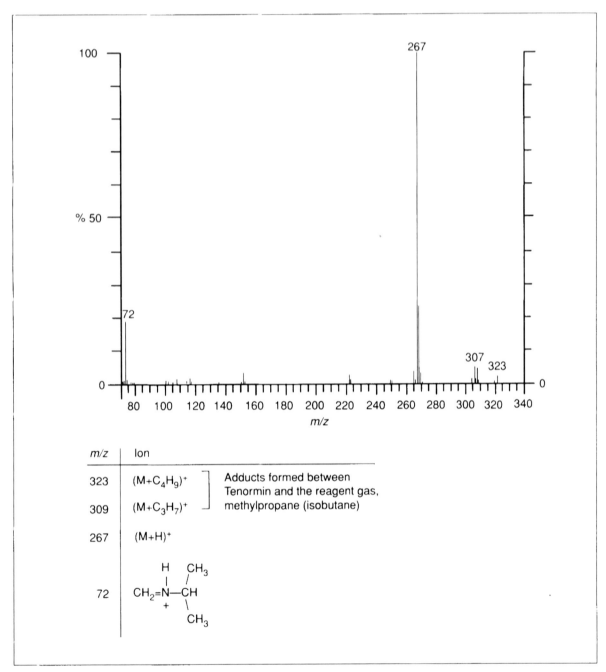

m/z	Ion	
323	$(M+C_4H_9)^+$	Adducts formed between Tenormin and the reagent gas, methylpropane (isobutane)
309	$(M+C_3H_7)^+$	
267	$(M+H)^+$	
72	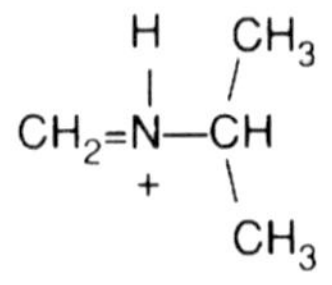	

Figure 11b Mass spectrum of Tenormin using chemical ionisation

THE ROYAL SOCIETY OF CHEMISTRY

Unilever

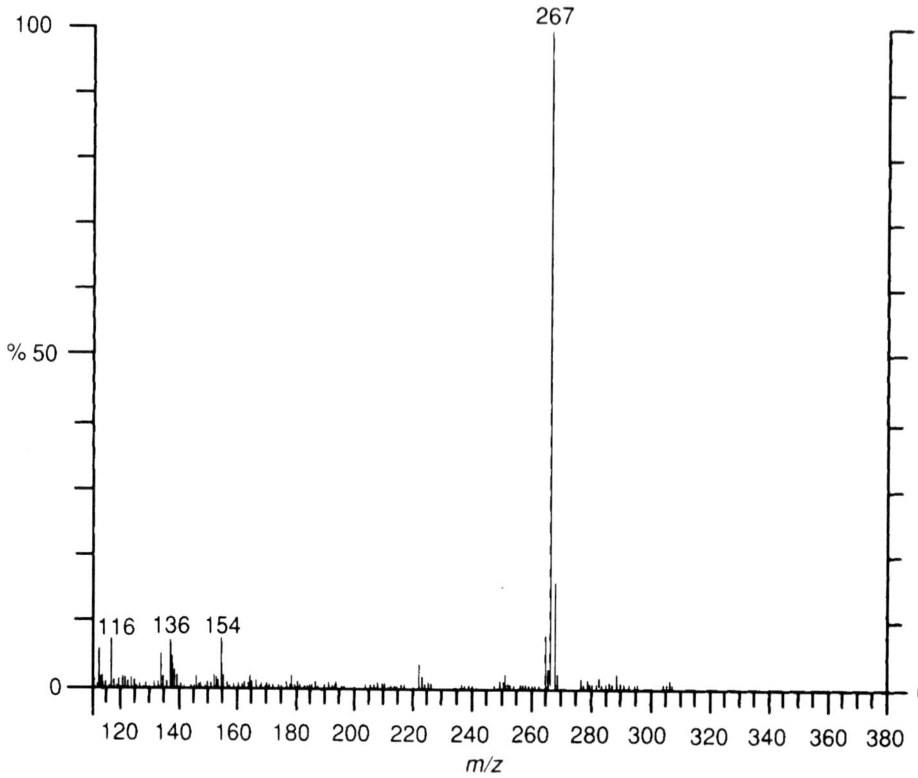

m/z	Ion	
267	$(M+H)^+$	
154	(Protonated 3-nitrophenyl methanol)$^+$	from the matrix
136	(Protonated 3-nitrophenyl methanol $- H_2O)^+$	from the matrix
116	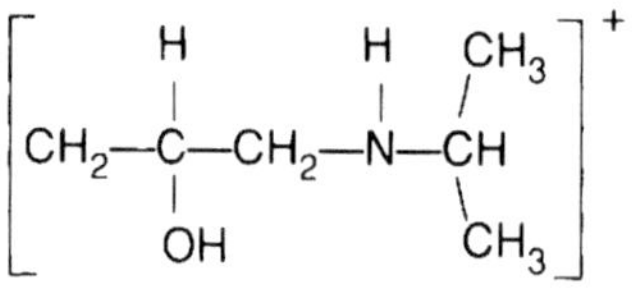	

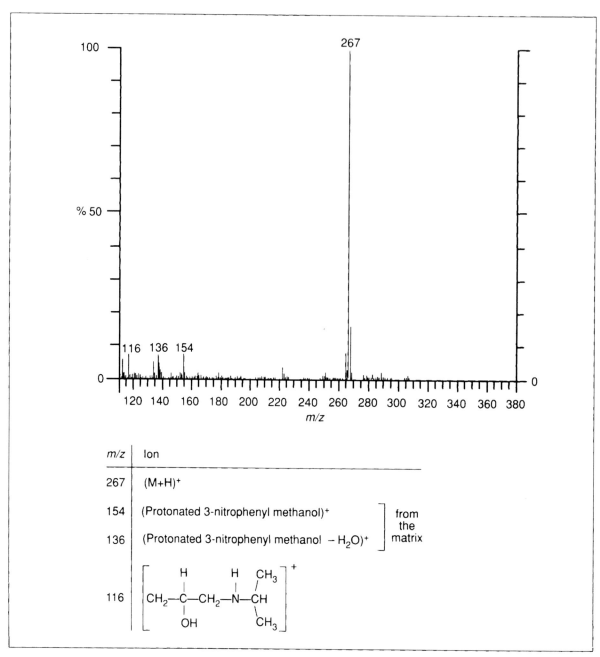

Figure 11c Mass spectrum of Tenormin using fast atom bombardment

Fast atom bombardment (FAB)

If the sample is susceptible to decomposition when heated, FAB is a useful technique because it involves no heating at all. A beam of atoms with high kinetic energy, usually xenon atoms with several keV energy, is used to strike a solution of the sample in a 'matrix' compound such as glycerol (propan-1,2,3-triol) or 3-nitrophenyl methanol (*m*-nitrobenzyl alcohol), on the end of a probe (*Fig. 12*).

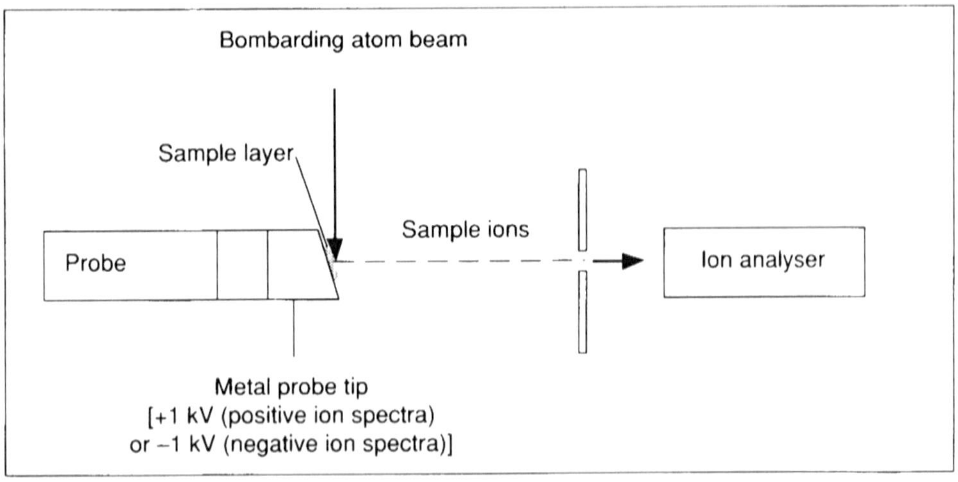

Figure 12 Fast atom bombardment – the ion source

The fast atoms are generated by accelerating xenon ions to 6–9 keV, then neutralising them as they pass xenon atoms at low pressure. Neutralisation takes place by electron transfer:

$$\underset{\text{fast ion}}{Xe^+} + Xe \longrightarrow \underset{\text{fast atom}}{Xe} + Xe^+$$

When the sample is struck by the fast atoms, it is desorbed from the surface of the probe by the transfer of momentum, usually as an ion. In common with chemical ionisation the sample molecule in FAB is usually detected as $(M+H)^+$ or as $(M-H)^-$. The ions produced are analysed in the mass spectrometer in the same way as ions produced by other methods. The FAB mass spectrum of Tenormin is shown in *Fig. 11c*.

Salts of the type AX can be examined by FAB, and the A^+ ions are detected (usually as AXA^+) in the positive ion mode, and the X^- ions in the negative ion mode.

Interpretation of the mass spectrum of a compound

With the aid of a computer it is possible to obtain information from the spectrum quickly – *eg* the mass of the molecular ion peak can be measured with sufficient precision to be able to assign the molecular formula directly. Libraries of mass spectra are commercially available on computer software, and the spectra of all the library compounds with the same molecular formula can be called up for comparison. The first spectrum in *Fig. 8* is of an unknown compound with a molecular ion peak corresponding to the formula $C_7H_6O_2$, and with it are the library spectra of five isomers of this formula. Although the computer can be used to compare one spectrum with another it is not necessary in this case. The peak at m/z = 105 (corresponding to the $C_6H_5CO^+$ ion) is sufficient to be able to eliminate all the isomers except benzoic acid. However, the spectrum of the unknown does not always correspond exactly to the library spectrum. This is entirely normal because the operating conditions under which the sample and the library compound were recorded might not have been exactly the same. Make, and to a greater extent, type of instrument, can influence the final spectrum. Unless there is a significant difference in the two spectra the major peaks should be sufficient to identify the compound.

If the spectrum is of a new or commercially sensitive compound then it will not appear in a commercial library, although the elemental composition can be determined. In these cases it is usual to confirm a suspected structure rather than try to determine the structure from first principles.

In confirming a suspected structure, or trying to elucidate a completely unknown structure, a rather different approach is used. The ratio of the two isotopes carbon-12 to carbon-13 (^{12}C:^{13}C) is 100:1.1, which is significantly greater than the isotopic ratios for hydrogen, nitrogen and oxygen. Therefore the isotopic peak at $(M+1)^+$ will be 1.1 per cent of the height of the molecular ion peak for every carbon atom present in the molecule. In the spectrum of benzoic acid *(Fig. 14)* the $(M+1)^+$ peak at m/z = 123 is approximately 8 per cent (7.7 per cent) of the height of the molecular ion peak. This is consistent with there being seven carbon atoms in the sample.

The presence of two other elements, chlorine and bromine, is usually easy to determine. The two isotopes of chlorine have mass numbers 35 and 37, and occur naturally in the ratio 76:24. Thus the appearance of two molecular ion peaks differing by 2 mass units, and in the approximate ratio 3:1 suggests the presence of a chlorine atom in the compound. Bromine has isotopes of relative masses 79 and 81 in approximately 1:1 ratio (50.5:49.5), so two molecular ion peaks of similar height differing by 2 mass units indicates the presence of a bromine atom. Compounds with more than one chlorine and/or bromine atom give the molecular ion and ion fragment peaks with the patterns shown in *Fig. 13*. All the peaks differ by 2 mass units.

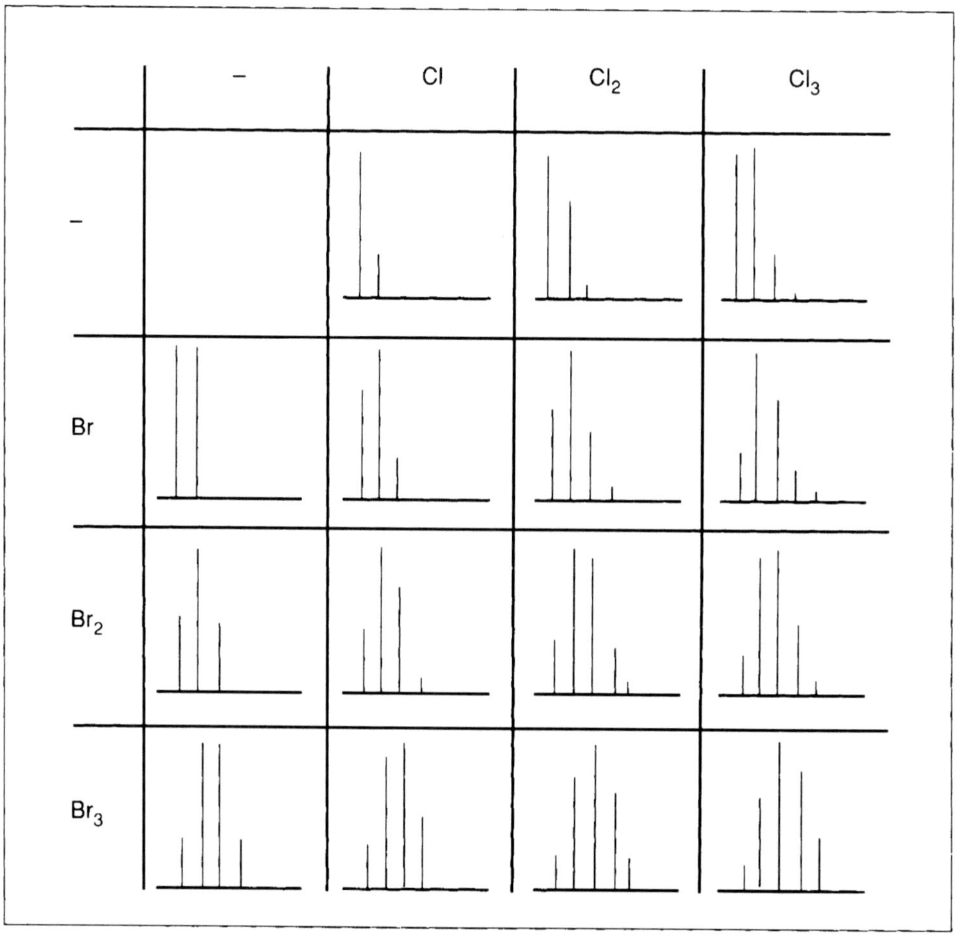

Figure 13 Peak patterns due to the presence of chlorine and bromine atoms in molecular ions and ion fragments

In the absence of chlorine and bromine atoms, if the relative mass of the molecular ion is odd, then probably an odd number of nitrogen atoms is present; and if the relative mass is even then either nitrogen is not present, or there is an even number of nitrogen atoms.

Once some idea of the molecular composition has been calculated an attempt can be made to determine the structure of the compound. This is not always easy because without some prior knowledge of the structure it is difficult to work out exactly what has been lost from the molecular ion. However, there are some groups that are commonly lost from the molecular ion and some of these are shown in Table 2.

For compounds containing a benzene ring, monosubstituted compounds might be expected to give a peak at *m/z* = 77, corresponding to $C_6H_5^+$. While such a peak is often observed, bond fission occurs more frequently one bond away from the benzene ring.

Figures 14 and *15* show the mass spectra of benzoic acid and methyl benzoate with the assignment of the major peaks.

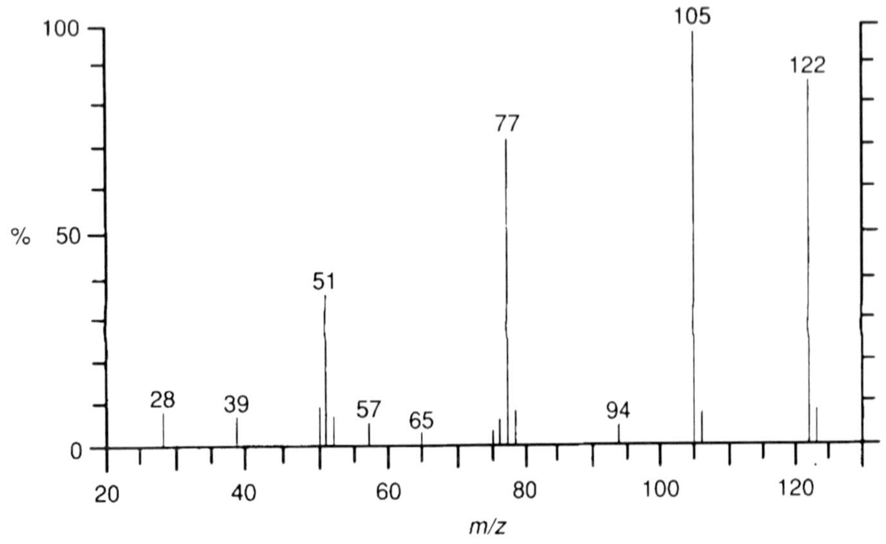

m/z	Ion	
122	M^+	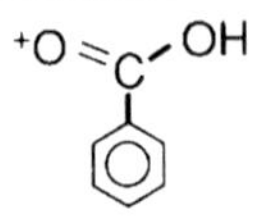
105	$(M\text{-}OH)^+$	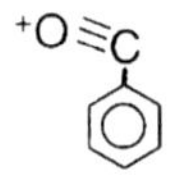
77	$(M\text{-}OH\text{-}CO)^+$	$C_6H_5^+$
51	$(M\text{-}OH\text{-}CO\text{-}C_2H_2)^+$	$C_4H_3^+$

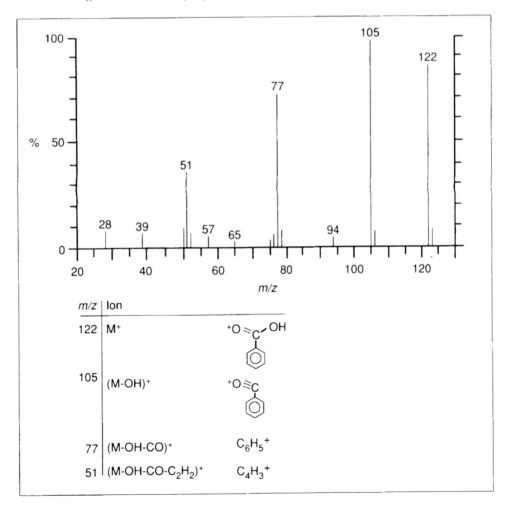

Figure 14 Mass spectrum of benzoic acid

THE ROYAL SOCIETY OF CHEMISTRY

Unilever

Table 2 Common losses from molecular ions

Ion	Groups commonly associated with the mass lost	Possible inference	
M − 16	O	$—NO_2$, $=\overset{+}{N}—\overset{-}{O}$, $>S{=}O$	
M − 16	NH_2	$ArSO_2NH_2$, $-CONH_2$	
M − 17	NH_3		
M − 17	OH	Alcohol	
M − 18	H_2O	Ketone, aldehyde, alcohol	
M − 28	CO	Quinone	
M − 28	C_2H_4	Aromatic ethyl ether	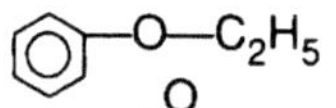
		Ethyl ester	$R—C(=O)—O—C_2H_5$
		Propyl ketone	$R—C(=O)—C_3H_7$
M − 29	CHO	Aldehyde	R—CHO
M − 29	C_2H_5	Ethyl ketone	$C_2H_5—CO—R$
M − 30	C_2H_6		
M − 30	OCH_2	Aromatic methyl ether	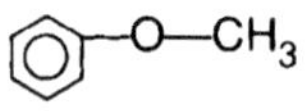
M − 30	NO	$Ar–NO_2$	
M − 31	OCH_3	Methyl ester	$R—C(=O)—O—CH_3$
M − 32	CH_3OH	Methyl ester	$R—C(=O)—O—CH_3$
M − 33	HS	Thiol	—SH
M − 34	H_2S	Thiol	—SH
M − 41	C_3H_5	Propyl ester	$R—C(=O)—O—C_3H_7$
M − 42	CH_2CO	Methyl ketone	$R—C(=O)—CH_3$
		Aromatic ethanoate	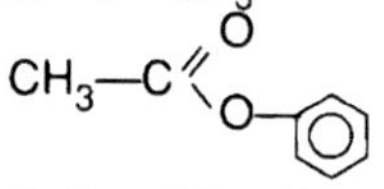
M − 42	C_3H_6	Butyl ketone	$C_4H_9—CO—R$
		Aromatic propyl ether	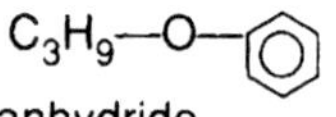
M − 44	CO_2	Carboxylic acid, ester, anhydride	
M − 45	CO_2H	Carboxylic acid	$R—C(=O)—OH$
M − 46	C_2H_5OH	Ethyl ester	$R—C(=O)—O—C_2H_5$
M − 46	NO_2	$Ar–NO_2$	
M − 48	SO	Aromatic sulphoxide	

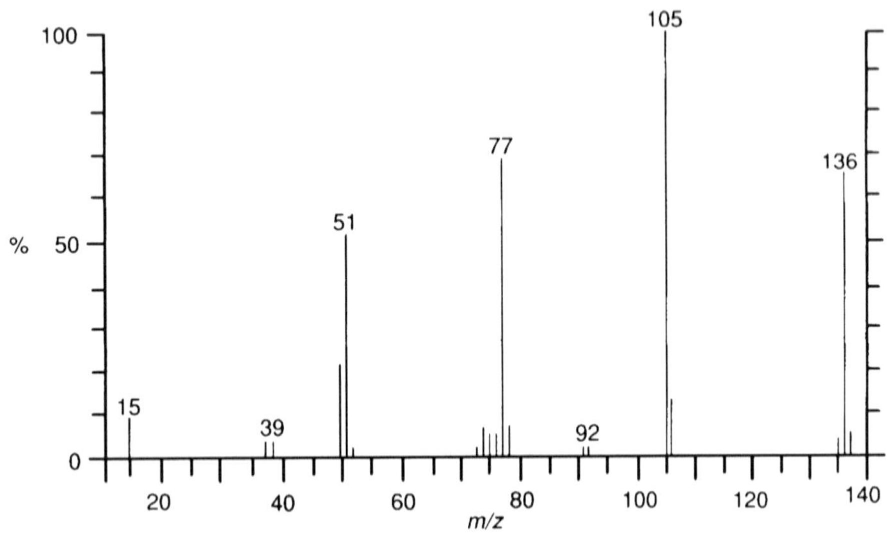

m/z	Ion	
136	M^+	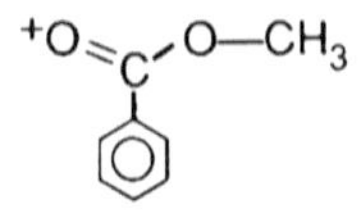
105	$(M\text{-}OCH_3)^+$	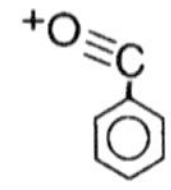
77	$(M\text{-}OCH_3\text{-}CO)^+$	$C_6H_5^+$
51	$(M\text{-}OCH_3\text{-}CO\text{-}C_2H_2)^+$	$C_4H_3^+$

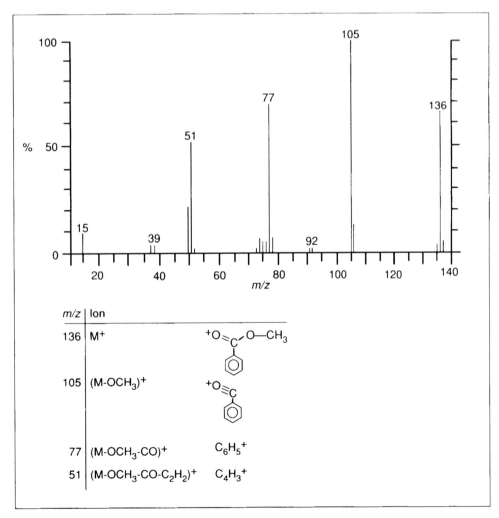

Figure 15 Mass spectrum of methyl benzoate

Applications of mass spectrometry

So far discussion of the application of mass spectrometry has been restricted to determining relative atomic/isotopic mass and identification of organic compounds. However, there are other applications.The mass spectrometer is sufficiently sensitive to measure the isotopic composition of substances accurately. Although the isotopic composition of elements is usually regarded as constant, there are some variations. Atmospheric oxygen, for instance, is richer in ^{18}O than fresh surface water. Hydrogen, carbon, oxygen and sulphur are just four elements whose isotopic compositions in different locations have been found to vary, and these variations have been studied using the mass spectrometer.

Mass spectrometer applications are also found in medicine and biochemistry. Many of the substances found in living systems are complex mixtures and are sensitive to fragmentation by using EI. Separation of individual compounds is done by using gas chromatography (GC) or high performance liquid chromatography (HPLC) and the mixture from the column (the eluate) is fed into a mass spectrometer (see chapter 5). Once inside the spectrometer soft ionisation methods are frequently used to ionise the sample. For example, FAB is often used to determine the residue sequence of peptides. Similarly, radioactively labelled compounds and compounds labelled with stable isotopes such as ^{13}C are used to determine the fragmentation

THE ROYAL SOCIETY OF CHEMISTRY

Unilever

mechanisms of organic molecules and extended studies have led to the identification of metabolic pathways. Labelled drugs allow pathways of drug metabolism to be studied on a picomole scale.

Examples of GC-MS studies (gas chromatography interfaced to a mass spectrometer) include:

1 drug abuse by athletes; oestrogens in the urine of pregnant females;

2 determination of prostaglandins in human semen;

3 the presence of ergotamine (an alkaloid) and other hallucinogens in blood;

4 the separation and identification of the compounds responsible for giving foodstuffs their aromas and tastes;

5 proteins labelled with ^{15}N-glycine;

6 the identification of bile acids in the gall bladder of an Egyptian mummy;

7 characterising tar from the Mary Rose;

8 the separation and detection of urinary acids thought to be connected with sudden infant death syndrome (SIDS) or cot death; and

9 the highly specific and sensitive quantitative procedure for the determination of residues or contaminants in foodstuffs thereby facilitiating increased safety.

Mass spectrometry is a powerful technique but it does have some disadvantages; for example: optical isomers cannot be distinguished from their spectra (although *cis/trans* isomers sometimes can); and although only small samples are required for analysis the technique is destructive, and the sample cannot be recovered.

Applications also exist in environmental analysis. Although concentration methods are sometimes required to supply the spectrometer with sufficient sample to analyse, GC-MS is sensitive enough to detect pollutants such as polychlorinated biphenyls (PCBs), which are found in surface water in concentrations on the nanogram per dm^3 (ng l^{-1} or 10^{-9} g l^{-1}) scale. It is important that they are detected because they do not degrade in nature and build up in the food chain. Some seals, for example, have been found to have several per cent PCBs in their total fat content.

Monitoring for dioxin can also be done by using mass spectrometry. Dioxin can be found in the effluent stacks of incinerators, and is a most undesirable substance in the environment because of its toxicity and stability.

Mass spectrometers have been adapted for use on board spaceships, where less attention has to be paid to the inclusion of vacuum pumps because the atmosphere is already at low pressure. Indeed, in some cases the atmosphere has been sampled using an open source. The exhaust gases of the space rocket must be avoided, of course, as must the emissions from manned modules. The Viking space craft that landed on Mars was equipped with GC-MS, and both the Martian atmosphere and soil were analysed (it was found that the partial pressures of water, nitrogen and carbon dioxide are very low, and that the soil contained a lot of chlorine and sulphur).

Biomarkers in the petrochemical industry

Biomarkers are organic compounds whose carbon skeletons provide an unambiguous link with a known natural product, and these are particularly useful in the petrochemical industry.

When the marine organisms that formed crude oil died, their bodies were subjected to extreme heat and pressure, resulting in the formation of the oil. In the chemically reducing environment present many of the natural products – *eg* steroids and terpenoids – were saturated. These saturated products are now present in crude oil, and serve as biomarkers. It is thus possible to identify the organisms from which the oil was formed, enabling the terrain at the time of formation to be deduced. Once the biomarkers from a particular oilfield have been identified it is also possible to determine where an oil sample originated. Thus, if an oil spillage occurs at sea and nobody admits responsibility, the biomarkers could be used to identify its source. It may then be a simple matter to find out which tankers were carrying oil from that field at the time of the incident, and so identify the culprit.

Identifying the biomarkers is done by separating the components of the crude oil using gas chromatography (see page 123) and feeding them directly into the chamber of a mass spectrometer. Once the components have been identified, by their retention times on the chromatography column, and their fragmentation patterns, the biomarkers can be detected on subsequent occasions by selective ion monitoring, in which only the characteristic peaks are looked for.

Geological dating

Naturally occurring potassium contains 0.01167 per cent of the radioactive isotope ^{40}K. One of its daughter products is ^{40}Ar. If a rock sample contains potassium, and the argon formed by the decay of ^{40}K is unable to diffuse out of the crystal, it is possible to determine the age of the rock sample.

The argon content of a sample is determined by using a mass spectrometer to analyse the gas released when the sample is heated to melting in a vacuum. It is assumed that none of the argon produced *in situ* diffuses out of the crystal lattice of the rock, and that there is no loss or gain of potassium – *ie* it is assumed that the sample has remained a closed system since its formation.

A tracer of ^{38}Ar is used to determine the ^{40}Ar in a method known as isotope dilution, in which the unknown has a reference sample added against which it can be measured. The total remaining potassium can be measured by using several methods, including flame photometry, atomic absorption spectrometry and isotope dilution.

It is possible that some atmospheric argon is trapped in geological specimens, and this will register on the mass spectrum once it has been released from the crystal lattice. Atmospheric argon consists predominantly of three isotopes, ^{36}Ar, ^{38}Ar and ^{40}Ar. The latter isotope will clearly interfere with the results expected from the rock sample, but this can be compensated for from a knowledge of the natural abundances of the isotopes – the ratio of ^{40}Ar:^{36}Ar is 295.5:1. Thus the peak at m/z = 40, due to the argon in the rock, can be calculated.

By using the half life of ^{40}K and the radiogenic ^{40}Ar value, the amount of potassium that has decayed can be determined. From a knowledge of the ratio of naturally occurring isotopes it is then possible to determine how much ^{40}K was originally present. From this the proportion that has decayed can be calculated, and hence the age of the sample can be found. This is known as potassium argon dating (*Fig. 16*).

Recent trends

Improved spectrometer technology over the past decade has made detection of smaller and smaller concentrations of materials possible. Electron multipliers in the detectors of spectrometers can now achieve a gain of 10^7 electrons per ion, and instruments are available that can routinely measure elemental compositions down to 10^{-12}g g^{-1} or 10^{-12}g cm^{-3} (pg g^{-1} or pg cm^{-3}) of sample. By interfacing other techniques with mass spectrometry it is possible to analyse solid, liquid and gaseous samples – often automatically.

Mass spectrometers employing a magnetic field are now being complemented, and in some cases superseded, by quadrupole spectrometers. The ions produced

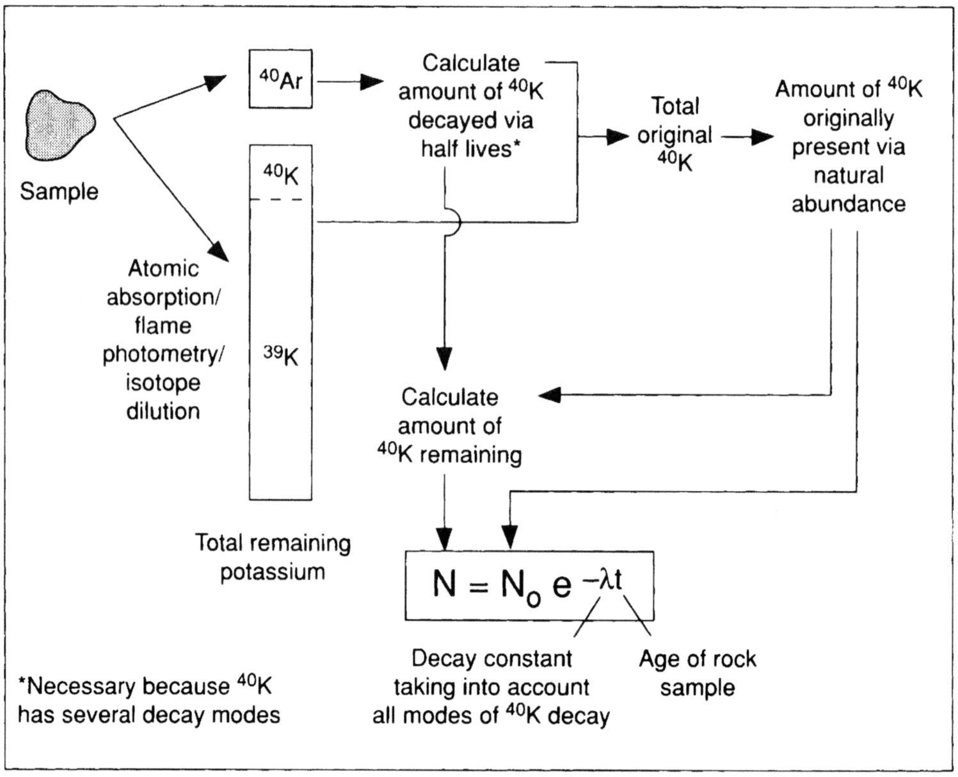

Figure 16 Schematic diagram of the potassium-argon dating method

travel down the gap between four rods through which a radiofrequency and a dc current are passed. The ion path in such spectrometers is helical. These have the advantage of being cheaper than magnetic sector instruments, but are not capable of such a high resolution. In theory, magnetic sector instruments will differentiate between ions of *m/z* values of 60 000 and 60 001, although in practice are likely to resolve 600.00 from 600.01 or 60.000 from 60.001. However, quadrupole spectrometers can only produce unit resolution (*ie* 60 and 61).

The size and cost of quadrupole instruments has decreased significantly in recent years, and benchtop models are now available. The lower potentials necessary in these instruments mean that they are much less prone to arcing inside the spectrometer. They are frequently interfaced to other techniques, such as gas chromatographs, where the presence of ions at particular nominal mass values is sought.

In the future this technique will probably be more accessible to non-specialists who need specific data quickly, perhaps without the detailed interpretation a skilled spectrometrist could give.

Unilever

THE ROYAL SOCIETY OF CHEMISTRY

Exercises

Below are the mass spectra of some organic compounds. Try to determine their structures. Interpretations are given.

Exercise 1

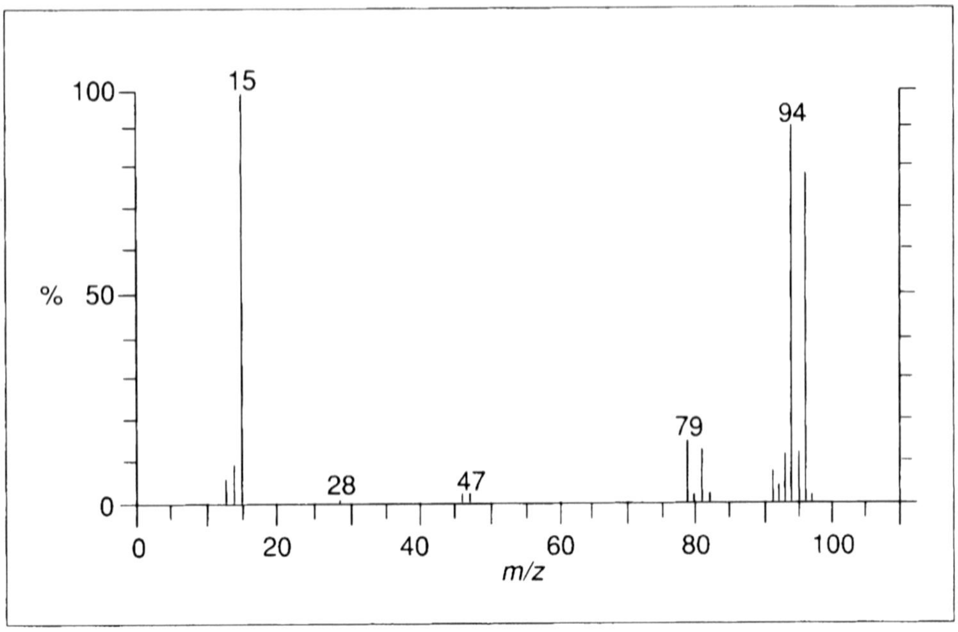

Figure 17

The peaks of similar abundance, at *m/z* =94 and 96, are likely to be isotope peaks. Their relative proportions and their mass difference of two would suggest Br. This is supported by the peaks at *m/z* = 79 and 81. Subtracting 79 from 94 or 81 from 96 leaves 15, which corresponds to CH_3. The peaks at *m/z* = 12-14 correspond to the loss of successive protons from this group, and the peaks at *m/z* = 91, 93 and 95 correspond to the loss of protons from the molecular ion. The peak at *m/z* = 94 is larger than the peak at 96 because it comprises $CH^{81}Br^+$ and $CH_3{}^{79}Br^+$, as well as the ^{79}Br isotope being very slightly more abundant. Therefore, we can conclude that the compound is bromomethane, CH_3Br.

The peak at *m/z* = 97 is due to the $^{13}CH_3{}^{81}Br^+$ ion.

THE ROYAL SOCIETY OF CHEMISTRY

Unilever

Exercise 2

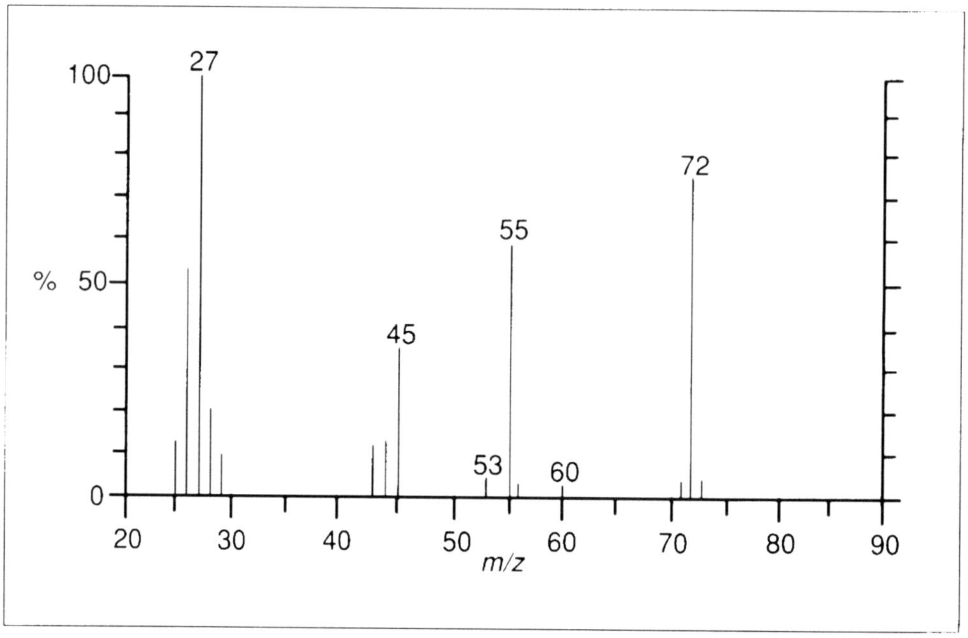

Figure 18

No isotopic cluster due to a halogen is present, and the peak at m/z = 73, *ie* $(M+1)^+$, suggests that 2.7 per cent ÷ 1.1 per cent = 2.5 carbon atoms are present. If three carbon atoms are present, 36 of the 72 mass units remain and either two oxygens, two nitrogens or one of each is likely to be present. One nitrogen is unlikely because the relative formula mass is even. Formulae which fit are $C_3H_8N_2$ and $C_3H_4O_2$. The absence of a peak at M-16 eliminates a primary amine (NH_2 group). Cyclic and secondary amine structures are also unlikely because the peaks in the spectrum are not abundant enough to suggest either CH_3–NH masses or losses. This leaves us with two oxygen atoms to try to account for. The peak at m/z = 55 is M-17, which would correspond to the loss of an OH group. The peak at m/z = 27 corresponds to C_2H_3, CH_2=CH, and its abundance suggests its relative stability. This leaves us with C_2H_3, C, OH, and O to piece together. The possibility of a carboxylic acid group is confirmed by the peak at m/z = 45 ($COOH^+$). This suggests that the compound is propenoic acid (acrylic acid), CH_2=CHCOOH.

Unilever

THE ROYAL SOCIETY OF CHEMISTRY

Exercise 3

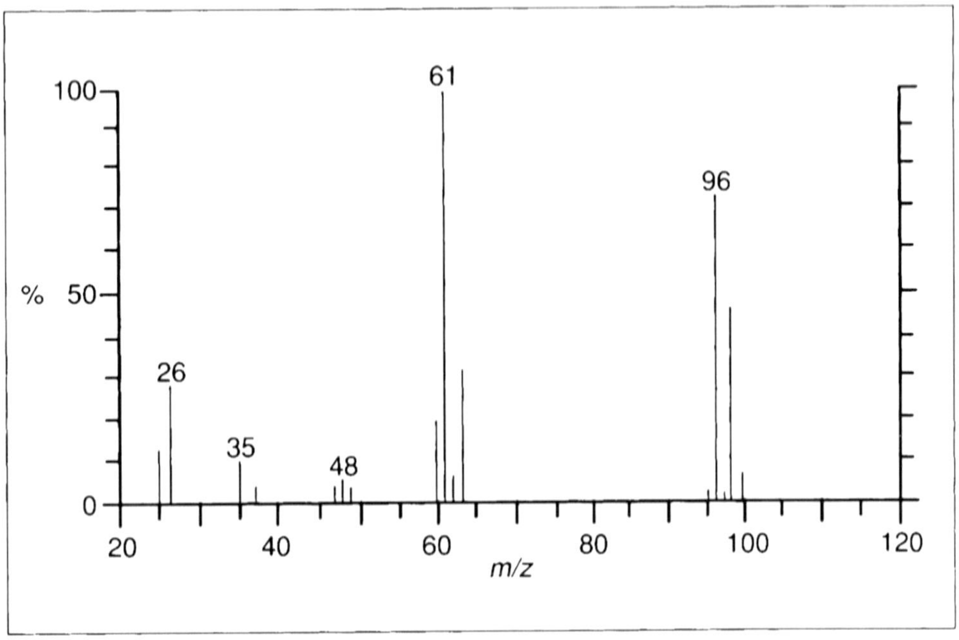

Figure 19

The abundance of the peaks at *m/z* = 96 and 98 lead us to consider whether the peak at 100 is truly the molecular ion peak, rather than an isotopic peak. A glance at *m/z* = 35 and *m/z* = 37 shows the presence of peaks in the ratio 3:1, strongly suggesting chlorine. The abundance of the peaks at *m/z* = 96, 98 and 100 in the approximate ratio 9:6:1 suggests that two chlorine atoms are present. The peaks at *m/z* =61 and *m/z* = 63 (formed by the loss of one chlorine atom) have heights in the ratio of 3:1 – typical of the pattern expected for a single chlorine atom remaining in the molecule. This accounts for all but 26 mass units. The group C_2H_2 fits this, so the compound must be dichloroethene, $C_2H_2Cl_2$, but without further information it is impossible to decide which isomer it is (actually it is *trans*-1,2-dichloroethene).

THE ROYAL SOCIETY OF CHEMISTRY

Unilever

Exercise 4

The two compounds are known to be 1-aminobutane, $CH_3CH_2CH_2CH_2NH_2$, (*A*) and *N,N*-dimethyl aminoethane, $CH_3CH_2N(CH_3)_2$,(*B*). Try to decide which spectrum corresponds to which isomer, and explain why the spectra appear as they are.

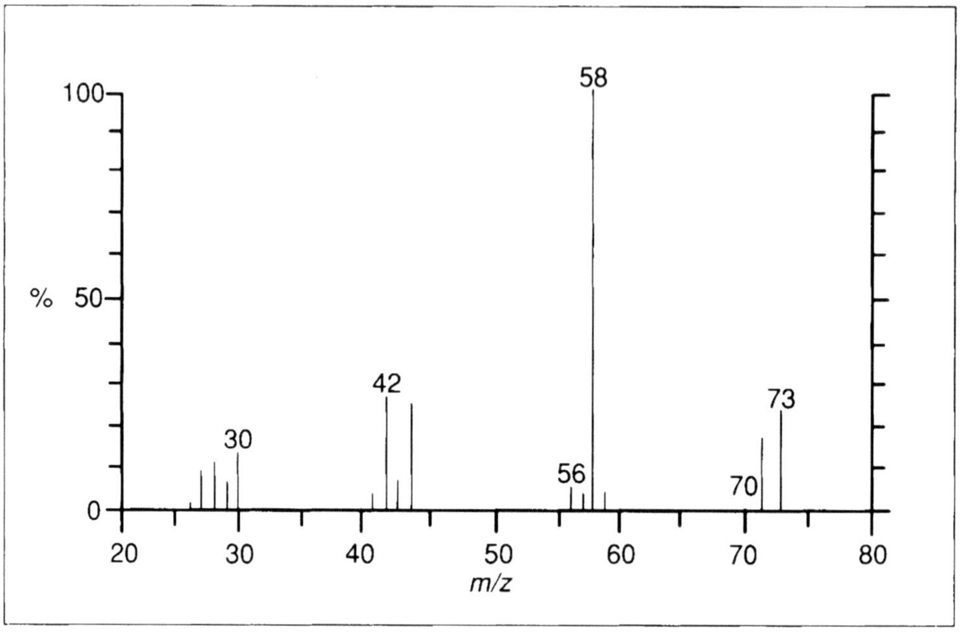

Figure 20

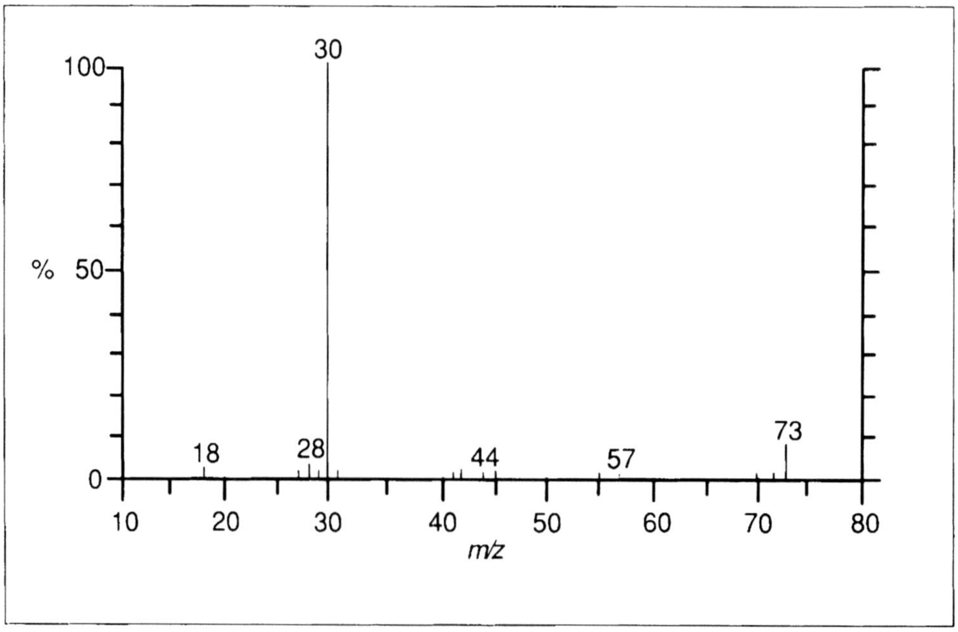

Figure 21

Although it might appear difficult to assign the spectra, it is not. What must be considered is the relative stabilities of the ions that might be formed. Basically, a positive charge on a highly substituted ion will be more stable than an ion with a straight chain because the effect of the positive charge can be reduced by the inductive effect of the side chain groups. *Figure 20* shows a major peak at (M-15) – *ie* at *m/z* = 58. This corresponds to loss of a CH_3 group because loss of an NH group is highly unlikely in the case of (A) and impossible in the case of (B). The remaining ions would be:

$$[CH_2CH_2CH_2NH_2]^+ \text{ and } [CH_2N(CH_3)_2]^+ \text{ or } [CH_3CH_2NCH_3]^+$$

(*1*) (*2*) (*3*)

Ion (*1*) would be less stable than ions (*2*) and (*3*), although it is not possible to say which of these two ions gives the peak on this information alone. A tentative guess at this stage would be that *Fig. 20* is the mass spectrum of the tertiary amine.

Figure 21 shows only one significant peak, at *m/z* = 30. The only sensible ions that can give this peak are $[CH_2NH_2]^+$, and $[CH_3NH]^+$. In neither case can (*B*) give these ions because it has no hydrogen atoms bonded to the nitrogen atom. Thus, the peak at *m/z* = 30 must be due to the $[CH_2NH_2]^+$ ion because there is no methyl group bonded to the nitrogen in (*A*) so *Fig. 21* is the mass spectrum of (*A*).

An experienced mass spectrometrist would be able to interpret the spectra directly from the knowledge that amines are susceptible to fragmentation one bond away from the C–N bond (the bond to the nitrogen) – *ie* (*B*) would fragment as follows:

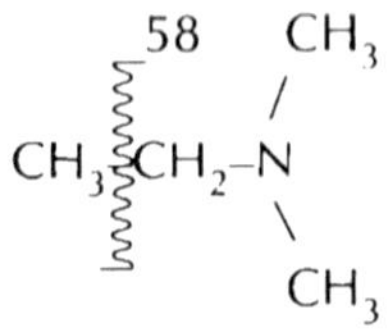

The peak at *m/z* = 58 in *Fig. 20* would then correspond to ion (*2*) above. Similarly (*A*) would fragment here:

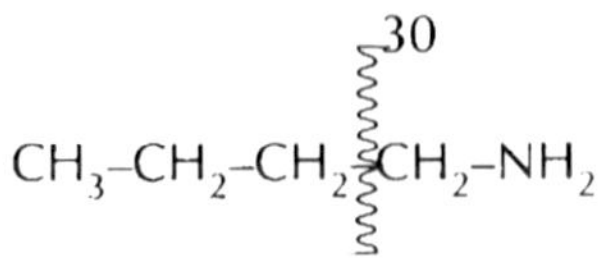

giving the peak at *m/z* = 30.

THE ROYAL SOCIETY OF CHEMISTRY

Unilever

2. Nuclear magnetic resonance spectroscopy

Nuclear magnetic resonance (NMR) spectroscopy gives information on the environment in which the nuclei of atoms are found in molecules and compounds. It is possible to derive an enormous amount of information from a single spectrum, and in many cases this will facilitate the determination of the structure of a molecule. Indeed, the NMR spectrum of a compound is frequently the first spectral information to be consulted.

The theory behind the technique is rather more complex than for mass spectrometry and infrared spectroscopy, but the interpretation of the spectra is probably no more difficult, if not easier, once a little familiarity has been gained.

The theory

A nucleus possessing a spin in the presence of an external magnetic field can align itself either with the external field (+) or against it (–) (*Fig. 1*).

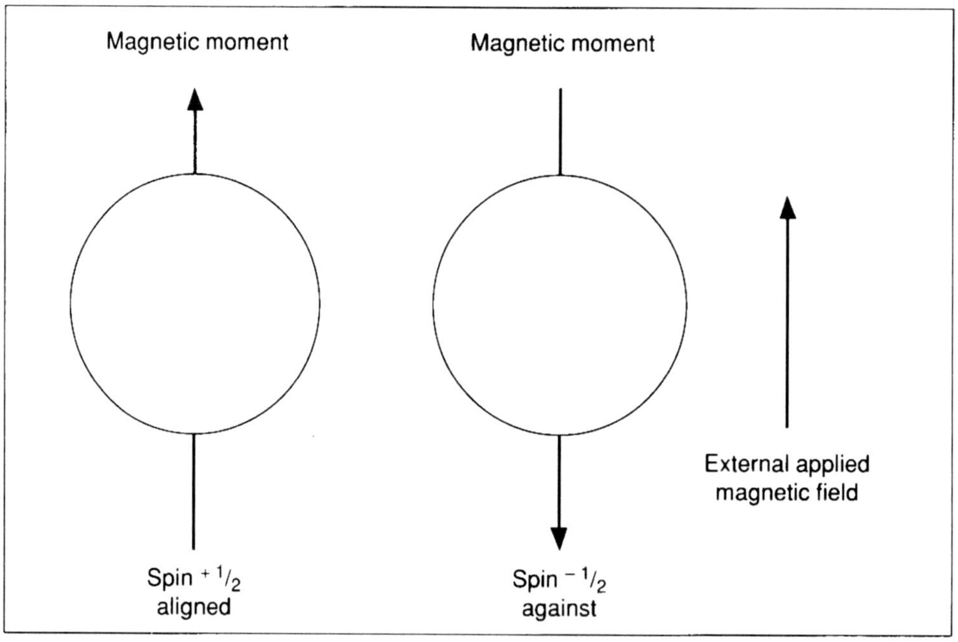

Figure 1 Nuclear magnets aligned with and against an external magnetic field

It is found that many nuclei spin about an axis. Because the nuclei are positively charged, this spin is associated with a circulation of electric charge. Circulating charges give rise to magnetic fields, so nuclei with spin also have a magnetic moment, rather like the magnet of a compass needle. When put in an external magnetic field the nuclei tend to turn (like compass needles in the earth's field) to a preferred orientation. Other, less favoured, orientations have higher energy. The nuclei obey quantum laws and for some nuclei said to have a spin quantum number of a $^{1}/_{2}$ only two orientations can be adopted. They are the most favoured and least favoured orientation (*Fig. 2*).

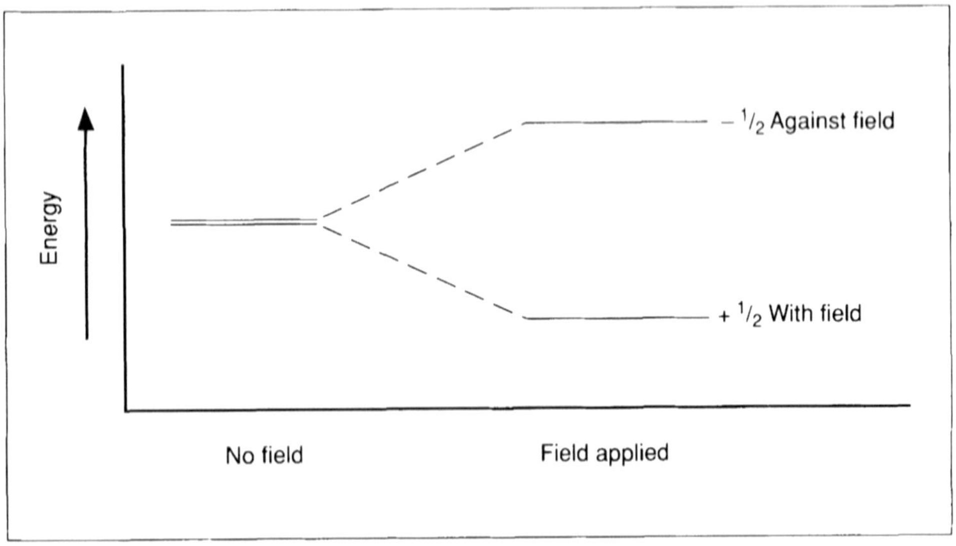

Figure 2 Splitting of energy levels in an external magnetic field

Exchange of energy between the nuclear spin and the thermal motion of the molecules containing them distributes the spins between the two energy levels in such a way that there are more nuclei in the lower than upper level.

Transitions between the two energy levels can occur if radiation of the correct frequency is absorbed.

The spin up (lower energy) state will have a higher population given by the ratio

$$\frac{N_{upper}}{N_{lower}} = e^{-\frac{\Delta E}{kT}}$$

where ΔE = difference between energy levels in joules
k = Boltzmann constant
T = temperature in kelvin
N = the number of nuclei at each energy level

Because ΔE is extremely small, the difference in populations will also be extremely small. For hydrogen nuclei it is approximately one in 10^5 for a ΔE of 6×10^{-24} J in a 2.35 tesla (T) external field. Transition from the lower to the upper state is possible by absorption of radiation of the correct frequency (*Fig. 3*).

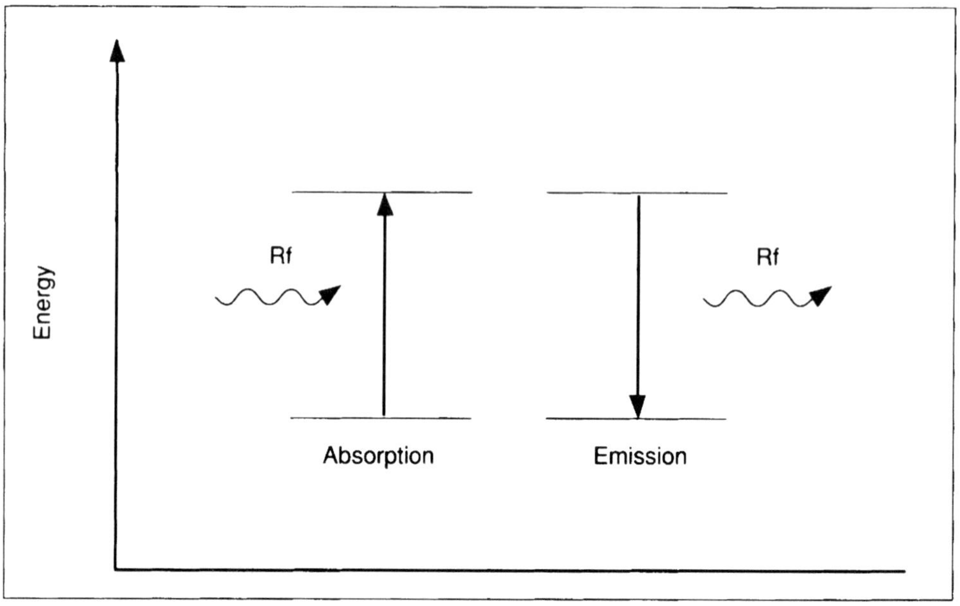

Figure 3 Absorption and emission of radiation

The radiation absorbed is in the radiofrequency region of the electromagnetic spectrum. For a 2.35 T magnetic field the radiation is typically in the range 5–100 MHz, and its precise frequency can be calculated using the formula

$$\nu = \frac{\gamma B_o}{2\pi}$$

where ν = frequency of the radiation absorbed in hertz
γ = a constant of proportionality called the magnetogyric ratio (the ratio of the magnetic moment to the angular or gyric moment) which differs according to the type of nucleus considered, and which is effectively a measure of the magnetic strength of the nucleus (units: radian T^{-1} s^{-1})
B_o = the strength of the applied magnetic field in tesla (1 T = 10^4 gauss)

The energy difference between the spin states is very small. Nuclei with a spin can be made to resonate between the spin states if:

1 a large enough external magnetic field is applied to ensure a significant difference between the energy states; and

2 radiation of the correct frequency is applied.

So far we have assumed that only two spin states are possible. This is true for all nuclei with spins of $^1/_2$ – *ie* nuclei whose spin states can be $+^1/_2$ or $-^1/_2$; these include ^1H, ^{13}C, ^{19}F, and ^{31}P. However, other spin states do exist – *eg* ^6Li (which can have spin states +1, 0 or -1) and ^{23}Na (which can have spin states $+^3/_2$, $+^1/_2$, $-^1/_2$, or $-^3/_2$) have further energy levels available between which transitions can occur. However, ^1H-NMR is the most widely used.

The symbol for the maximum nuclear spin is *I*, and for *I*=1 three energy levels are available – *eg* ^{14}N – (*Fig. 4*), and for nuclei with *I*=$^3/_2$ four levels are available – *eg*

^{33}S – (*Fig. 5*). Generally for spin *I*, (2*I*+1) energy levels are available. Nuclei with no spin (*I*=0) – *eg* ^{4}He, ^{12}C, and ^{16}O – are inactive because only one energy level is available.

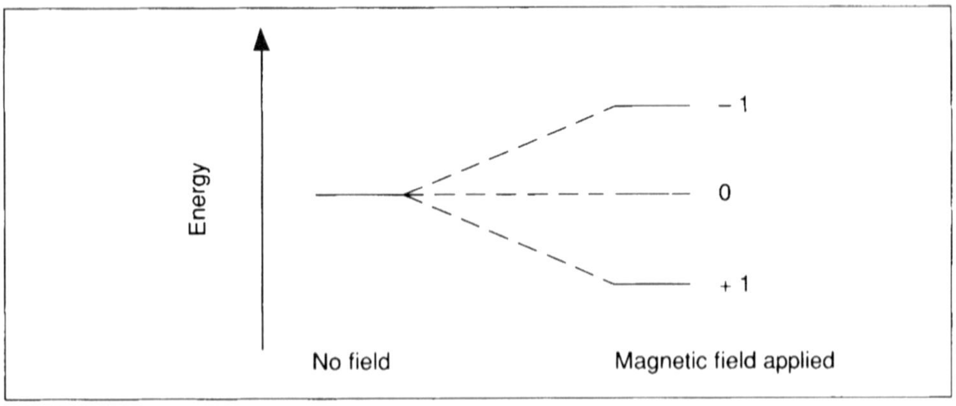

Figure 4 Energy levels available for nuclei with spin = 1

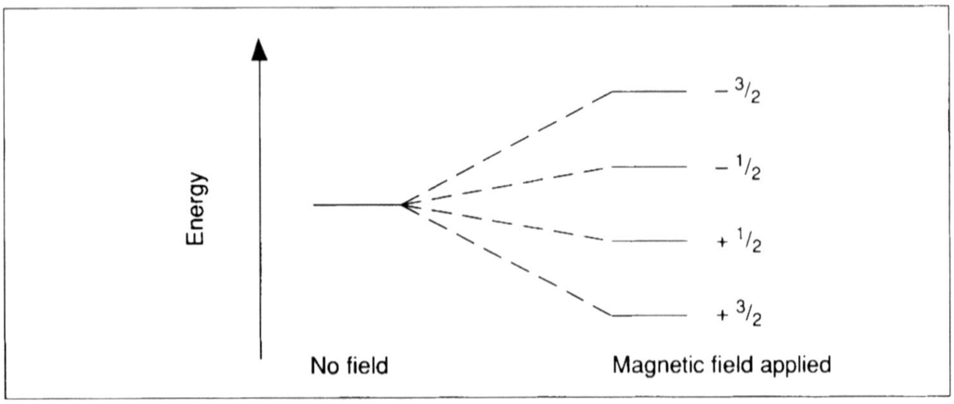

Figure 5 Energy levels available for nuclei with spin = 3/2

There are no authoritative rules that enable the precise spin of all nuclei to be predicted, but some generalisations can be made. These relate the atomic and mass numbers with the observed nuclear spins.

Atomic number	**Mass number**	**Nuclear spin (*I*)**
Even or odd	Odd	1/2, 3/2, 5/2 ...
Even	Even	0
Odd	Even	1, 2, 3 ...

When the frequency of the radiation supplied corresponds to the energy difference between levels the population of the higher energy state increases as radiation is absorbed. The equilibrium population distribution is re-established by spin-lattice relaxation processes whereby the energy previously absorbed is shared with either the surroundings (spin-lattice relaxation) or with other nuclei (spin-spin relaxation). Spin-lattice relaxation processes are often quicker in liquid samples than solid samples because of the greater molecular mobility in the liquid phase. Most solutions have relaxation times in the range 10^2–10^{-4} s, the majority of ^1H and ^{13}C nuclei taking a fraction of a second, whereas solid samples can take several minutes. Relaxation processes can be speeded up by the presence of a paramagnetic material – *eg* molecular oxygen or chromium(III) 2,4-pentandionate (acetylacetonate).

In the analysis of an organic sample it is the ^1H- or proton-NMR spectrum that is usually most useful because hydrogen atoms are present in such large numbers, bonded in a variety of environments. However, from the theory presented so far one would expect that all hydrogen atoms would resonate at the same frequency – *ie* at 100 MHz in a 2.35 T field.

When a molecule is placed in a magnetic field, the electrons surrounding the nuclei behave like perfectly conducting shells, and weak electric currents are induced in them. The currents flow in such a way as to produce a magnetic field which opposes the applied field; the nuclei at the centre, therefore, experience a fractionally smaller field than the applied external field (*Fig. 6*).

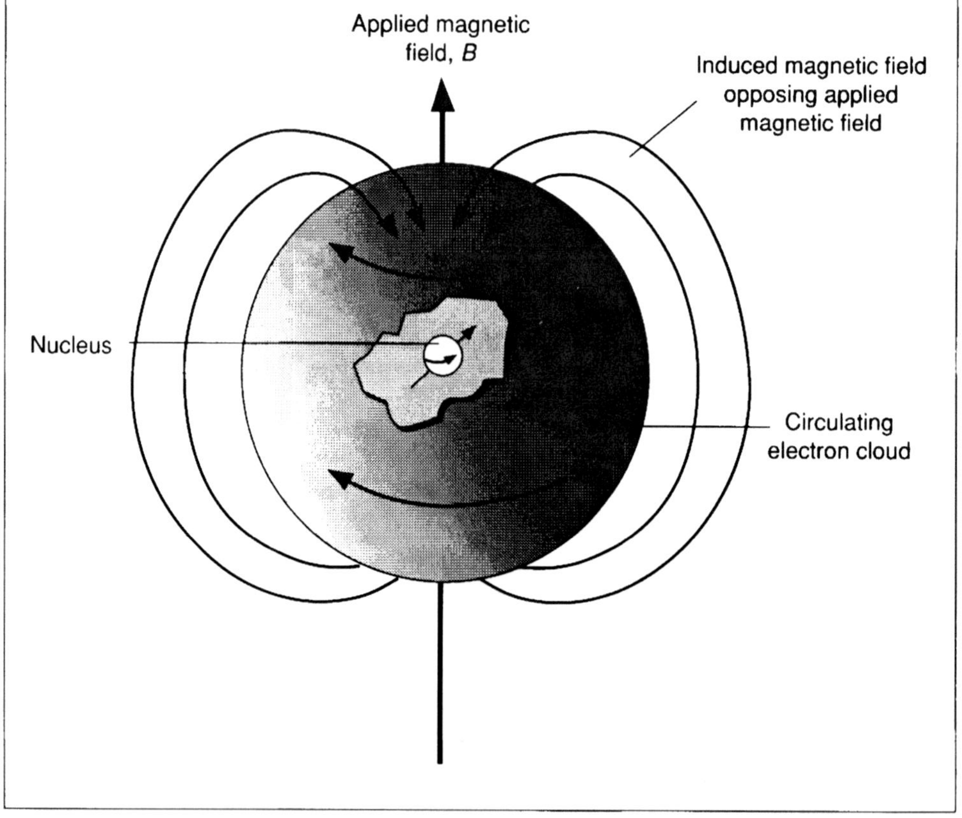

Figure 6 Shielding of an isolated nucleus by circulation of the surrounding electron cloud

Because the electron distribution around chemically different hydrogen atoms in a molecule are not the same, the induced fields vary slightly. Conseqently, the nuclei experience different magnetic fields in the same external field. The effect is, however, very small; for hydrogen atoms it is only a few parts per million. Fortunately the line widths are very small and it is still possible to measure these so-called 'chemical shifts'. The different environments of the protons in ethanol are shown by its low resolution spectrum (*Fig. 7*). Notice that the signals occur at fields differing by a few parts per million. It is for this reason that NMR spectrometers require magnetic fields that are stable and homogeneous to a few parts in 10^8.

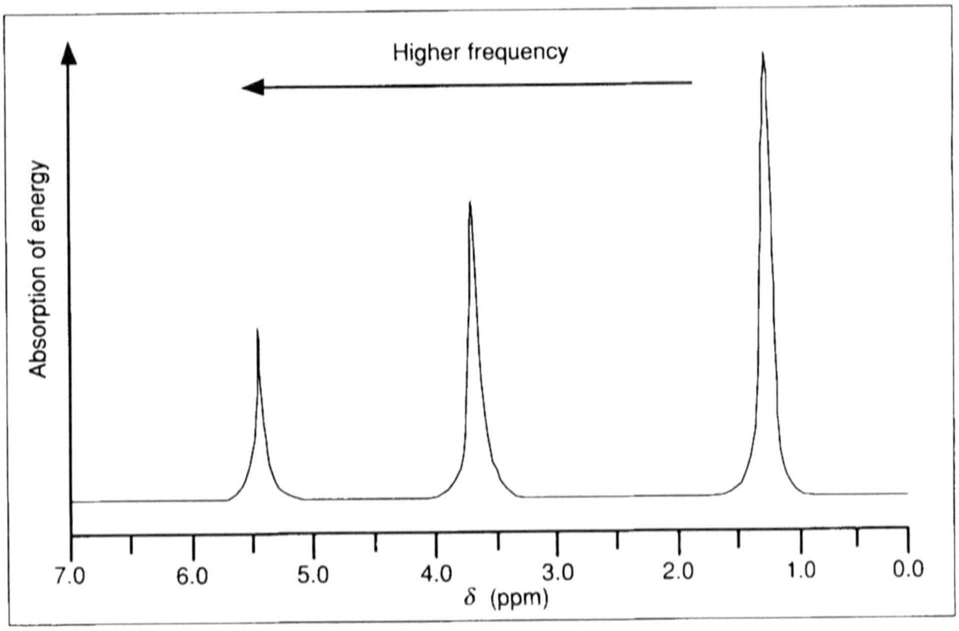

Figure 7 Low resolution spectrum of ethanol (an explanation of the scale used is given on page 36).

There are two variables that can be altered when recording an NMR spectrum:

1 the magnetic field can be kept constant and the range of radiofrequencies scanned, or

2 the radiofrequency can be kept constant and the magnetic field scanned.

A few simple instruments scan the magnetic field. A radiofrequency detector is set at right angles to the radiofrequency transmitter inducing resonance, and a recorder charts the absorption of energy as a function of the applied field or frequency (*Fig.8*). However, more advanced Fourier Transform NMR machines are the most common type of spectrometer in use today (see page 47).

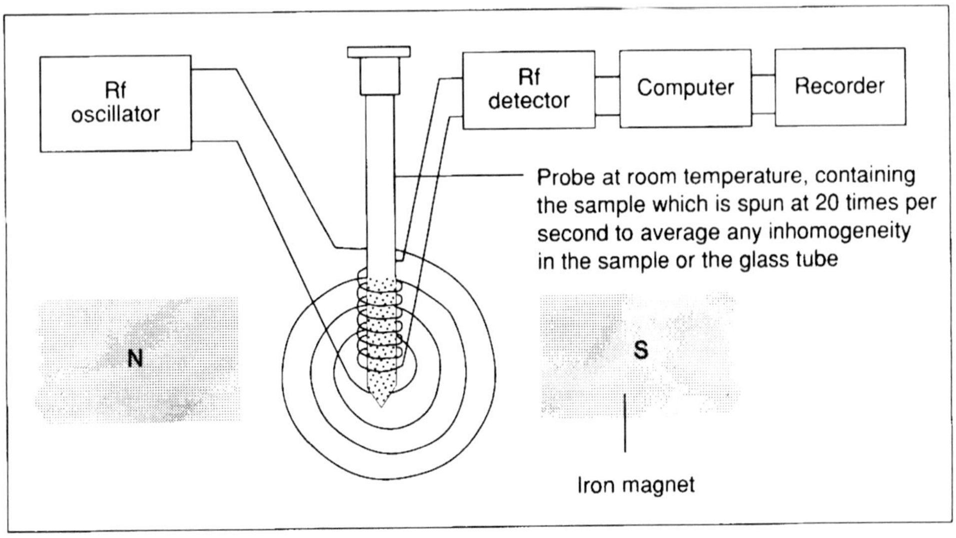

Figure 8 Diagram of a simple NMR spectrometer

The NMR spectrum of ethanol (*Fig. 7*) shows that the absorption of energy is displayed against neither magnetic field nor frequency. Instead, it is on a scale (with no units) that increases from right to left. Peaks on this scale, δ, have the same value no matter what the magnetic field or frequency range of the instrument used because the chemical shift is induced by the applied field and is proportional to it. Values on the scale can be derived from measurements in either hertz (frequency) or tesla (magnetic field), and are always measured relative to a standard that gives a reference peak at one end of the scale. The reference material that is usually used is tetramethylsilane, TMS ($Si(CH_3)_4$).

δ can be calculated using:

$$\delta = \frac{B_{TMS} - B_{sample}}{B_{TMS}} \times 10^6 \text{ ppm}$$

or

$$\delta = \frac{\nu_{TMS} - \nu_{sample}}{\nu_{TMS}} \times 10^6 \text{ ppm}$$

where B = magnetic field strength at resonance
ν = radiofrequency at resonance

Resonances are always expressed in terms of chemical shift, measured in parts per million (ppm), so that results are reproducible no matter what machine the spectra are run on, and no matter what applied magnetic field is used (machines are available that run at different magnetic fields).

By definition the δ value of TMS is zero. Most organic proton resonances are then on a scale of 0-10 on the lower field/higher frequency side of zero. The relationship between field, frequency and shielding is shown in *Fig. 9*.

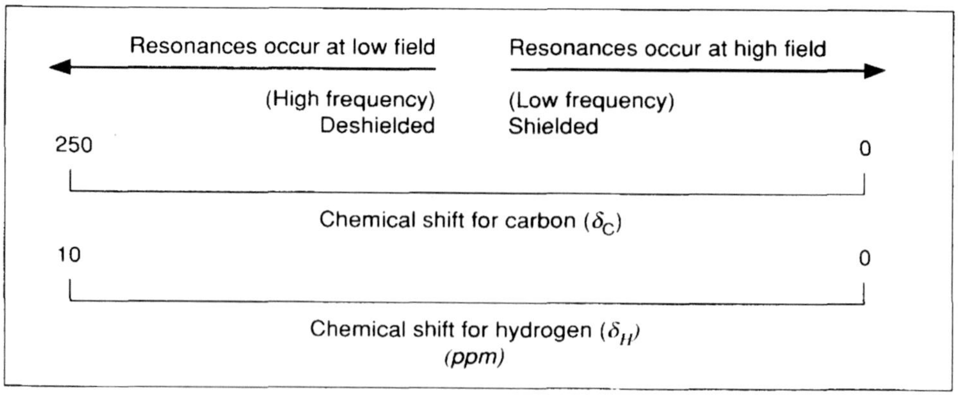

Figure 9 The relationship between field, frequency and nuclear shielding for proton and ^{13}C-NMR

Tetramethylsilane has a number of useful features:

1 it is non-toxic and inert;

2 it gives a signal that resonates well away from almost all other organic hydrogen resonances because the protons are so well shielded, and do not interfere with the spectrum;

3 because there are 12 protons in the same environment they all resonate at the same frequency so the single peak is intense and easily recognised; and

4 the boiling point of TMS is fairly low so it can be boiled off if the sample is required for anything else.

The high resolving power of modern spectrometers enables spectra to be produced which display far more information than that shown in *Fig. 7*. The high resolution spectrum of ethanol is shown in *Fig. 10*. Each set of peaks is centred on a δ value known as the chemical shift. The δ values vary according to the chemical environment and the δ value gives an indication of the degree of shielding experienced by the proton or protons. Tables of chemical shifts are available for protons in different environments (Tables 1 and 2).

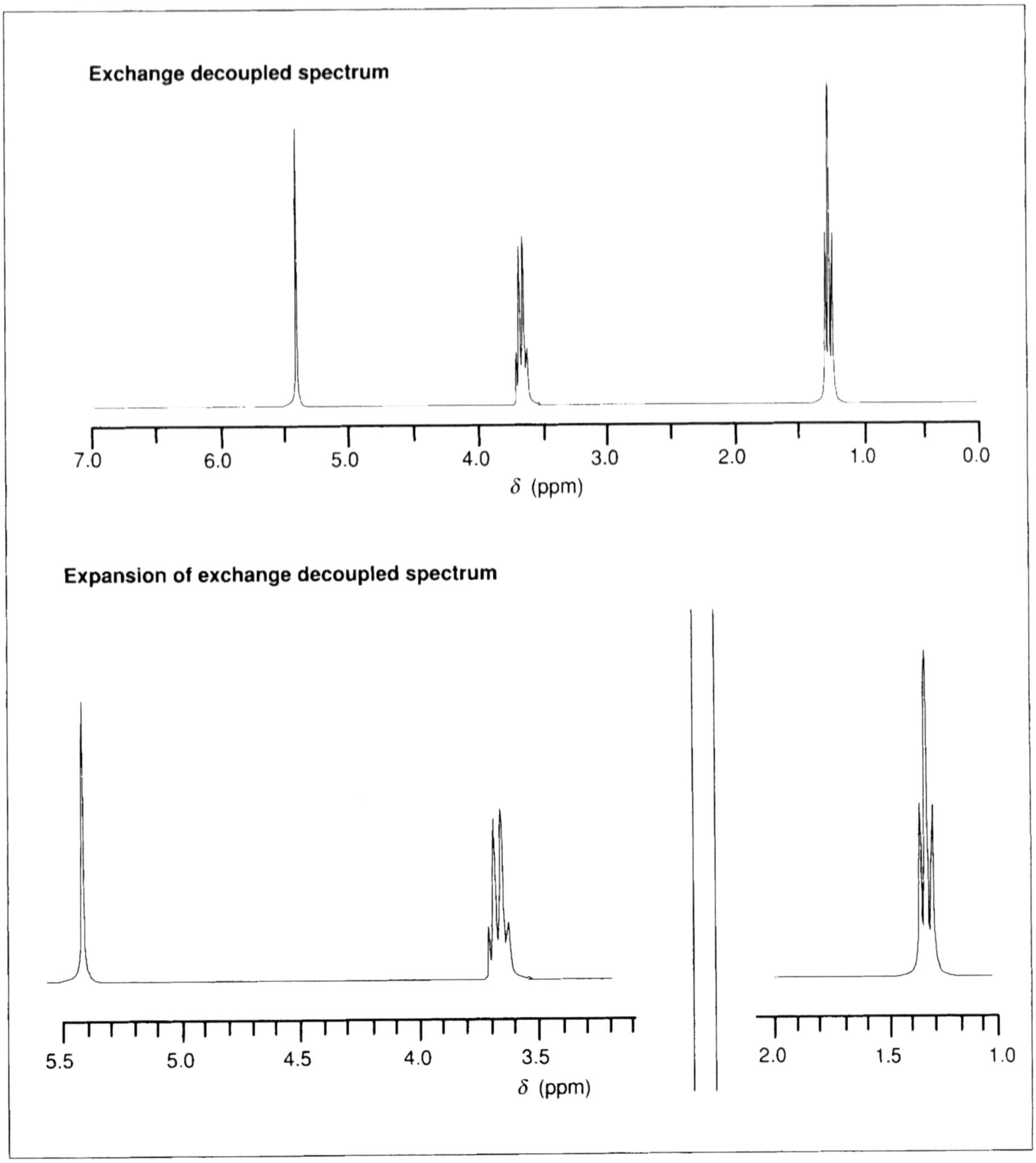

Figure 10 High resolution proton NMR spectrum of ethanol

Chemical shift is dependent on the electronegativity of the atoms in molecules. Multiple bonding can also cause extra shielding or deshielding compared with a single bond and this causes a change in chemical shift. For example, an ethynic bond C≡C will shield adjacent protons because the circulation of the triple bond electrons reduces the apparent magnetic field. The circulation of the delocalised (π) electrons in a benzene ring deshields the protons bonded to the ring because the induced magnetic field (due to the circulation of the electrons) reinforces the applied magnetic field in the region occupied by the protons (*Fig. 11*).

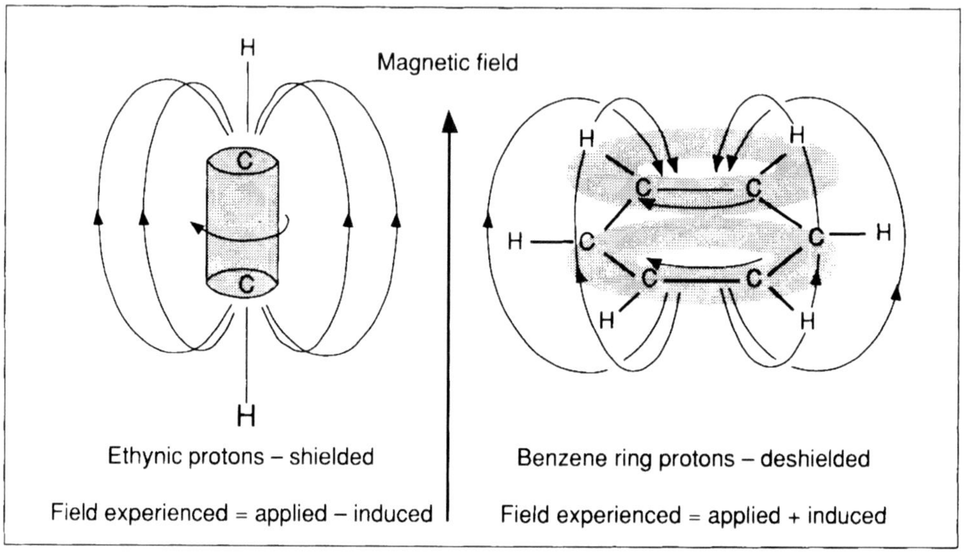

Figure 11 Shielding and deshielding of protons in ethyne and benzene

Table 1 Proton chemical shifts in aliphatic environments

These are typical values, and can vary slightly in different solvents and if induced magnetic fields within a molecule are stronger in one direction than the other (anisotropy).

Methyl protons	δ	Methylene protons	δ	Methine protons	δ
CH_3–R	0.7–1.6	RCH_2–R	1.4	CH–R	1.5
CH_3–Ar	2.3	RCH_2–Ar	2.3–2.7	CH–Ar	3.0
CH_3–C≡N	2.0	RCH_2–C≡N	2.3	CH–C≡N	2.7
CH_3–C(=O)–R	2.2				
CH_3–C(=O)–O–R	2.0	RCH_2–C(=O)–R	2.4	CH–C(=O)–R	2.7
CH_3–C(=O)–Ar	2.6			CH–C(=O)–Ar	3.3
CH_3–C(=O)–O–Ar	2.4	RCH_2–C(=O)–Ar	2.9	CH–N–C(=O)–R	4.0
		$ArCH_2$C(=O)R	3.7		
CH_3–N–R	2.3	RCH_2–N	2.5	CH–OH	3.9
CH_3–N–Ar	3.0	RCH_2–N–C(=O)–R	3.2	CH–O–R	3.7
CH_3–N–C(=O)–R	2.9	RCH_2–Cl	3.6	CH–O–Ar	4.5
CH_3–O–R	3.3	RCH_2–Br	3.5	CH–O–C(=O)–R	4.8
CH_3–O–Ar	3.8	RCH_2–I	3.2	CH–Cl	4.2
CH_3–O–C(=O)–R	3.7	RCH_2–OH	3.6	CH–Br	4.3
CH_3–O–C(=O)–Ar	4.0–4.2	RCH_2–O–R	3.4	CH–I	4.3
		RCH_2–O–Ar	4.3		
		RCH_2–O–C(=O)–R	4.1		
		$ArCH_2$–O–C(=O)–R	4.9		

Table 2 Calculated proton chemical shifts in aromatic environments

Aromatic proton shifts can be calculated using the equation

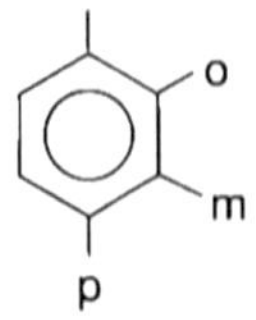

$$\delta_H = 7.27 + \sum_i z_i \qquad i = o, m, p$$

eg the protons in bromobenzene would have their peaks at:

$$\delta_{ortho} = 7.27 + 0.18 = 7.35$$
$$\delta_{meta} = 7.27 - 0.08 = 7.19$$
$$\delta_{para} = 7.27 - 0.04 = 7.23$$

For more complicated systems there will be as many adjustments to the chemical shift of 7.27 for each proton as there are substituents on the ring – *eg* a proton might be *ortho* to a methyl group (- 0.20) and *meta* to a nitro group (+ 0.26).

	Zo	*Zm*	*Zp*
CH_3–	–0.20	–0.12	–0.22
CH_3CH_2–	–0.14	–0.06	–0.17
$(CH_3)_2CH$–	–0.13	–0.08	–0.18
$CH_3C(=O)$–	0.62	0.14	0.21
$H_2NC(=O)$	0.61	0.10	0.17
HOC(=O)	0.85	0.18	0.27
$CH_3OC(=O)$	0.71	0.1	0.21
H_2N–	–0.75	–0.25	–0.65
$CH_3C(=O)NH$–	0.12	–0.07	–0.28
O_2N–	0.95	0.26	0.38
N≡C–	0.36	0.18	0.28
HO–	–0.56	–0.12	–0.45
CH_3O–	–0.48	–0.09	–0.44
$CH_3C(=O)O$–	–0.25	0.03	–0.13
Br–	0.18	–0.08	–0.04
Cl–	0.03	–0.02	–0.09
I–	0.39	–0.21	0.00
–S–	–0.08	–0.10	–0.22
–S(=O)–	0.3	0.1	0.2
$–SO_2$–	0.76	0.35	0.45

Spin-spin coupling

In a molecule the nucleus of an atom, A, can induce in the electrons of the chemical bonds attached to it a very weak magnetic moment. This moment affects the magnetic field at a neighbouring atom, B's, nucleus. It would increase the field when the atom A's nucleus is pointing one way and decrease the field when it points the other way. This interaction is known as coupling and this causes peaks to be split into a number of lines. Protons can usually interact with other protons that are up to three bonds away, *ie*

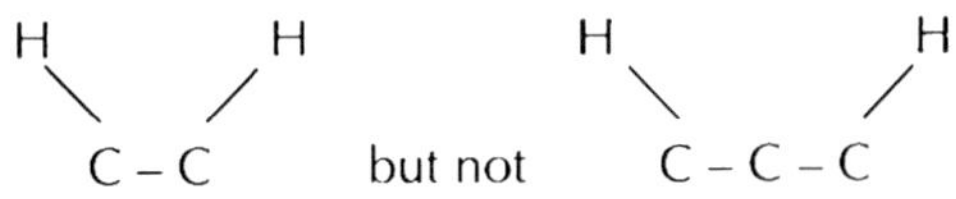

Protons with the same chemical shift do not show coupling with each other – *ie* coupling is not seen between protons in the same chemical environment (*nb* coupling over greater distances does exist but is much smaller.)

Thus in the spectrum of iodoethane, CH_3CH_2I, the CH_3 protons will interact with each of the CH_2 protons as follows. There are three energy states available to the two protons, depending on whether their spins are up or down, the combinations being:

1 both aligned with the field;

2 one aligned with the field and one against it (2 combinations); and

3 both aligned against the field.

ie

↑↑ ↑↓ ↓↓
↑↓

The methyl protons can therefore interact with each of these three energy states, at slightly different frequencies. Because there are two combinations having one nucleus spin up and one spin down this peak has twice the intensity (area under the peak) of the other two. This triplet of peaks is centred on a chemical shift of 1.8. (This value is downfield [larger chemical shift] of the range for methyl groups bonded to other alkyl groups because of the deshielding effect of iodine.)

Similarly the two CH_2 protons can couple (interact) with the three methyl protons in four ways:

↑↑↑ ↑↑↓ ↑↓↓ ↓↓↓
↑↓↑ ↓↑↓
↓↑↑ ↓↓↑

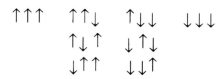

This gives a quartet of peaks with intensities in the ratio 1:3:3:1 according to the combination of spins. It is centred on the chemical shift expected of the CH_2 protons bonded to an alkyl group and an iodine atom (δ = 3.2). The spectrum of iodoethane is shown in *Fig. 12*.

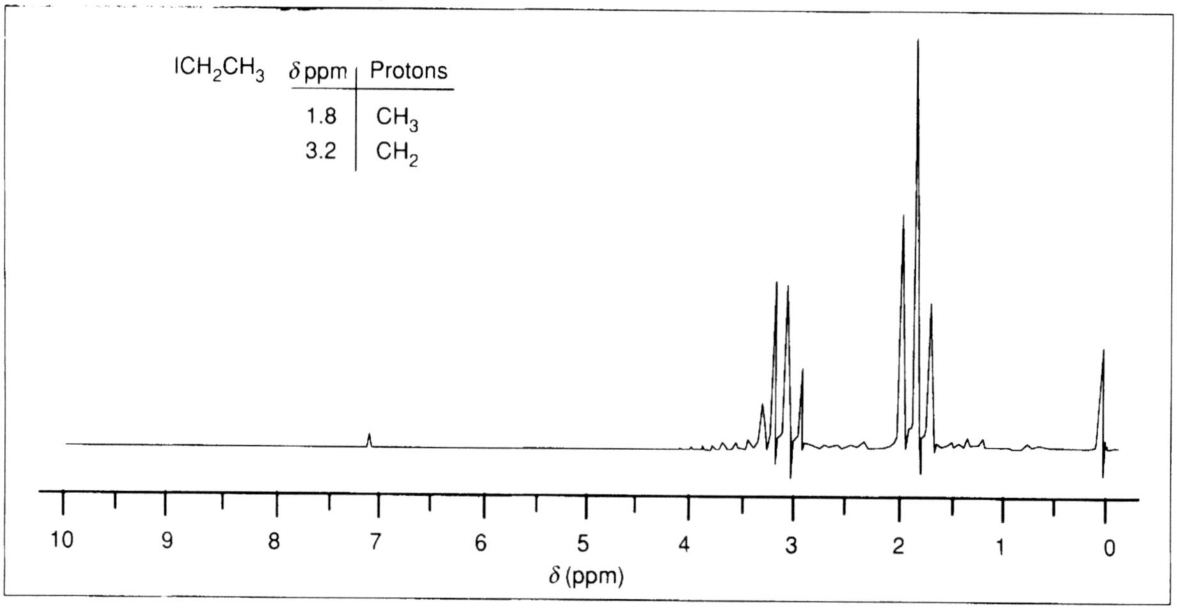

Figure 12 High resolution proton NMR spectrum of iodoethane

In general, if there are *n* protons three bonds away from the resonating group, the absorption will be split into a multiplet of *n*+1 lines. Their expected intensities can be predicted using Pascal's triangle:

Coupling with 0 protons:	1
Coupling with 1 proton:	1 1
Coupling with 2 protons:	1 2 1
Coupling with 3 protons:	1 3 3 1
Coupling with 4 protons:	1 4 6 4 1

Within a multiplet, the spectral lines are separated by differences which are constant for particular types of interaction between resonating nuclei (spin-spin coupling), and tables of these constants are available – *eg* Table 3. These coupling constants have the symbol *J*, and are always described in terms of frequency, whether the spectrum is obtained by scanning the frequency or the magnetic field.

Such *J* values are constant if expressed as a frequency – *ie* they are independent of magnetic field.

Table 3 Typical spin-spin coupling constants, *J* (Hz)

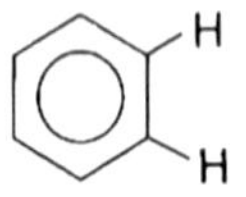	*J(ortho)* = 6–9 (1,2)
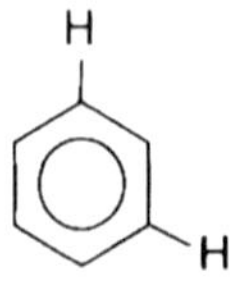	*J(meta)* = 1–3 (1,3)
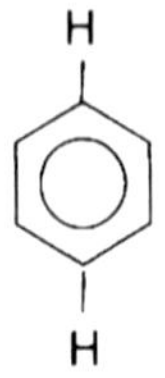	*J(para)* = 0–1 (1,4)
CH_2–CH_2	*J* = 5–8
CH_3–CH_2	*J* = 6–8
CH_3–CH	*J* = 5–7
CH–CH	*J* = 0–8

If two groups of spectral lines have similar chemical shifts then the relative line heights may not be in the predicted ratio. The intensities of the line on the sides of the peaks that are closest together can be greater than predicted. This is known as 'roofing' – *eg Fig. 13* – and this can cause confusion of an otherwise straightforward spectrum.

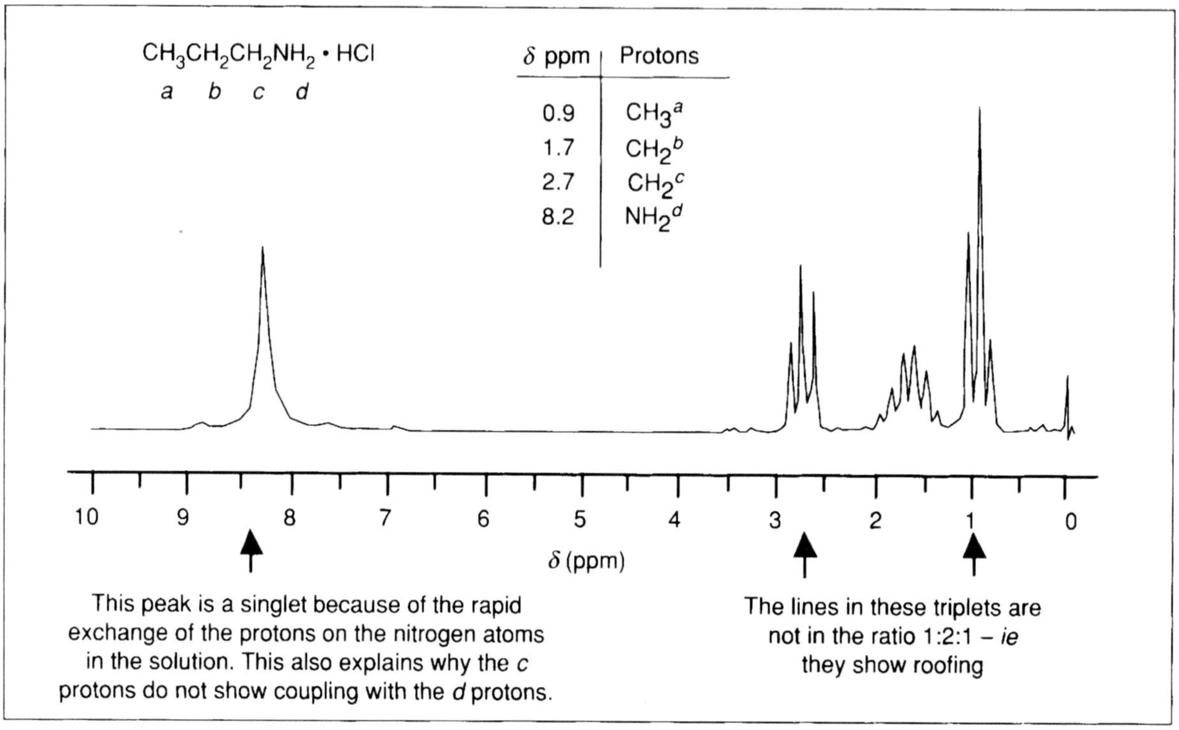
$CH_3CH_2CH_2NH_2 \cdot HCl$
a b c d
δ ppm | Protons
0.9 | $CH_3{}^a$
1.7 | $CH_2{}^b$
2.7 | $CH_2{}^c$
8.2 | $NH_2{}^d$
10 9 8 7 6 5 4 3 2 1 0
δ (ppm)
This peak is a singlet because of the rapid exchange of the protons on the nitrogen atoms in the solution. This also explains why the c protons do not show coupling with the d protons.
The lines in these triplets are not in the ratio 1:2:1 – ie they show roofing

Figure 13 Proton NMR spectrum of propanamine hydrochloride showing skewing, or 'roofing', of multiplet lines

In the hypothetical case of two doublets separated by a large chemical shift compared with their coupling constant the 'roofing' effect is small, but as the chemical shifts get closer together the spectrum becomes more and more distorted until the chemical shifts are identical and only a singlet is seen (*Fig. 14*).

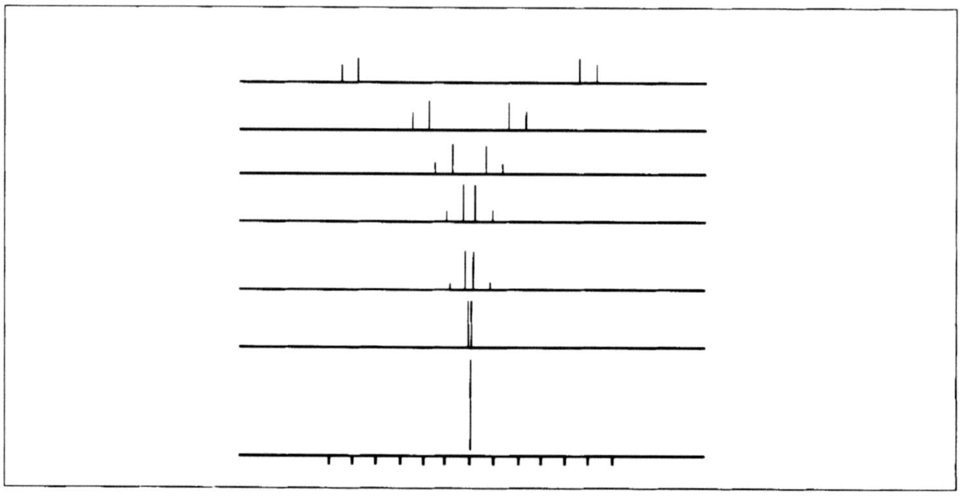

Figure 14 The effect on a spectrum of bringing sets of peaks closer together

So if the resonating protons couple with more than one set of protons more complex splitting patterns will be observed – *eg* the CH_2 protons in propanal (CH_3CH_2CHO), $\delta = 2.5$, couple with the methyl protons to give a quartet, and with the aldehydic proton to give a doublet at the same time *ie* this part of the spectrum therefore appears as a quartet of doublets (*Fig. 15*).

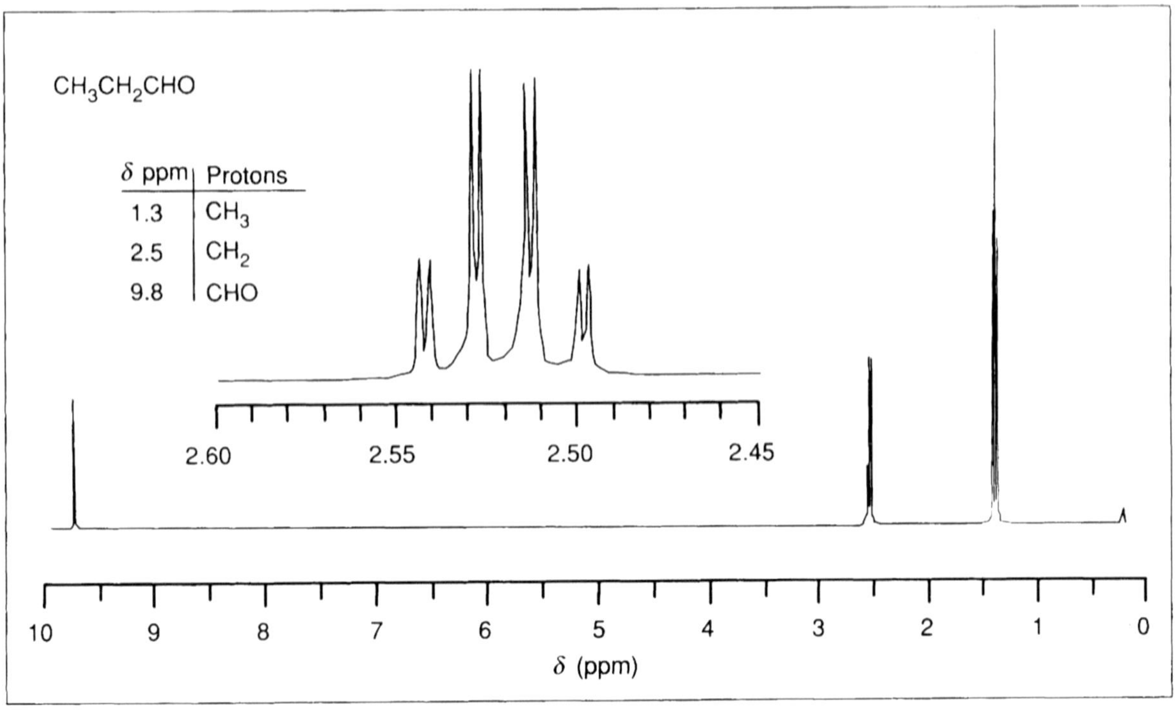

Figure 15 Proton NMR spectrum of propanal showing a quartet of doublets for the CH_2 protons

The three protons in a methyl group show coupling only with other protons, and not with each other – this makes spectral interpretation much simpler. However, if because of stereochemistry, chemical environments are not identical, slightly different patterns can be observed – *eg cis* and *trans*-3-phenyl-2-propenoic (cinnamic) acid ($C_6H_5CH{=}CHCOOH$) have different *J* values for the coupling between the protons attached to the double bond carbons. The coupling constant of *cis* protons is generally smaller (7-12 Hz) than that of *trans* protons (13-30 Hz). *Figure 16* shows the NMR spectrum of *trans*-3-phenyl-2-propenoic (cinnamic) acid, with the coupling constant calculation.

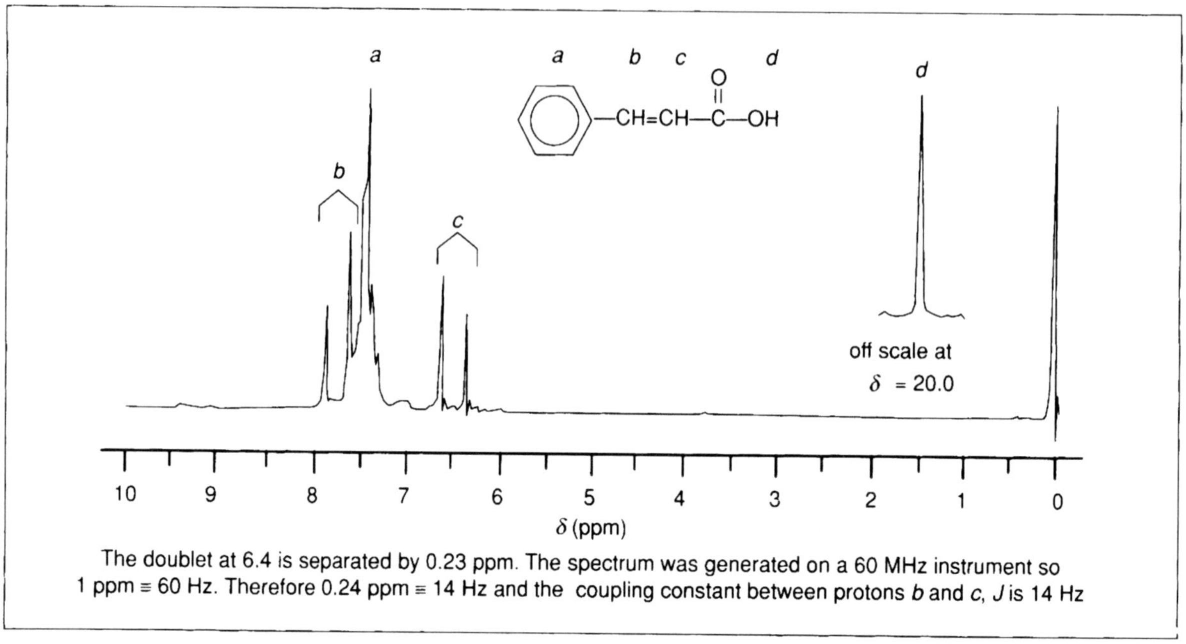

Figure 16 NMR spectrum of *trans*-3-phenyl-2-propenoic (cinnamic) acid

The coupling patterns for aromatic protons depend on the positions of ring substituents. *Figure 17* shows the spectral patterns expected from substituted ring protons, and *Fig. 18* shows the spectrum of 1-bromo-4-iodobenzene, which illustrates the 1,4- disubstituted pattern.

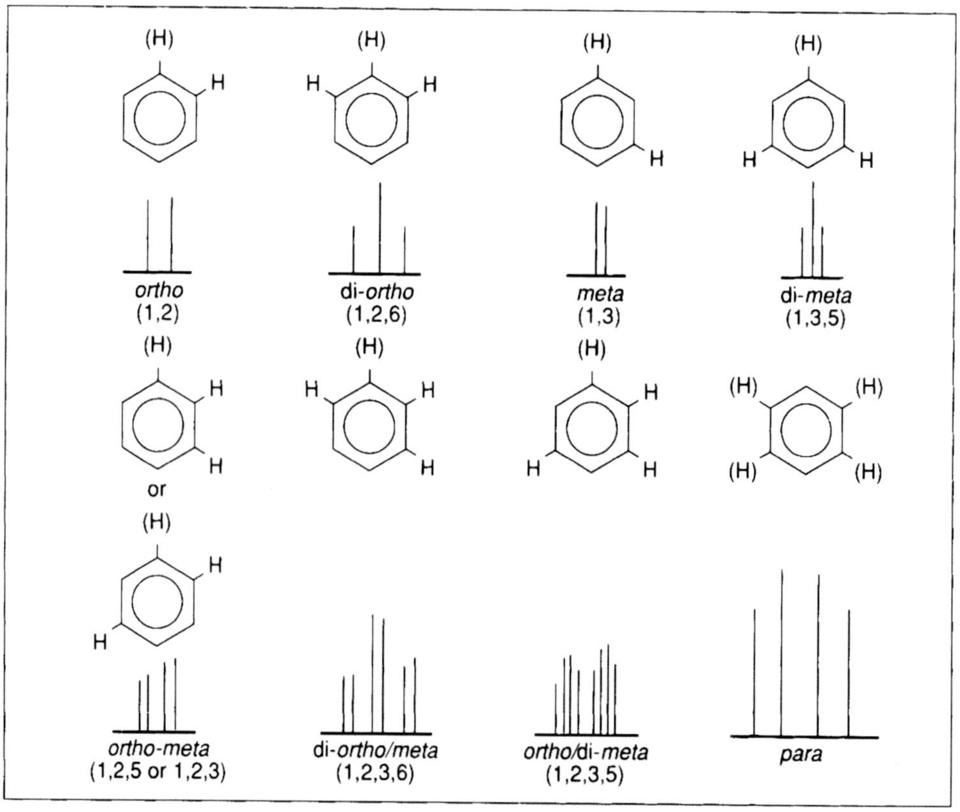

Figure 17 NMR aromatic coupling patterns as shown by the proton in brackets (terms refer to position of protons)

Unilever

THE ROYAL SOCIETY OF CHEMISTRY

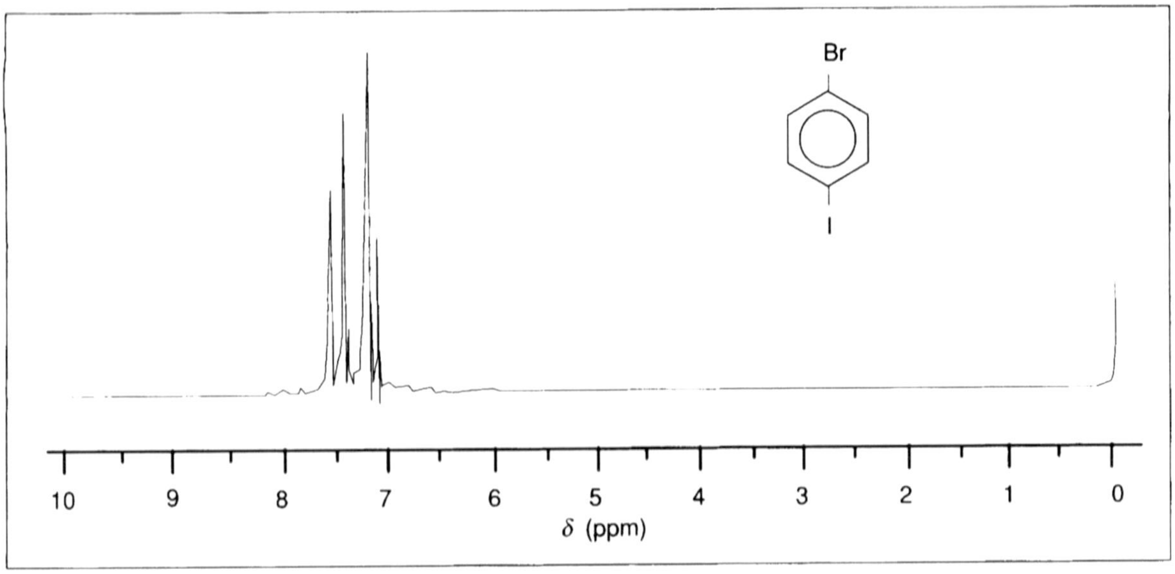

Figure 18 NMR spectrum of 1-bromo-4-iodobenzene showing 1,4-disubstituted aromatic coupling

Integration of peaks

The amount of energy absorbed at each frequency/magnetic field strength is proportional to the number of protons absorbing. Consequently, the area under each set of spectral lines is proportional to the relative number of protons absorbing. Many instruments will give this information directly by also plotting an integration curve on the spectrum. By measuring the height of the integration curve at each set of peaks the ratio of protons absorbing can be determined. The NMR spectrum of propyl ethanoate shown in *Fig. 19*, illustrates how the integration curve is used.

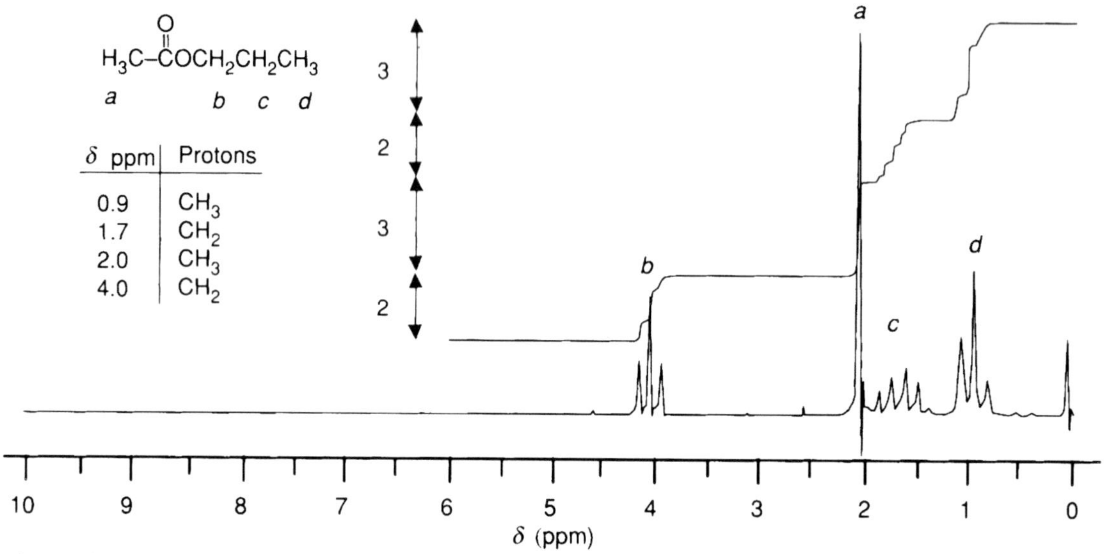

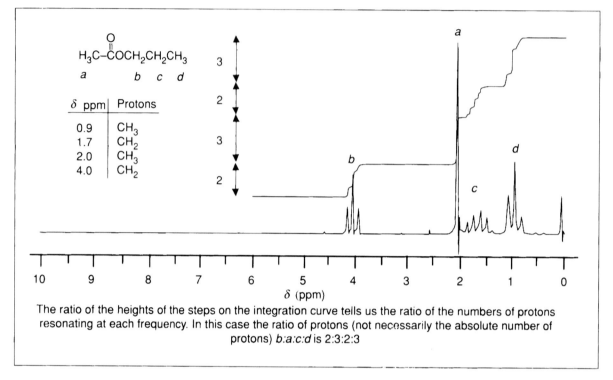

The ratio of the heights of the steps on the integration curve tells us the ratio of the numbers of protons resonating at each frequency. In this case the ratio of protons (not necessarily the absolute number of protons) *b:a:c:d* is 2:3:2:3

Figure 19 NMR spectrum of propyl ethanoate showing the integration curve and how it is used to obtain information on the number of protons resonating at each frequency

The latest machines produce a numerical computer printout which gives the area of each peak without the need to measure the height of the integration trace.

An unexpectedly simple peak appears in the spectrum of compounds containing an OH group. In the spectrum of ethanol *(Fig. 10)*, a single peak appears for the resonance of the hydroxy proton, at a chemical shift of 5.4. A triplet would normally be expected because the proton couples with the two CH_2 protons. However, the hydroxy proton is rapidly lost and replaced by other protons from hydroxy groups in its vicinity (*ie* exchanges), so no coupling is observed and only a single peak is seen in the spectrum. No coupling is observed because the hydroxy proton spends too short a time on each molecule. For instance if $J = 7$ Hz, to observe coupling the hydroxy proton must remain on a molecule in the order of 1/7 of a second. The CH_2 protons would be expected to give a quartet of doublets but the lack of coupling with the hydroxy proton means that only a quartet is observed. In *Fig. 20* the spectrum shows all of the coupling because no exchange of the OH proton occurs. These peaks are seen if the spectrum is run by putting the ethanol in a solvent that prevents proton exchange – *eg* DMSO (dimethyl sulphoxide).

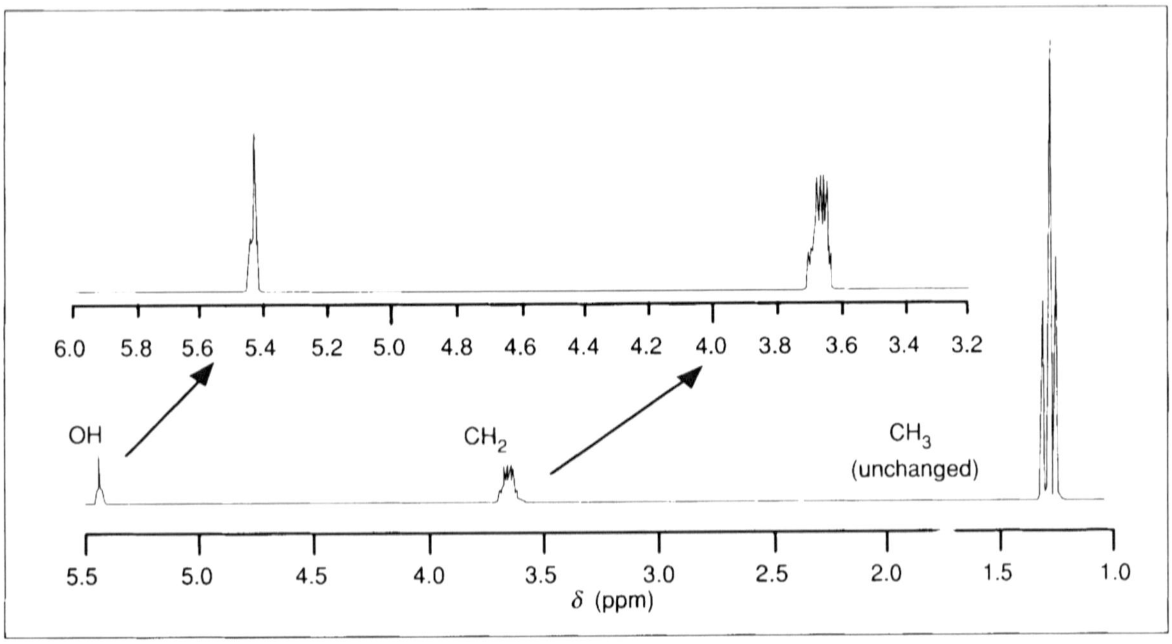

Figure 20 Non-exchange decoupled NMR peaks of ethanol

Practical considerations

About 20 mg of the sample is dissolved in 0.4 cm^3 of a solvent which has no hydrogen or is deuterated – *eg* CCl_4, $CDCl_3$, C_6D_6, d_6-DMSO (hexadeuterodimethyl sulphoxide, $(CD_3)_2SO$), or D_2O. The choice of solvent depends on the relative solubility of the sample in the various solvents. The reference material, usually tetramethylsilane (TMS) is added, and the solution is placed in a precision ground glass tube of 5 mm diameter to a depth of 2 or 3 cm. (2.5 mm and 10 mm diameter tubes can be used when only a small volume of solution is available or a large amount of sample has to be used.) The sample tube is then lowered into a probe, at room temperature (*Fig. 8*). The probe in older machines has both a transmitter and receiver coil connected to it, however, in modern Fourier Transform (FT) machines one coil performs both functions. In both machines the probe is immersed in the magnetic field.

The field in FT machines is often provided by a superconducting magnet that consists of alloy coils maintained at liquid helium temperature (4 K) but the magnet is insulated so that the sample tube remains at room temperature. The magnet retains its field strength once it has been energised with an electric current, provided the coils do not warm up or the energising current is lost.

The effect of slight variations in the magnetic field is minimised by spinning the sample at 20-30 revolutions per second and mounting 'shim coils' which can be used to reduce any inhomogeneity, in the magnet bore. This can be done manually or by computer control because the field has to be consistent to roughly one part in 10^9.

Older spectrometers relied on iron magnets and scanned the range of fields, and the receiver coils gave the spectrum directly. However, modern FT NMR spectrometers have a constant magnetic field and the range of frequencies is transmitted simultaneously for a few microseconds (typically 5 μs for ^{1}H and 10 μs for ^{13}C spectra). All the protons in the sample are excited, and each sends out radiofrequencies of the type shown in *Fig. 21*, as they relax. This is analogous to what happens when a bell is hit. A sound characteristic of the bell is given out and dies away with time. The time taken for complete decay depends on the relaxation time of the system.

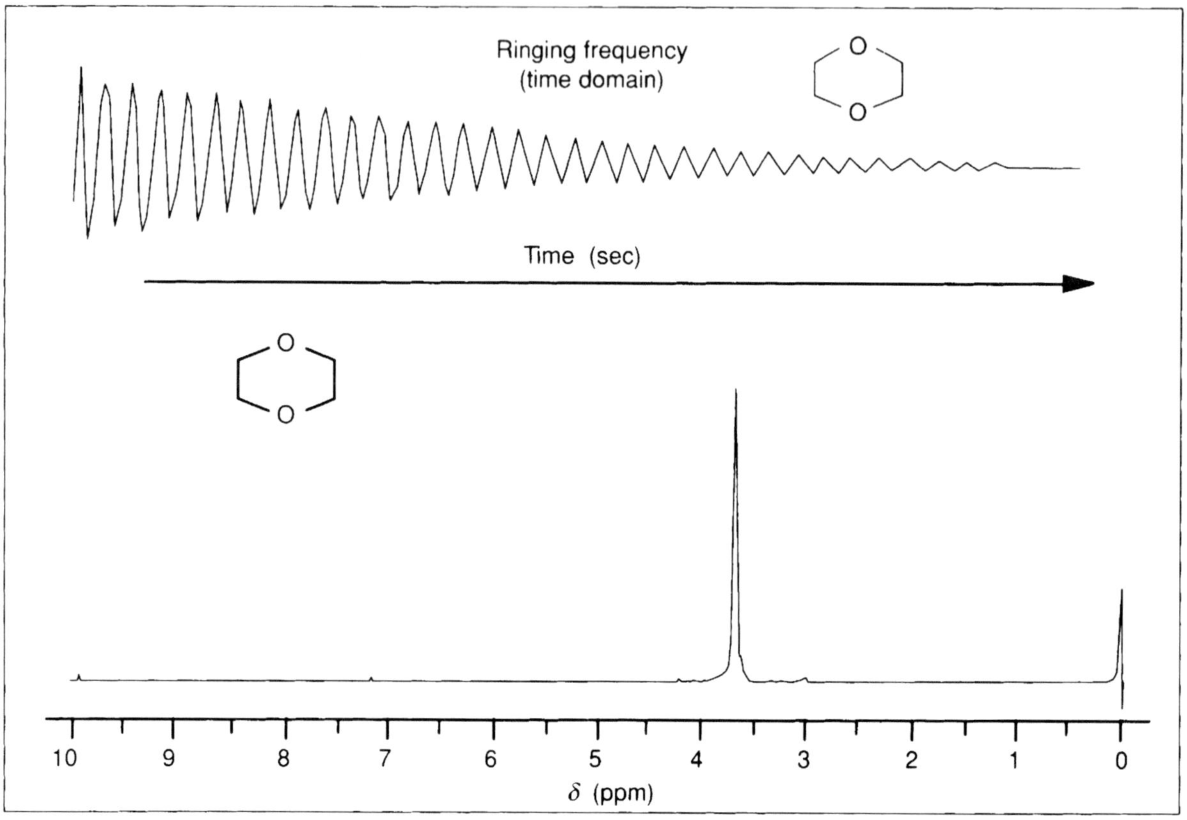

Figure 21 Time domain and final spectrum of dioxane

The amplitude of the decay curve is dependent on the degree of absorption, and all of the ringing frequencies transmitted will be superimposed on each other. A computer calculates the frequency (chemical shift) and intensity of each absorption and the spectrum is produced by a mathematical process known as Fourier transformation. The whole technique is known as pulsed NMR and typically the pulse is transmitted once a second until sufficient information has been collected. For ^{1}H-NMR spectra a 20 mg sample might require only one scan, while 1 mg might need 100–200 scans to collect enough data, depending on the relative mass of the compound and the strength of the magnetic field used. The individual scans can be added to produce the final spectrum. Pulsed NMR-FT spectra have the advantage that the background noise becomes less significant as more scans are added, because it is easier to register scans and put one scan on top of another to average them.

More advanced techniques

Spectra are often complicated by protons coupling with groups on both sides of them, and might be difficult to interpret. For instance, the C_b protons in propan-1-ol,

$$\overset{a}{CH_3}\overset{b}{CH_2}\overset{c}{CH_2}OH$$

couple with those of C_a and C_c. It is possible to simplify the spectrum by saturating the sample with radiation of the frequency at which one of the groups resonates. If the resonant frequency of the methyl protons (C_a) is transmitted while the data for the spectrum are being collected the protons on C_b show no coupling with the C_a protons. A triplet is then observed, and not a triplet of quartets (*Fig. 22*). In general, by removing the coupling of a known group it is possible to simplify the spectrum and decide precisely which other groups certain protons couple to. The technique is known as spin decoupling.

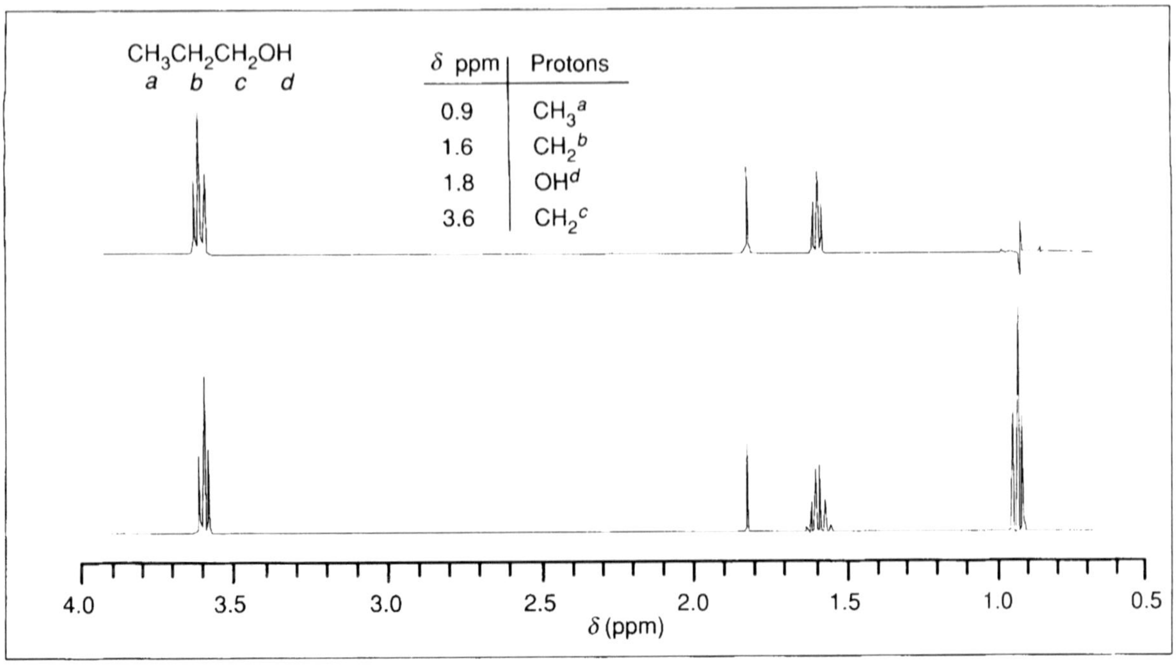

Figure 22 NMR and spin decoupled NMR spectra of $CH_3CH_2CH_2OH$

COSY

Spin decoupling has now largely been replaced by COSY (COrrelated SpectroscopY). In this method the spectrum is drawn horizontally and vertically and a contour or a three dimensional plot is constructed. This has the advantage over spin decoupling in that it shows all the coupling relationships. The full spectrum can be seen along the diagonal, and the coupling relationships are seen off-diagonal – *eg Fig. 23* where the NMR and COSY spectra of ethyl 4-methylbenzoate are shown.

Other NMR spectra

Although a number of nuclei other than ^1H will give NMR spectra, the most useful one is ^{13}C. The low abundance of this isotope (1.1 per cent) means that the probability of two ^{13}C atoms being bonded to each other is very small, so coupling is not observed and even if it were, the coupled signal would be too small to detect. ^{13}C-^1H coupling is possible, but this complicates the spectrum, so spin decoupling is frequently used to simplify it. Consequently the ^{13}C spectrum usually appears as a series of single peaks, each peak signifying the presence of a carbon atom in a

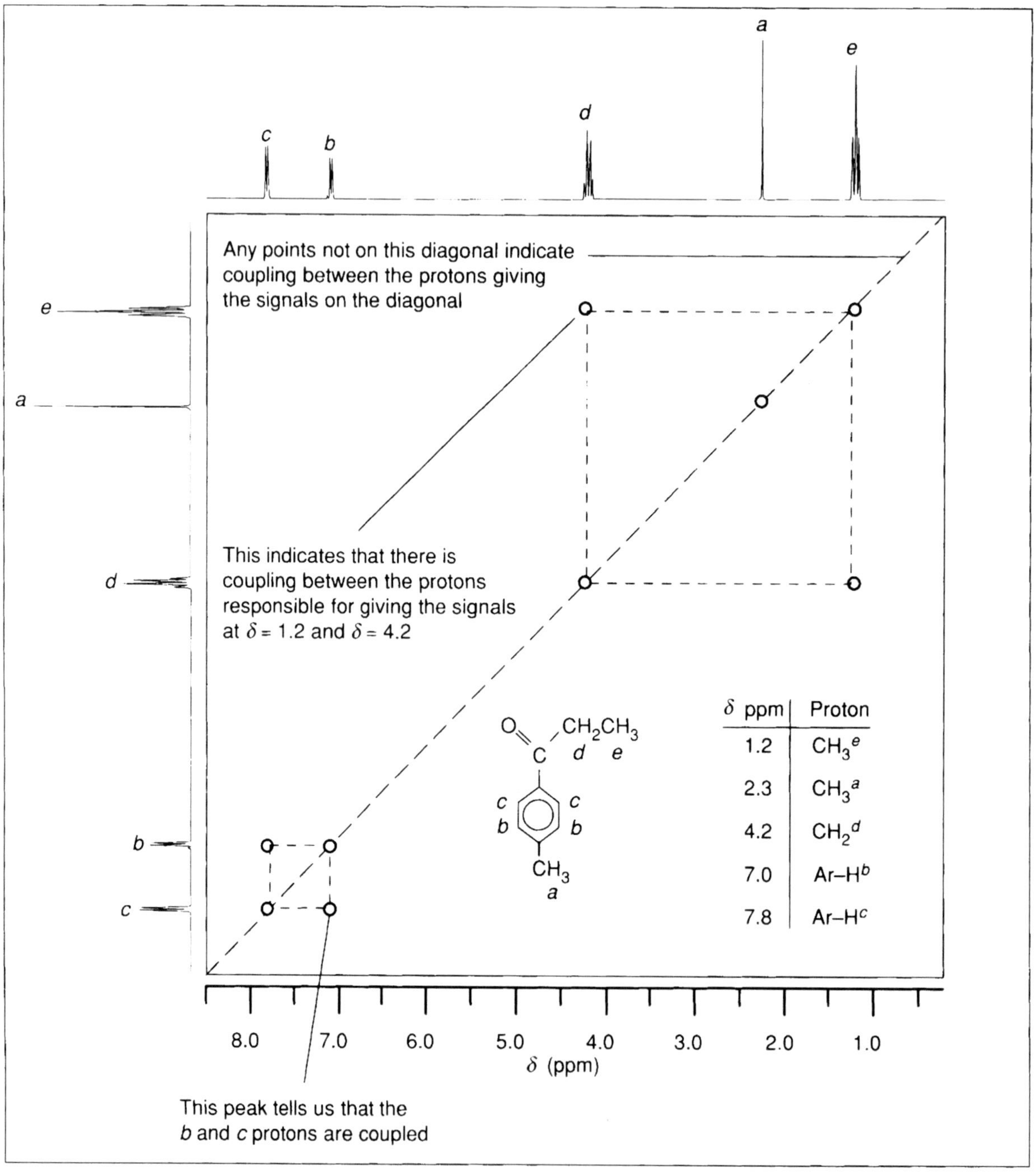

Figure 23 NMR and COSY spectra of ethyl 4-methylbenzoate

different chemical environment. In this type of NMR spectroscopy a useful integration peak may not be obtained because the signal strength is dependent on the environment of the carbon atoms. Therefore, these factors have to be taken into account when determining the number of equivalent carbon atoms in a particular position.

While proton NMR chemical shifts are in the 0-10 ppm range, ^{13}C chemical shifts are mostly in the 0-250 ppm range. The frequency at which ^{13}C resonates is also different – in a 2.35 T magnetic field protons resonate at 100 MHz, but ^{13}C resonates at 25.14 MHz. The ^{13}C spectrum of ethyl ethanoate is shown in *Fig. 24.* One similarity between the two techniques is that tetramethylsilane (TMS) is a useful reference material for ^{13}C NMR spectroscopy too, and by definition gives a chemical shift of zero.

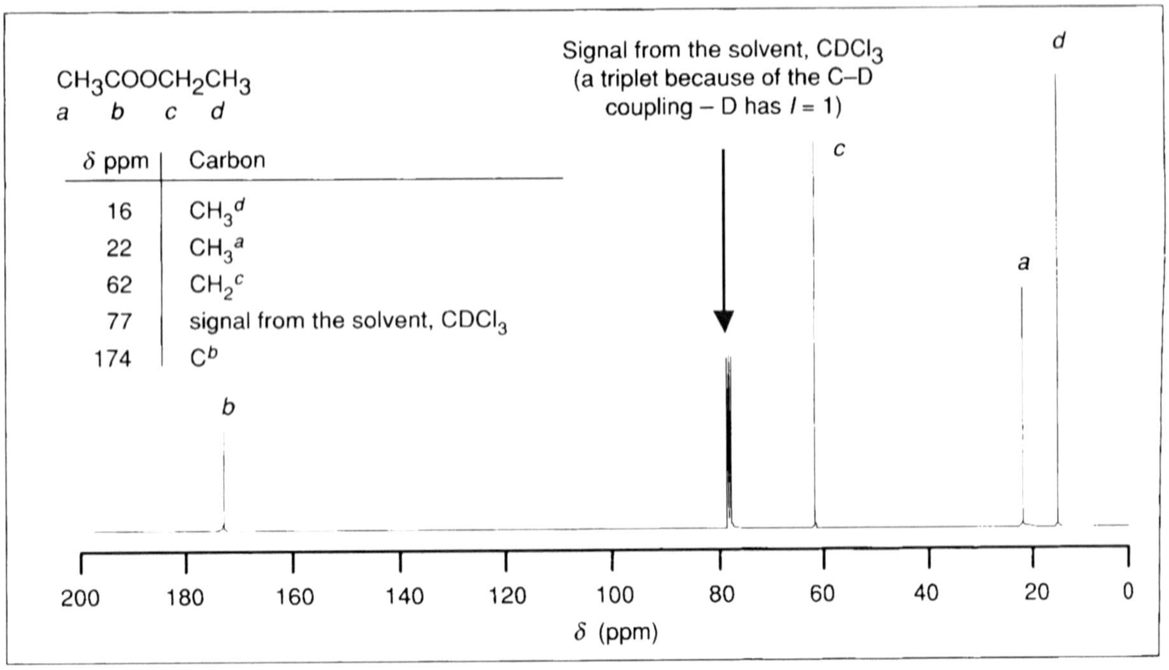

δ ppm	Carbon
16	CH_3^d
22	CH_3^a
62	CH_2^c
77	signal from the solvent, $CDCl_3$
174	C^b

Figure 24 ^{13}C-NMR spectrum of ethyl ethanoate

Applications of NMR

NMR is commonly used for structure determination, but some other important uses do exist. Relaxation NMR spectroscopy can be used to evaluate the proportions of solid and liquid phase components in fatty foodstuffs such as margarines and low fat spreads. A plot of the signal intensity from the protons (recorded by the radiofrequency detector) with time is generated, and appears as one curve superimposed on another *Fig. 25*.

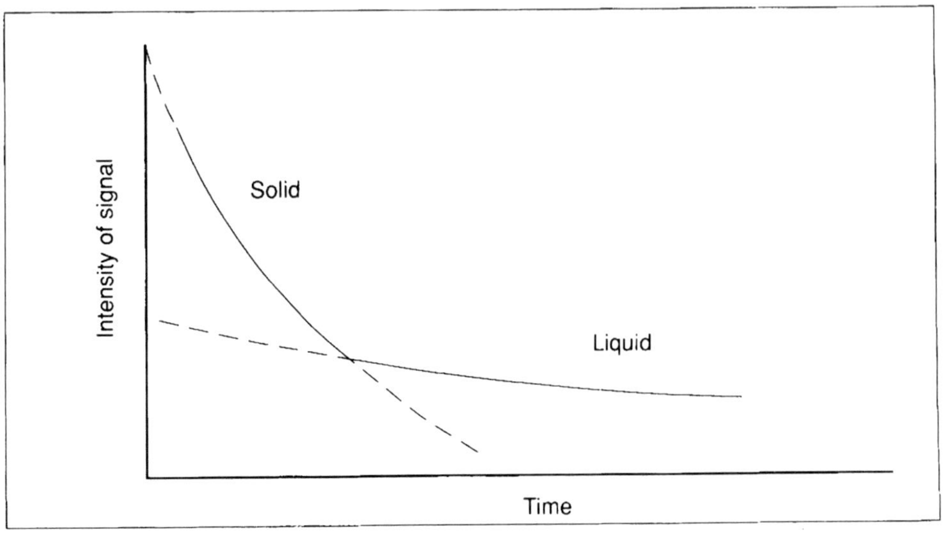

Figure 25 Intensity of NMR signal with time from a fatty foodstuff

This technique relies on the spin-spin (T_2) relaxation times of protons in the liquid phase being longer than those in the solid phase. Consequently, when the two curves are extrapolated back they give intensity values which are proportional to the relative amounts of protons in the solid and liquid phases. If the temperature of the sample is changed, the relative proportions may change, and if the experiment is repeated over a range of temperatures, a melting profile can be obtained.

The application of NMR in medicine is becoming increasingly common, from simple dynamic studies to complex diagnosis of tissue abnormalities. Pioneering work has been done at a number of institutions, including Hammersmith Hospital and Aberdeen and Oxford Universities. The studies described here were carried out in the NMR Imaging Facility at Queen Mary and Westfield College, London.

^{31}P-NMR of blood and cell fluids enables the kinetics of pH control in diabetics to be monitored. An absence of insulin can lead to harmfully high cell acidity levels, and an infusion of sodium hydrogencarbonate can help to re-establish the body's normal cell pH. This can be monitored by measuring the chemical shift separation between the organic and inorganic phosphorus signals. This is because the inorganic phosphorus NMR signal is highly pH dependent and can move as much as 1 ppm per pH unit. The area under each peak can also be used to gain data on the relative concentrations of each biochemical (*eg* ATP), and hence the metabolic status of the tissue.

Using ^{1}H-NMR in body scanning has become quite common. The intensity of ^{1}H-NMR signals depends both on the protons' density and their relaxation times. Consequently protons in water, proteins, lipids, carbohydrates and most body sites would be expected to give different signals. However, the main species, in the liquid state, with proton densities high enough to give an appreciable signal are water and lipids. The environments of the resonating nuclei give them different relaxation times, and hence different signals. Consequently different organs in the body can be differentiated.

Magnetic resonance images, as they are called (the 'nuclear' is dropped to avoid any association with nuclear radiation) look similar to X-ray images (*Figs. 26-28*). Magnetic resonance images can be acquired from a limb, the head or the whole body. Images from soft tissue can be acquired in any plane, and these complement the information from hard tissue data (such as X-rays). The scanning takes approximately 20 min, so the subject has to remain still with the part of their anatomy being scanned inside a large bore magnet. With animals this is achieved by light anaesthesia.

There are no known side effects associated with this technique, which means that subjects can be scanned regularly, including the young and the frail, to monitor any changes in condition.

This technique is now available in many clinics and has been used to diagnose successfully and monitor a variety of conditions such as cancer, hydrocephalus and multiple sclerosis.

Unilever

THE ROYAL SOCIETY OF

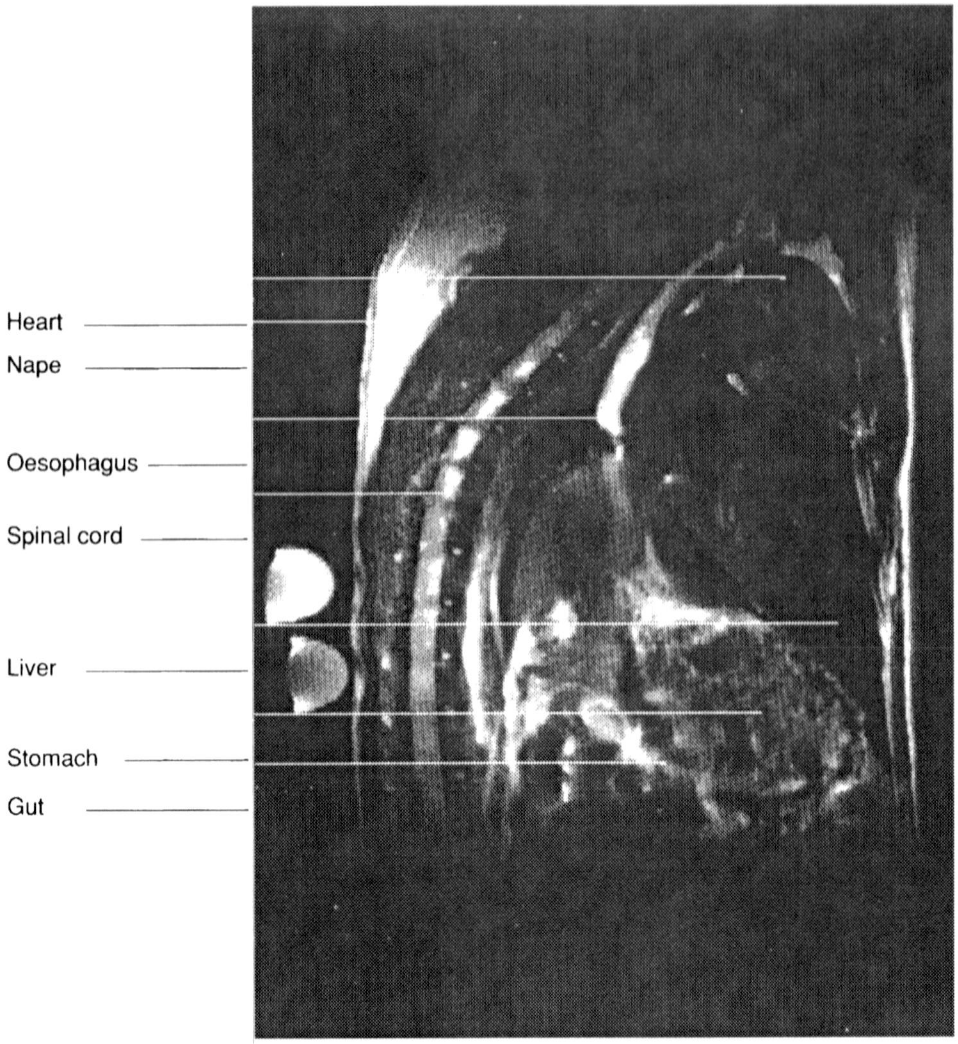

Figure 26 Saggital slice ^{1}H MR image of a live, anaesthetised guinea pig from a slice of 1.5 mm thickness with an in-plane resolution of 100 nm. Note the clear delineation of soft tissue structures such as the spinal cord, oesophagus, stomach and liver as well as heart, gut and subcutaneous fat at the nape of the neck

THE ROYAL SOCIETY OF CHEMISTRY

Unilever

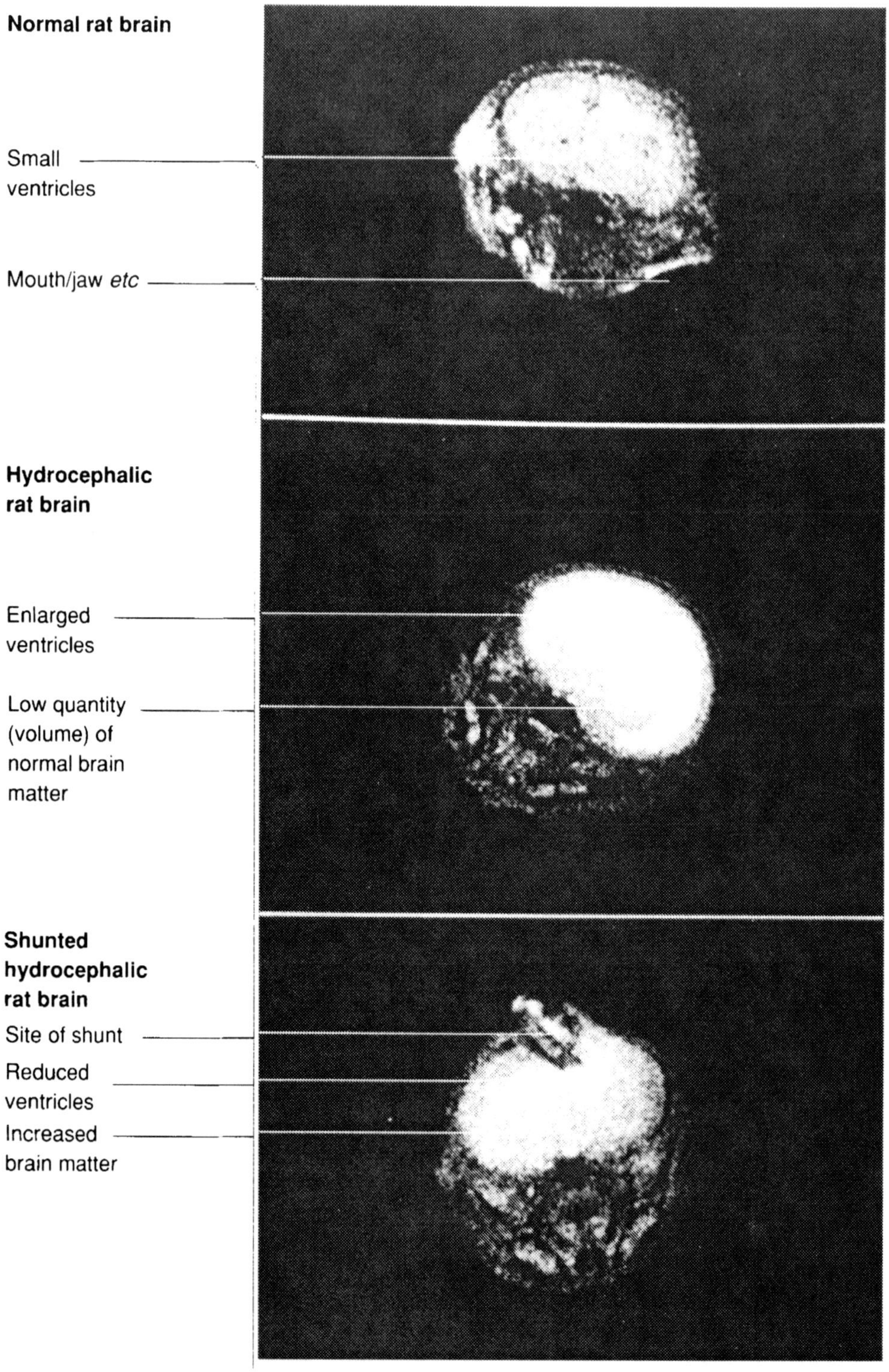

Figure 27 Coronal slice ^{1}H MR images of rats born with congenital hydrocephalus (water on the brain) from a 2 mm slice thickness. Top – normal animal with small ventricles. Middle – litter mate of affected animal with enlarged ventricles depicting build up of fluid. Bottom – another animal from the same litter that also suffered with the same disease but has been shunted in a similar manner to the techniques currently used in affected children. Note the reduced ventricle size and the increase of brain tissue relative to the diseased but untreated animal

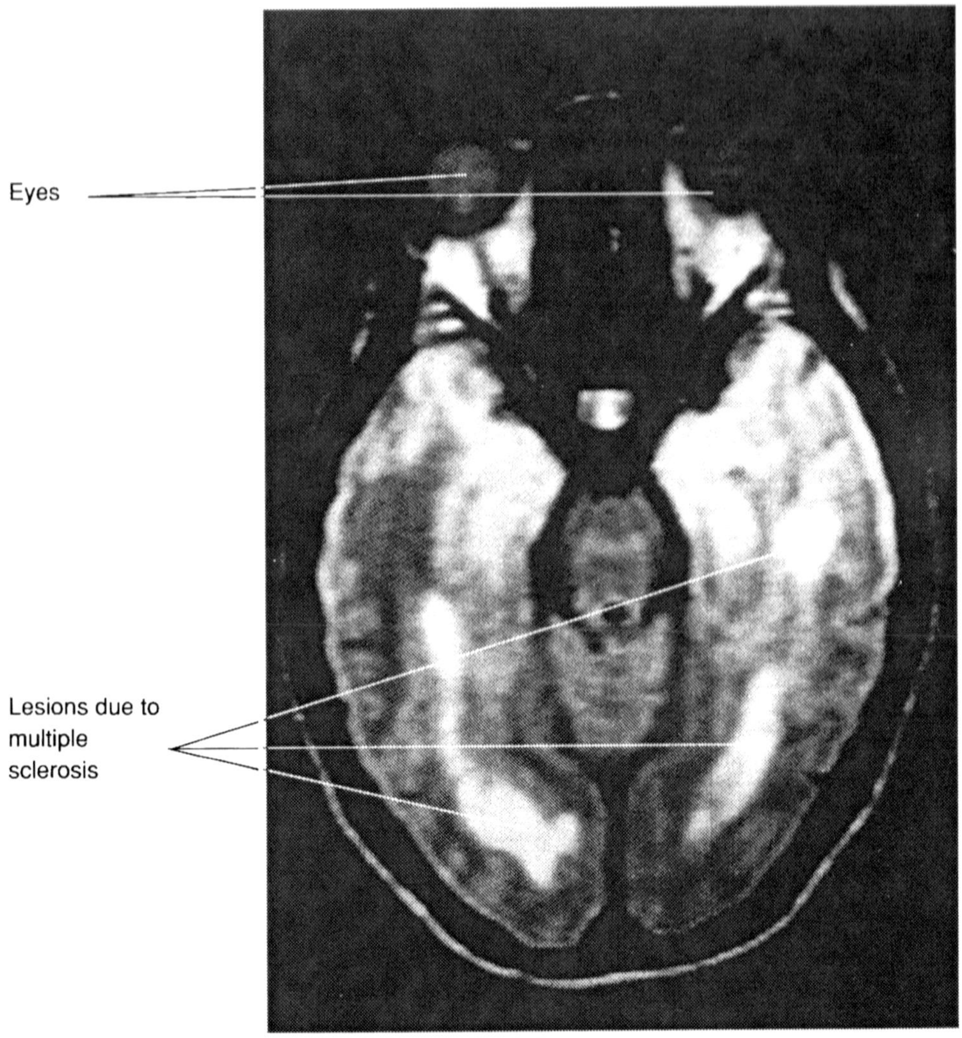

Figure 28 Transverse slice ^1H MR image of 5 mm thickness from human brain at the level of the eyes. The bright, white areas in both hemispheres relate to oedema formation in demyelinating lesions due to multiple sclerosis

Exercises

By using Table 1 (page 38) and any other information given, it should be possible to determine the structures of each of the unknowns in the examples given below.

Exercise 1

A hydrocarbon, liquid at room temperature. The empirical formula of the compound is C_8H_{10}.

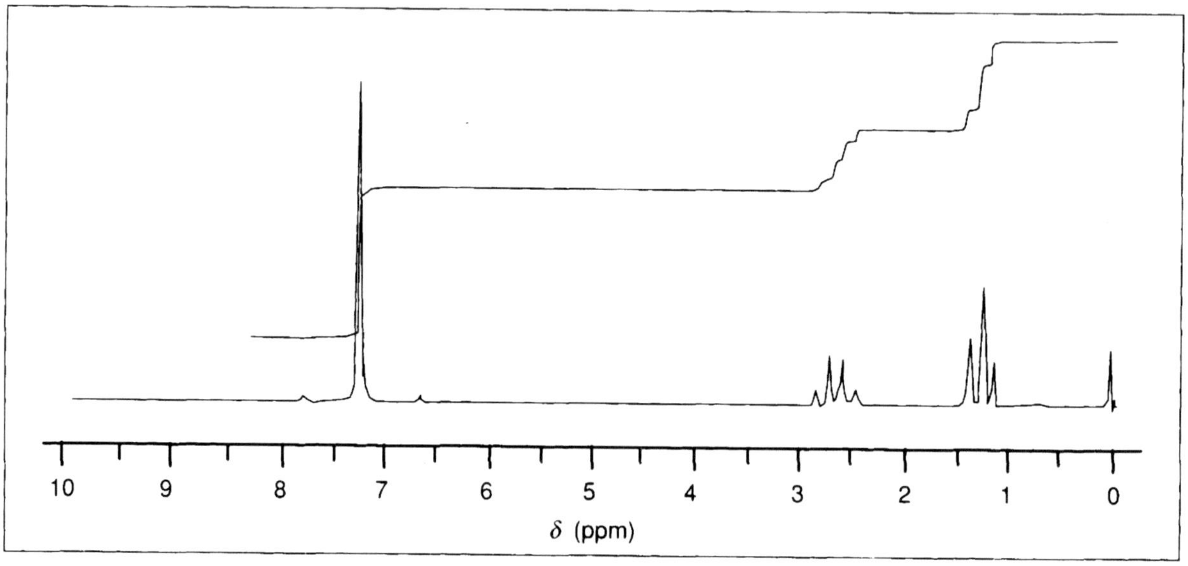

Figure 29

Chemical shift (δ)	Multiplicity (no of lines)	Integration
1.2	Triplet	3
2.6	Quartet	2
7.2	Singlet	5

The singlet at chemical shift (δ) = 7.2 indicates a benzene ring, and the fact that the integrated peak gives a ratio of 5 suggests that it is monosubstituted. The quartet at δ = 2.6 tells us that whatever is resonating at this value is coupling with three other protons. The integration curve reveals that two protons are resonating, so it would be reasonable to assume that a CH_2 group is involved. From Table 1 the only reasonable possibility is R-CH_2-Ar.

The triplet at δ =1.2 is due to protons coupling with two other protons – the CH_2 protons – and the integration curve reveals that there are three protons resonating at this shift – *ie* a CH_3 group is present.

Putting all this information together, the structure is ethylbenzene.

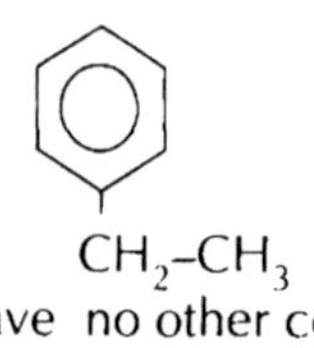

Ethyl groups that have no other coupling always give a triplet (the CH_3 protons) and a quartet (the CH_2 protons). However, the chemical shifts of these two peaks change with the electronic environment.

Unilever

THE ROYAL SOCIETY OF CHEMISTRY

Exercise 2

A colourless mobile liquid at room temperature, with a sweet smell, having the empirical formulae $C_4H_8O_2$.

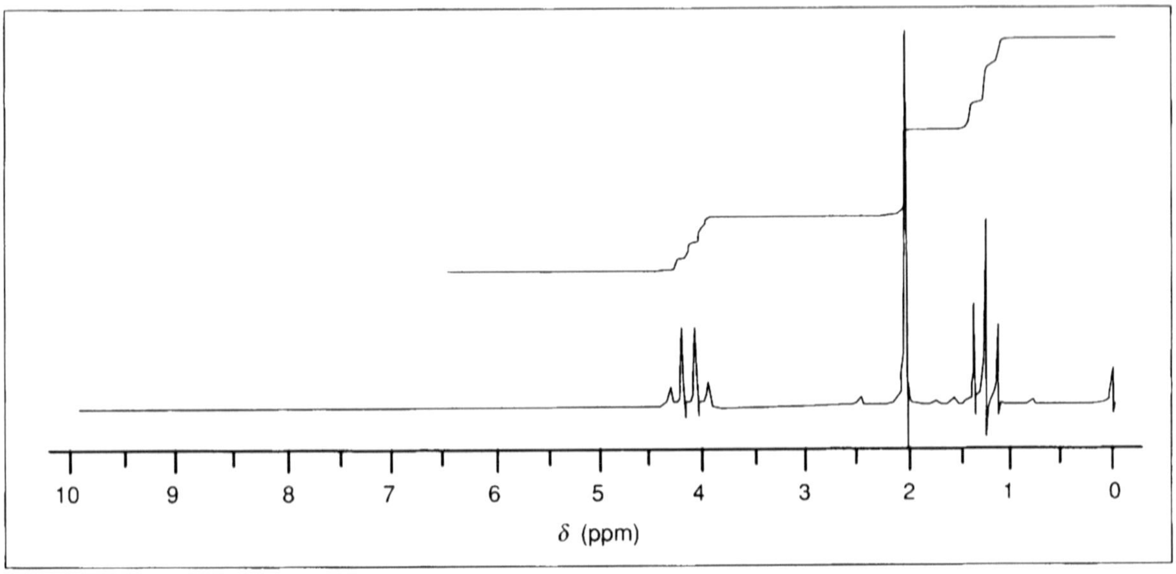

Figure 30

Chemical shift (δ)	**Multiplicity (no of lines)**	**Integration**
1.2	Triplet	3
2.0	Singlet	3
4.1	Quartet	2

From Table 1 the only realistic resonances at $\delta = 4.1$ are due to $C\underline{H}_3$–O–CO–Ar and R–$C\underline{H}_2$–O–CO–R′. Because the peak is split into a quartet, the resonating protons must be coupling with three other protons, and as the integration value is 2 the most sensible assignment would be R–$C\underline{H}_2$–O–CO–R′, where R must be CH_3. This would also account for the triplet at $\delta = 1.2$, where the methyl protons couple with the two CH_2 protons. The integration curve supports this, revealing that three protons are resonating at this value. Further evidence for dismissing the first structure suggested is that there are no aromatic protons present in the molecule – there is no peak in the range expected from benzene ring protons (δ = *ca* 6.5 – 8.2).

So far we have the structure CH_3–CH_2–O–CO–R′, an ester. The integration of the singlet at $\delta = 2.0$ reveals that three protons are present, suggesting a methyl group. This is supported by the value in Table 1.

Thus, *Fig. 30* is the ^{1}H-NMR spectrum of CH_3–CH_2–O–CO–CH_3, *ie* ethyl ethanoate.

THE ROYAL SOCIETY OF CHEMISTRY

Unilever

Exercise 3

This compound has a relative formula mass of 122, and is a solid at room temperature. It contains carbon, hydrogen and oxygen as the only chemical elements.

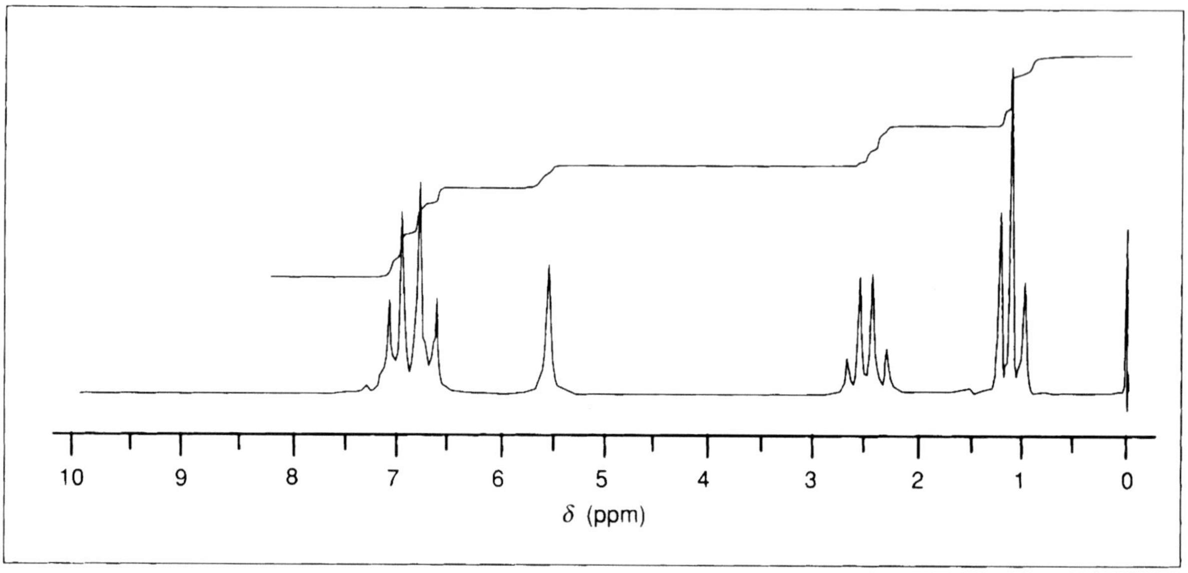

Figure 31

Chemical shift (δ)	Multiplicity (no of lines)	Integration
1.2	Triplet	3
2.5	Quartet	2
5.5	Singlet	1
6.8	Quartet	4

From Table 1, the three proton triplet centred on $\delta = 1.2$ is due to a methyl group attached to and coupled with a CH_2 group. The latter group appears at $\delta = 2.5$.

The CH_2 protons must also couple with the methyl protons to give a quartet, as is seen at $\delta = 2.5$. The shift value also suggests that the CH_2 unit is bonded to a benzene ring.

The multiplet at $\delta = 6.8$, with integration curve data showing four protons, indicates a benzene ring. The splitting pattern is that expected for a 1,4- disubstituted ring. So far we have

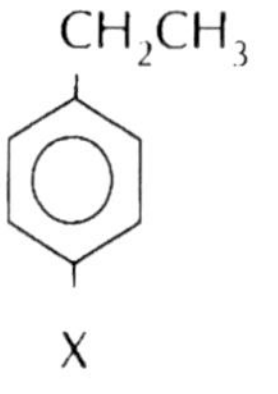

The relative mass of the components identified so far is $C_6H_4 + C_2H_5 = 105$. This leaves 17 mass units unaccounted for, and the only reasonable unit that corresponds to this mass is OH.

Thus if X is OH the molecule must be 4-ethylphenol,

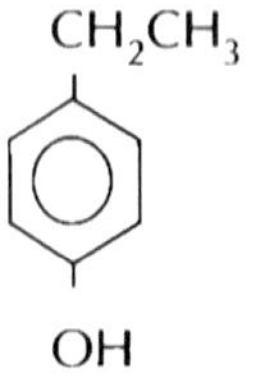

Assigning the singlet at chemical shift = 5.5 would have been difficult in the absence of any other information because hydroxy protons can resonate over a range of shifts, depending on factors such as solvent, pH and temperature as well as the chemical environment of the proton.

Exercise 4

A solid at room temperature, containing carbon, hydrogen and oxygen only. The structure of this compound can be determined from its spectrum (*Fig. 32*).

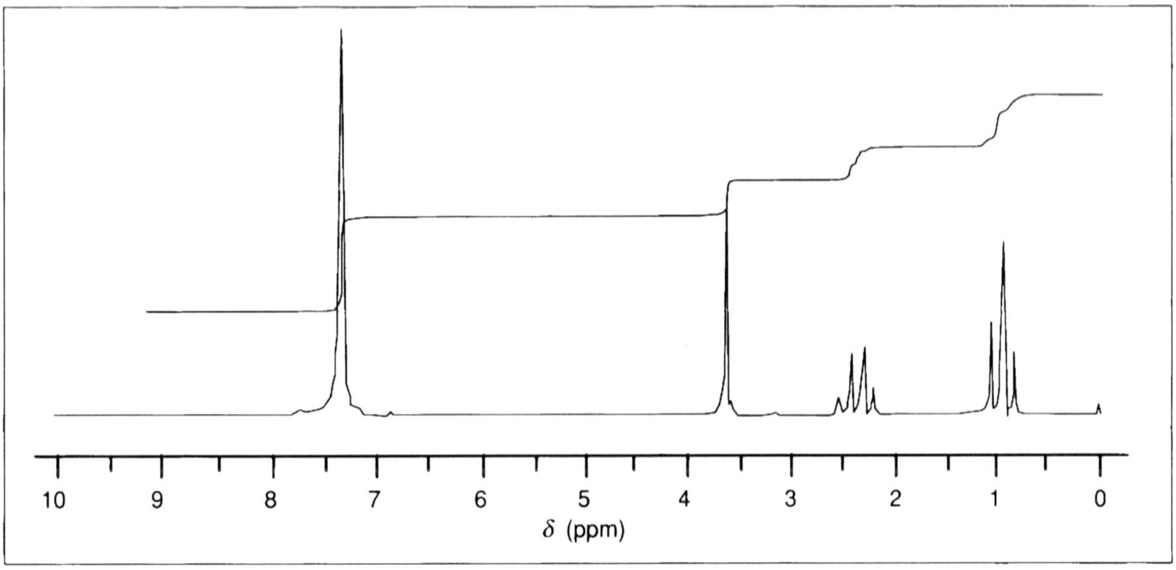

Figure 32

Chemical shift (δ)	**Multiplicity (no of lines)**	**Integration**
0.9	Triplet	3
2.35	Quartet	2
3.6	Singlet	2
7.2	Singlet	5

The singlet with integration ratio 5, at $\delta = 7.2$ indicates a monosubstituted benzene ring. A convenient way of determining the nature of the side chain is to start with the quartet at $\delta = 2.35$. Whatever is resonating at this value is coupling with three protons (from the multiplicity), and there are two protons (from the integration curve) resonating. A group that would conform to this is a CH_2 group bonded to a CH_3 group. The chemical shift should give some information about what else is bonded to the CH_2 unit. From Table 1 the possibilities are:

CH_3–CH_2–Ar or CH_3–CH_2–CO-R

The first possibility can be discounted because if the molecule is ethylbenzene there would be no peak $\delta = 3.6$. This peak could be from another side chain, but in that case the ring protons would give a more complicated splitting pattern and the integration curve would suggest only four (and not five) ring protons.

Therefore, the ethyl group must be bonded to a carbonyl carbon atom. However, the structure is still not solved, because the R function on CH_3–CH_2–CO–R has not been determined. The only peak not assigned so far is the singlet at $\delta = 3.6$ – representing two protons. These protons must be adjacent to groups that they cannot couple with, otherwise the peak would not be a singlet. One group is obviously the carbonyl and the other will be the benzene ring. The final structure is therefore 1-phenyl-2-butanone.

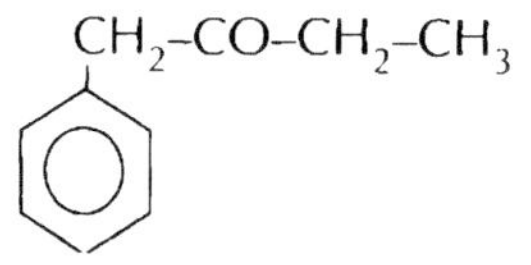

It is possible that there is a carbonyl or an ester function between the CH_2 group and the benzene ring, but this would shift the resonance of the ring protons further downfield (see Table 2 for values).

Exercise 5

This compound has percentage composition C 73.2 per cent, H 7.3 per cent and O 19.5 per cent by mass. It has two oxygen atoms per molecule and its ^{1}H-NMR spectrum is shown in *Fig. 33*.

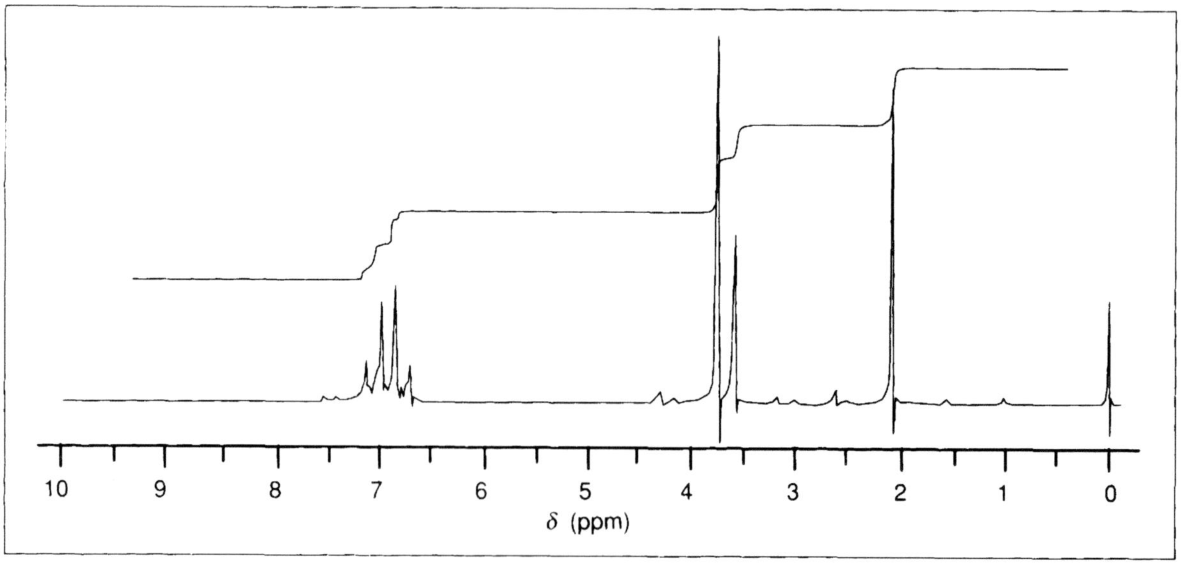

Figure 33

Percentage composition data gives the empirical formula C_5H_6O. Each molecule contains two oxygen atoms, so the molecular formula must be $C_{10}H_{12}O_2$. The information can be obtained from the integration of the NMR peaks because the ratios add up to 12.

Chemical shift (δ)	Multiplicity (no of lines)	Integration
2.1	Singlet	3
3.55	Singlet	2
3.7	Singlet	3
6.85	Quartet	4

Apart from the quartet at $\delta = 6.85$ all the peaks in this spectrum are singlets. Consequently, apart from the four ring protons (the quartet) no other protons couple with each other. The splitting pattern of the ring protons suggests that it is 1,4-disubstituted.

The signal from the protons at $\delta = 2.1$ is likely to be from one of the following:

CH_3–CN, or CH_3–CO–R, or CH_3–CO–O–R

The nitrile can be dismissed because nitrogen is not present in the compound, but the other two are possibilities. If the unknown is an ester, it must have at least one CH_2 unit between the ester function and the benzene ring (the chemical shift is for an alkyl ester and not an aromatic ester). Thus a signal would be expected at $\delta = 4.9$ from the CH_2 protons in Ar–CH_2–O–CO–R (Table 1). This is not observed, so the compound must contain CH_3–CO–R.

From Table 1, the singlet from the two protons at $\delta = 3.55$ could be due to either Ar–CH_2–CO–R or R–CH_2–OH. The latter is not possible because R–CH_2– involves coupling between the protons, and this is not observed.

Combining the information so far, we have:

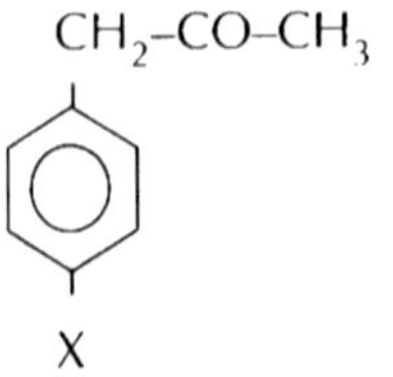

and have accounted for C_9H_9O. This leaves CH_3O unaccounted for. Two possibilities are: OCH_3 and CH_2OH. The latter would give two NMR signals, and we only have one peak at $\delta = 3.7$ representing three protons which is unassigned. This corresponds to OCH_3.

Thus the structure of the compound is 4-methoxyphenylpropanone.

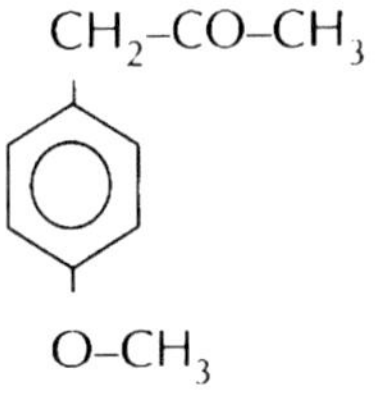

THE ROYAL SOCIETY OF CHEMISTRY

Unilever

3. Infrared spectroscopy

Infrared spectroscopy can be used to detect the functional groups present in a sample, and information can sometimes be obtained about the proximity of one group to another. In some cases it is also possible to give data on the amount of sample present.

The theory

The atoms in molecules are not static, but vibrate about their equilibrium positions, even in the solid state. Each atom vibrates with a frequency which depends on its mass and the length and strength of any bonds it has formed. Molecular vibrations are stimulated by bonds absorbing radiation of the same frequency as the natural frequency of vibration of the bond (*ie* in the range $1.20 \times 10^{13} - 1.20 \times 10^{14}$ Hz) which is in the infrared region of the electromagnetic spectrum. So that the numbers are more manageable absorption is usually quoted in wavenumbers (units cm^{-1}, the reciprocal of wavelength in cm – *ie* the number of wavelengths that make up one centimetre). The relationship between frequency, wavelength, wavenumber and energy is shown in Table 1.

Table 1 Relationship between the frequency, wavelength, wavenumber, and energy of infrared radiation

Frequency (ν) (Hz)	**Wavelength (λ) (m)**	**Wavenumber (cm^{-1})**	**Energy ($kJ\ mol^{-1}$)**
1.20×10^{13}	2.50×10^{-5}	400	4.79×10^{3}
1.20×10^{14}	2.50×10^{-6}	4000	4.79×10^{4}

Frequency, wavelength and energy are interrelated:

$$c = \nu\lambda$$

where c = velocity of light ($3.00 \times 10^{8}\ ms^{-1}$)
ν = frequency in Hz
λ = wavelength in m

and

$$E = h\nu L \text{ for a mole of photons}$$

where E = energy
h = Planck's constant (6.63×10^{-34} Js)
L = Avogadro constant ($6.02 \times 10^{23}\ mol^{-1}$)

For most purposes it is assumed that each vibration occurs independently of all others around it (and that the atoms behave as simple harmonic oscillators). A variety of vibrations are possible. Some are shown in *Fig. 1*. The centre of mass of the molecule remains constant during these vibrations.

Unilever

THE ROYAL SOCIETY OF CHEMISTRY

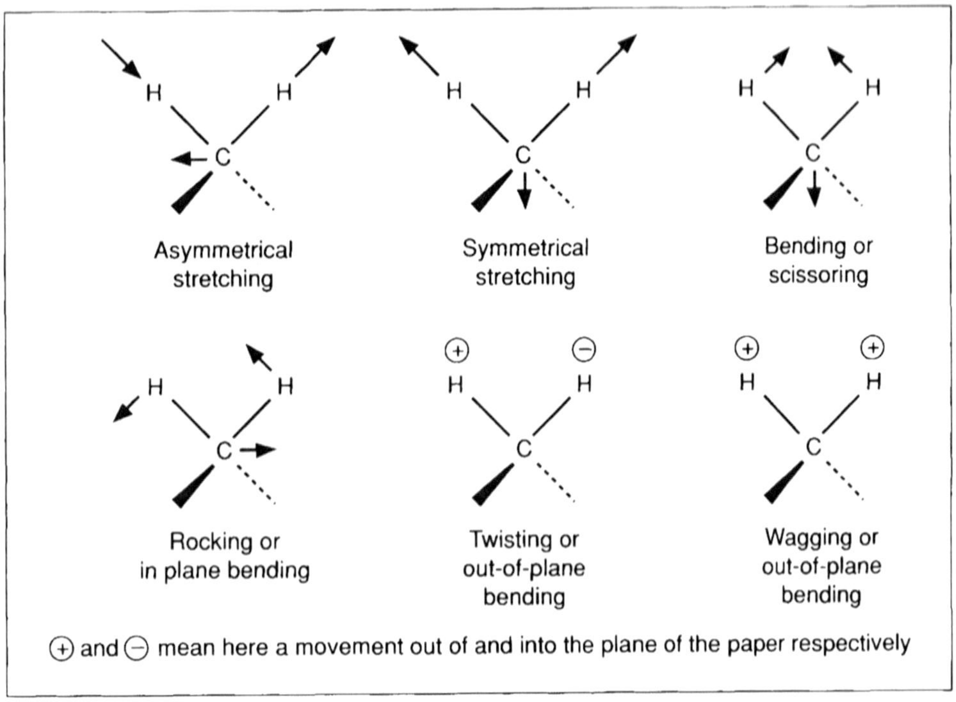

Figure 1 Some modes of vibration

The spectrometer

Conventional infrared spectrometers (known as dispersive infrared spectrometers) have two beams of radiation, one passing through the sample, the other passing through a reference cell (*Fig. 2*).

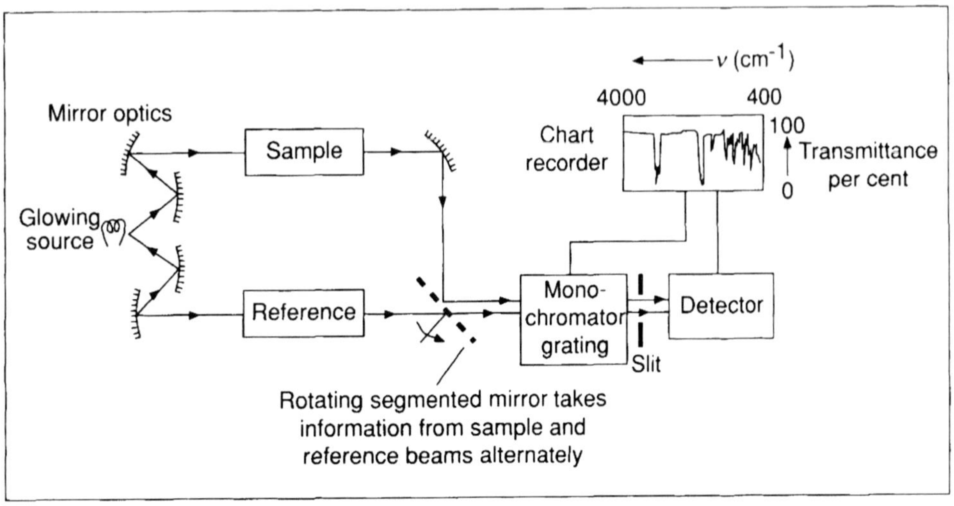

Figure 2 The infrared spectrometer

How it works

The infrared source used can be a simple coil of nichrome wire, or a more complex water cooled rod of silicon carbide. Radiation across the frequency range is passed through the sample. As a particular frequency is absorbed by the sample less radiation is transmitted, and the detector compares the intensity passing through the sample with the intensity passing through the reference – the reference can be air. However, for solutions an identical cell containing the solvent is used as the reference so that the instrument does not record the absorption of the solvent as coming from the sample.

Detecting the radiation passing through the sample or reference cell is usually done by either a photomultiplier or a photodiode (which converts photons of radiation into tiny electrical currents); or a semiconducting cell (that emits electrons when radiation is incident on it) followed by an electron multiplier similar to those used in mass spectrometers (see page 8). Photocells containing triglycine sulphate (TGS) are commonly used, though a mercury cadmium telluride (MCT) cell is often used if fast scans using a very sensitive detector, are required (these operate at liquid nitrogen temperatures). In both cases the spectrum is generated by comparison of the currents produced by the sample and the reference beams.

The data system (a computer) records the spectrum as transmittance (the amount of radiation passing through the sample) against decreasing wavenumber (*ie* increasing wavelength – *eg Fig. 3*).

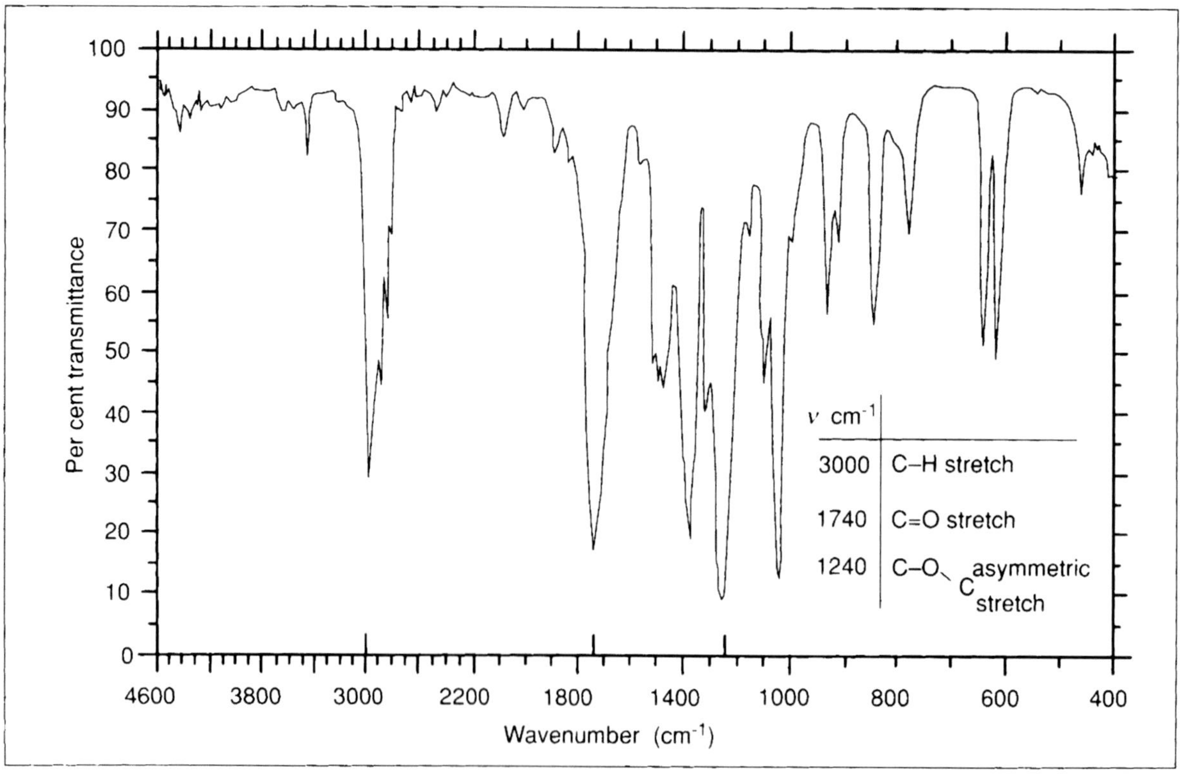

Figure 3 Infrared spectrum of ethyl ethanoate

Fourier transform infrared (FTIR)

Conventional infrared spectra are obtained by passing radiation through the sample and reference, and using a monochromator to select radiation of only one frequency at a time (monochromatic radiation). In FTIR only one beam is used and this means that all the required frequencies pass through the instrument simultaneously. A computer is used to interpret the resulting information (by a mathematical treatment known as Fourier transformation) and plot the spectrum.

The spectrum obtained by using this technique is sometimes slightly different from conventional scanning, but any variation is usually insignificant.

FTIR has a number of advantages over conventional infrared:

1 the whole spectrum can be run in a matter of a few seconds, rather than several minutes;

2 the faster scan speed means that the spectra of compounds leaving a chromatography column can be obtained as they exit the column without collecting them first;

3 the sensitivity of the technique is greater because the 'background noise' is at a much lower level;

4 the spectrum of a solvent or a known impurity can be removed from the observed spectrum because the information is initially converted to a digital signal – a computer can subtract one spectrum from another; and

5 a small sample can yield a spectrum by adding the information from several scans to produce a single spectrum.

Sample preparation

Infrared spectroscopy is not as sensitive as ultraviolet/visible spectroscopy because the energies involved in the vibration of atoms are far smaller than the energies of electronic transitions. However, very small amounts of sample will give good, reproducible, spectra if prepared correctly.

Gases

These are introduced into a special cell, typically 10 cm in length, although longer ones are available for gases with very low partial pressures.

Liquids

Liquids are used simply in the form of a thin film, kept in place by two potassium bromide discs made from single crystals (these are quite expensive). A drop of the liquid is placed on one disc and the second is placed on top and this spreads the sample into a thin film. Sodium chloride cannot be used across the whole frequency range because it absorbs radiation below 625 cm^{-1} and the data in the range 400 – 625 cm^{-1} would be lost. For many situations this is not a problem but where it is potassium bromide can be used instead of sodium chloride.

Solutions can also be held between discs, but if the concentration is very low a greater path length is required. A cell made of potassium bromide must be used. These are available with path lengths between 0.01 mm and 0.5 mm. Clearly, aqueous solutions cannot be used to clean potassium bromide discs or cells, otherwise they will dissolve, so the discs are cleaned with tetrachloromethane, followed by polishing to achieve a flat surface. In extreme cases washing with ethanol can be used so that the small water content in the alcohol dissolves the stained surface layer of the disc. Spectra of aqueous solutions are usually obtained by using special cells (*ie* zinc sulphide or zinc selenide) and the strong absorption peaks due to the water are disregarded or subtracted by computer.

Many organic solvents absorb in the infrared region, so reference cells of the same path length containing the pure solvent are put into the reference beam. Absorption bands of some solvents are shown in *Fig. 4*.

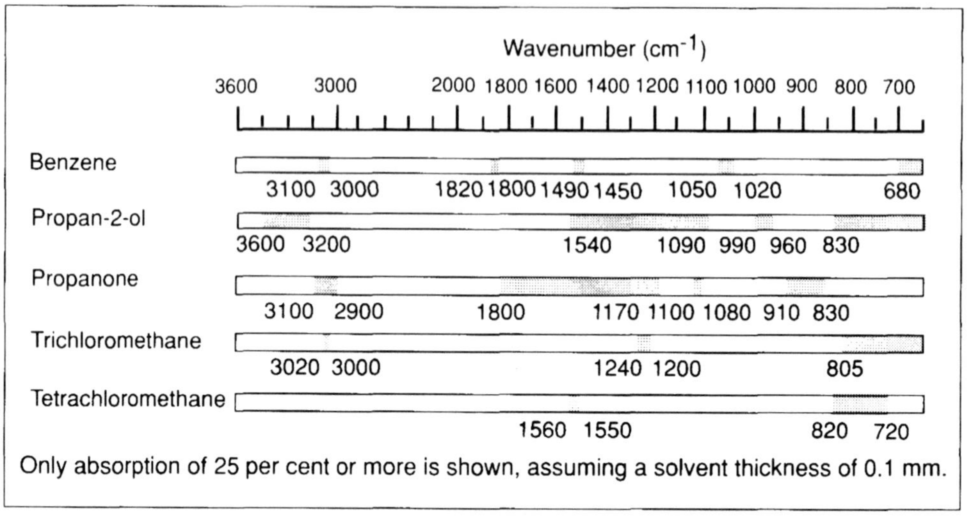

Figure 4 Absorption characteristics of some solvents

The solvent may have an effect on the spectrum. Hydrogen bonding of N-H and O-H groups is particularly responsible for changes in absorption frequency. Hydrogen bonding, for instance, causes bonds to broaden and shift to lower frequencies, see page 78.

Solids

A simple way of obtaining the infrared spectrum of a solid is to prepare a solution. Alternatively, the spectrum of the solid can be determined by first producing a disc or a mull.

A disc is made by grinding *ca* 1 mg of the solid with 100-250 mg of potassium bromide powder using a mortar and pestle. The very fine powder is then put into a circular die, and placed under a mechanical pressure of 1–7 x 10^8 Nm^{-2} (15 000–100 000 pounds per square inch) under vacuum. The pressure is maintained for up to six minutes. The resulting transparent disc is used to record the spectrum. It is essential to produce a very fine powder, otherwise the radiation will simply reflect off the surface of the particles.

Making a mull involves grinding the solid to a fine powder, and then adding a liquid – *eg* Nujol (a long chain hydrocarbon) to produce a paste with a consistency like that of toothpaste. Other liquids can be used if Nujol masks the C–H bands of the sample, and these include hexachlorobutadiene, perfluorokerosene and chlorofluorocarbon grease. The mull is then placed between potassium bromide or sodium chloride discs. This method has the advantage that if the mull layer is too thick the discs can be pressed closer together reducing the thickness and producing a better spectrum – *ie* if the absorptions are too intense.

Vibrations absorbing infrared radiation

Only vibrations resulting in the change of a dipole moment, and having resonant frequencies in the infrared region of the spectrum, will absorb infrared radiation. Consequently, simple gas molecules such as H_2, Cl_2, and O_2 do not have infrared spectra (because they do not have dipoles).

Sulphur dioxide and carbon dioxide both potentially have three ways of absorbing energy and vibrating (*Fig. 5*).

The symmetrical stretch of carbon dioxide does not result in any change of the dipole of the molecule. Thus it would not be expected to absorb radiation, and this

Unilever

THE ROYAL SOCIETY OF CHEMISTRY

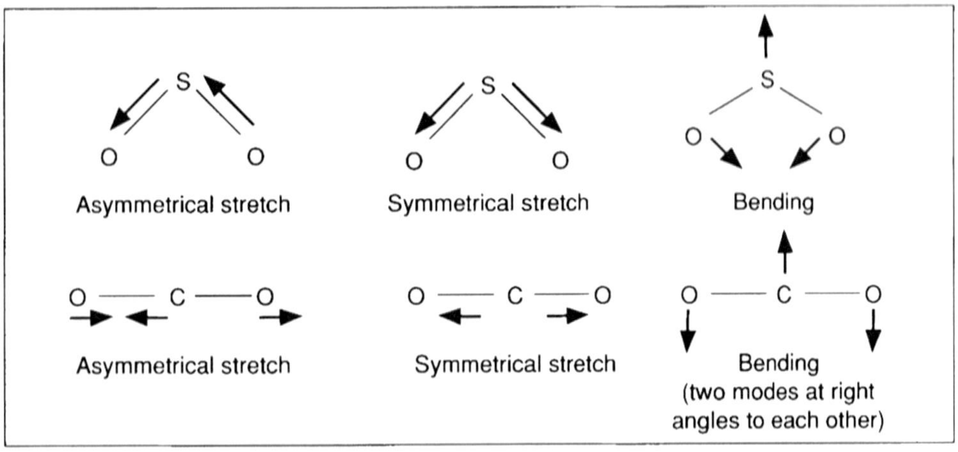

Figure 5 Vibrational modes of sulphur dioxide and carbon dioxide

mode is inactive. All the other modes of both molecules do change the molecular dipole, and absorb infrared radiation – *ie* they are active modes (*Figs. 6* and *7*).

THE ROYAL SOCIETY OF CHEMISTRY

Unilever

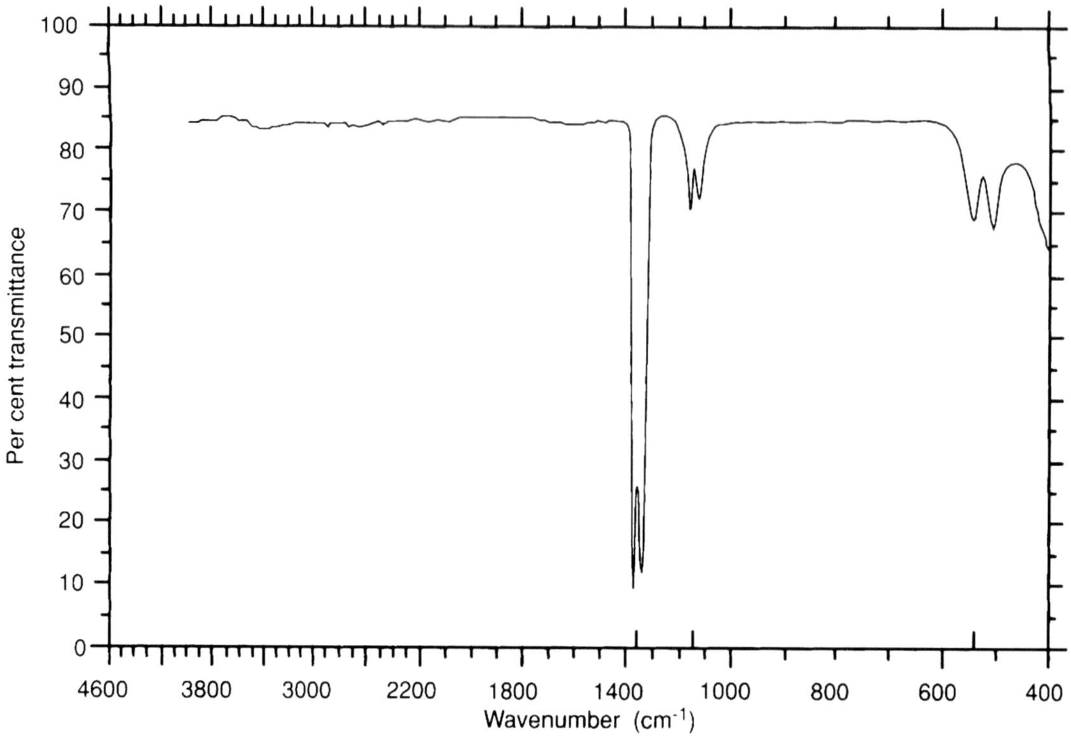

ν cm^{-1}	
1360	Asymmetrical stretch
1150	Symmetrical stretch
540	Bending

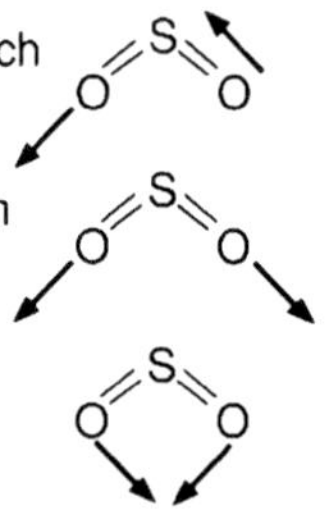

Figure 6 Infrared spectrum of sulphur dioxide

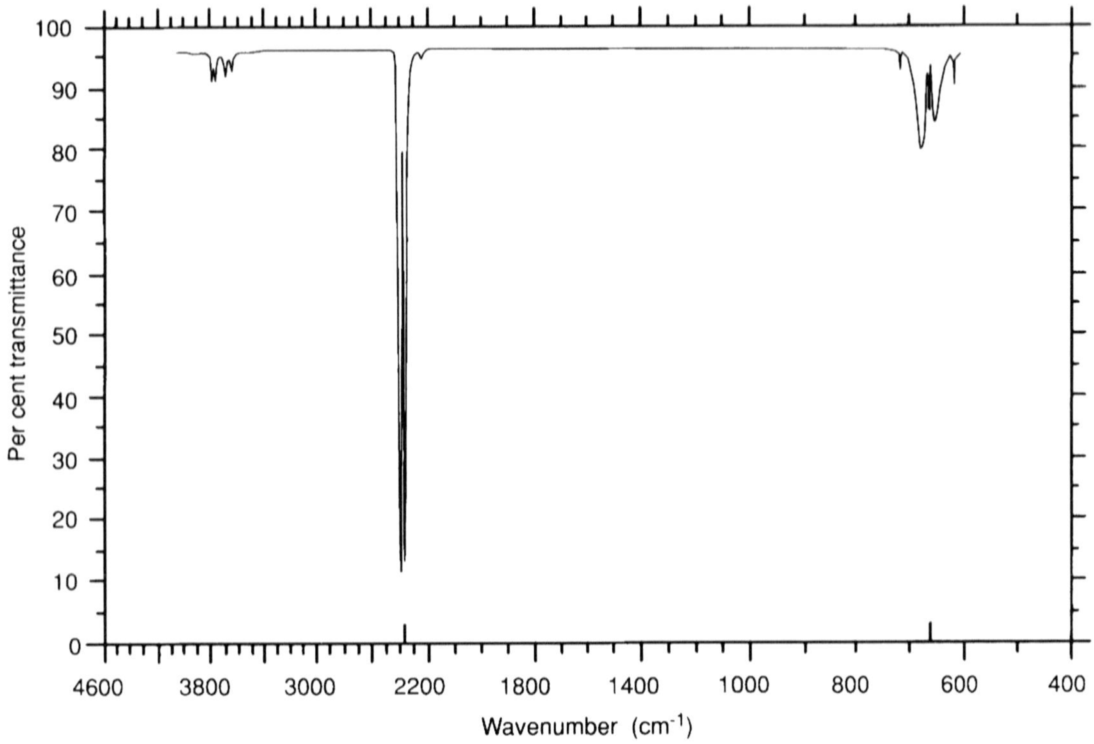

ν cm^{-1}		
2360	Asymmetrical stretch	O=C=O
670	Bending	O=C=O

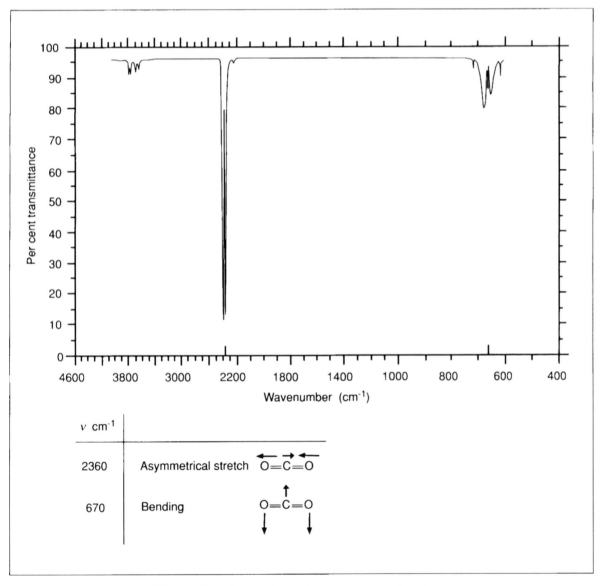

Figure 7 Infrared spectrum of carbon dioxide

Interpretation of spectra

Simple molecules such as carbon dioxide and sulphur dioxide have spectra that are relatively easy to interpret. However, as the complexity of the sample increases it becomes more difficult to assign each absorption to a particular vibrational mode. This is not a great problem because although the exact description of the vibration is difficult, it is possible to assign particular peaks to the vibrations of functional groups. An example of this is the C=O bond stretching in organic molecules, which always occurs in the range 1640–1815 cm^{-1} (Table 2).

The infrared spectrum of propanal illustrates this, because the strong absorption at 1730 cm^{-1} falls into the range expected for an aliphatic aldehyde (1740–1720 cm^{-1}, *Fig. 8*).

THE ROYAL SOCIETY OF CHEMISTRY

Unilever

Table 2 Characteristic absorptions of carbonyl groups

		Wavenumber (cm^{-1})
RCOCl	acyl chloride	1815–1790
RCOOR'	aliphatic ester	1750–1730
RCHO	aliphatic aldehyde	1740–1720
RCOOH	aliphatic acid	1725–1700
RCOR'	aliphatic ketone	1725–1700
ArCHO	aromatic aldehyde	1715–1695
ArCOR	aromatic ketone	1700–1680
ArCOAr	diaromatic ketone	1670–1650
$RCONH_2$	aliphatic amide	1680–1640

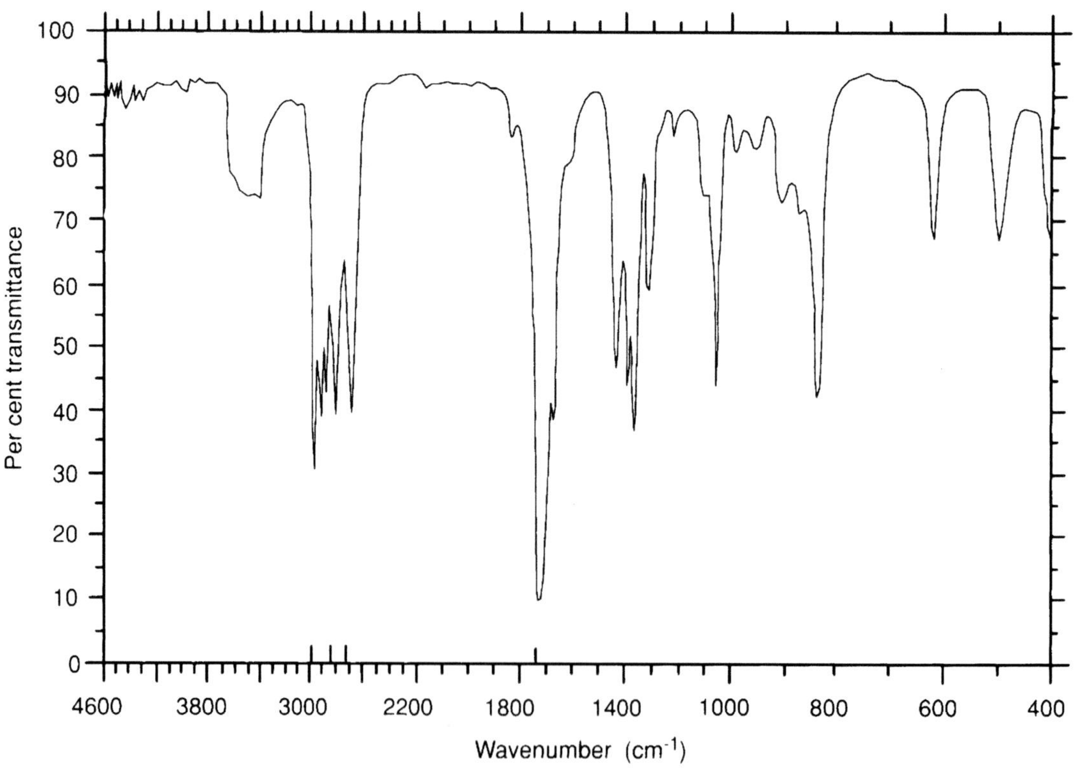

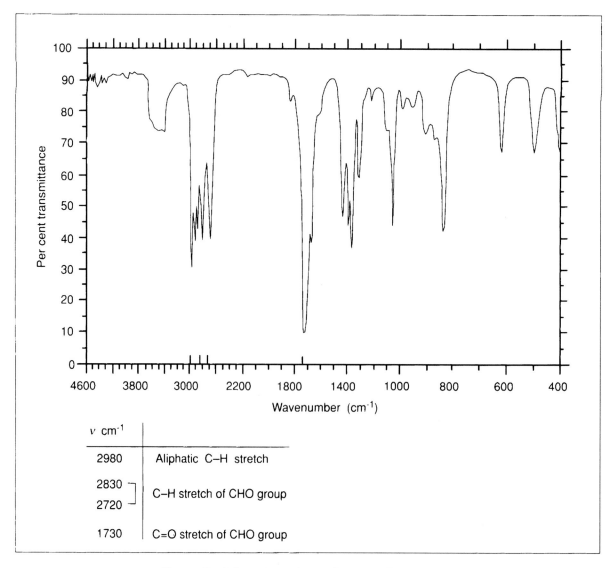

ν cm^{-1}	
2980	Aliphatic C–H stretch
2830 2720	C–H stretch of CHO group
1730	C=O stretch of CHO group

Figure 8 Infrared spectrum of propanal

Because particular types of vibration always occur at a similar frequency it is possible to build up a table of characteristic absorption frequencies. So, when the infrared spectrum of a compound of unknown structure is presented, it is probable that sufficient information can be derived from the spectrum to identify the functional groups present. In most cases more information is normally required to determine the structure.

The fact that functional groups give absorptions at particular frequencies can be used to show the purity of a sample. Contamination owing to solvent residues or by-products will show absorptions not observed in the pure compound. This is used extensively in industry, where it is possible to find that the same compound is marketed by a number of manufacturers but the infrared spectra of the products are slightly different. This happens in the pharmaceutical industry, particularly, where the infrared spectrum of a drug or its formulation is often included in its patent.

It is quite common for correlation tables to include data on the usual intensity and/or band width of absorption peaks as well as their wavenumbers (Table 3).

THE ROYAL SOCIETY OF CHEMISTRY

Unilever

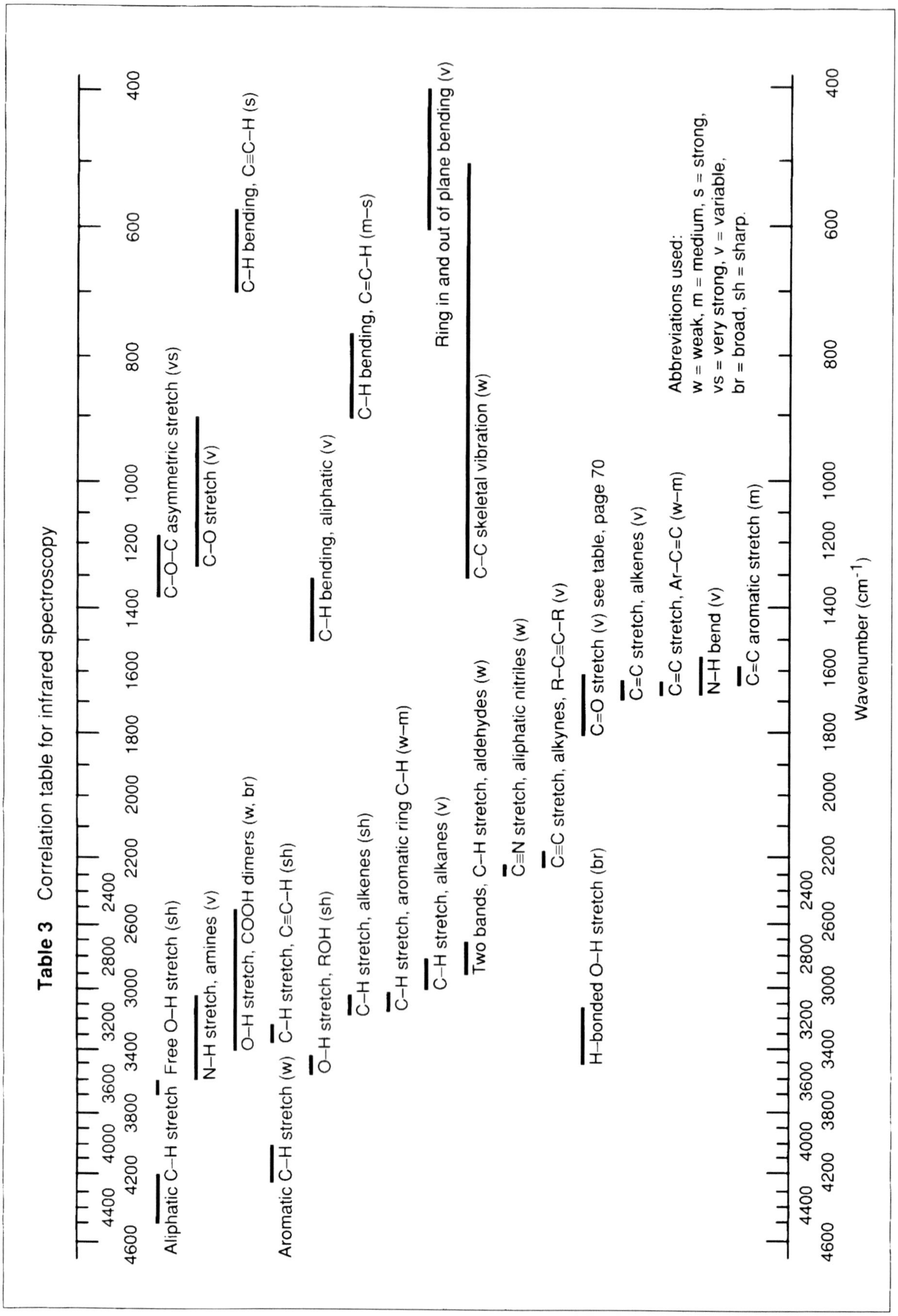

Unilever

THE ROYAL SOCIETY OF CHEMISTRY

The infrared spectrum can conveniently be split into four regions for interpretation:

4000–2500 cm^{-1}: absorption of single bonds to hydrogen – *eg* C–H, O–H, N–H

2500–2000 cm^{-1}: absorption of triple bonds – *eg* C≡C and C≡N

2000–1500 cm^{-1}: absorption of double bonds – *eg* C=C, C=O

1500–400 cm^{-1}: absorption owing to other bond deformations – *eg* rotating, scissoring and some bending.

There are some exceptions – *eg* N–H bending is observed at 1550-1620 cm^{-1}.

4000–2500 cm^{-1}

The high frequency is explained by the low mass of the hydrogen atom. The spectrum is altered if ^2H is present – *ie* the C–H stretch of $CHCl_3$ is at a higher frequency than that of $CDCl_3$ (*Figs 9* and *10*).

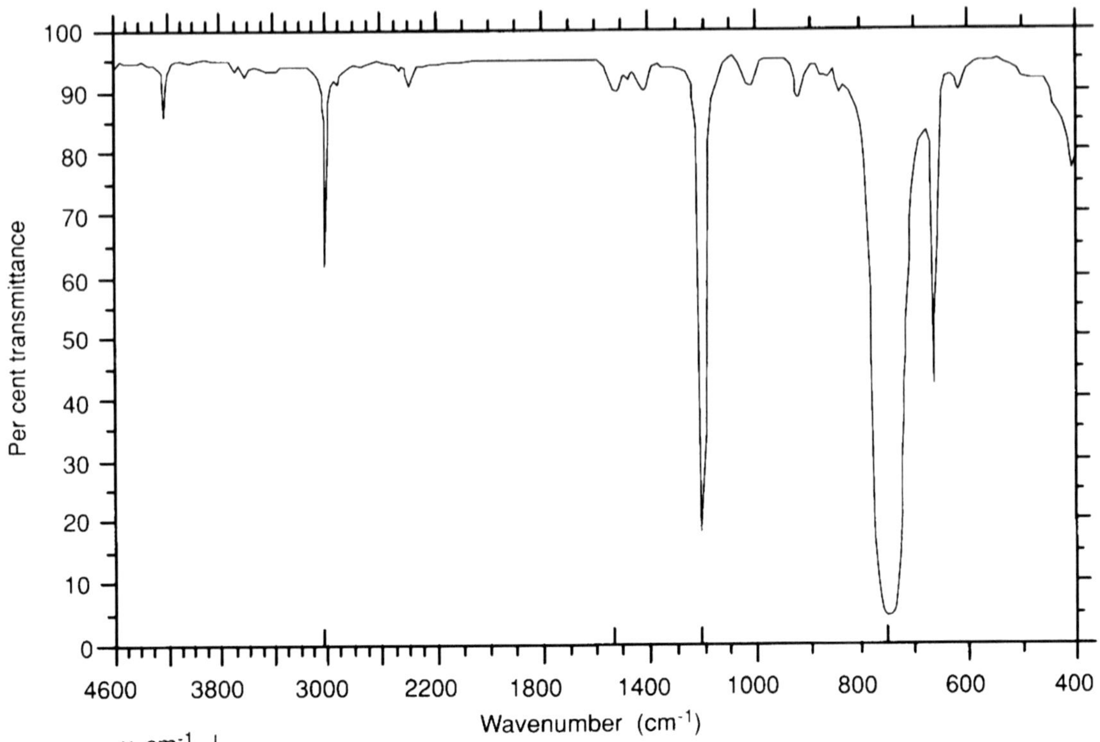

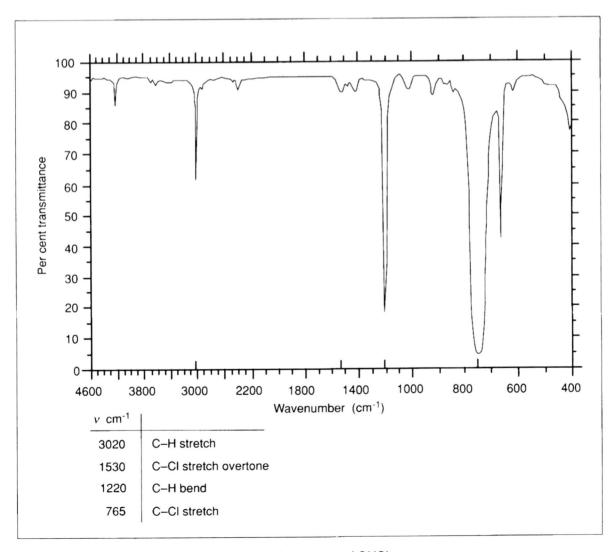

ν cm^{-1}	
3020	C–H stretch
1530	C–Cl stretch overtone
1220	C–H bend
765	C–Cl stretch

Figure 9 Infrared spectrum of $CHCl_3$

THE ROYAL SOCIETY OF CHEMISTRY

Unilever

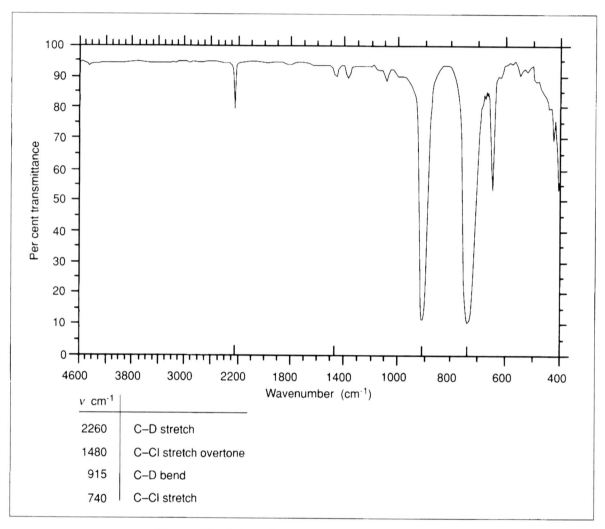

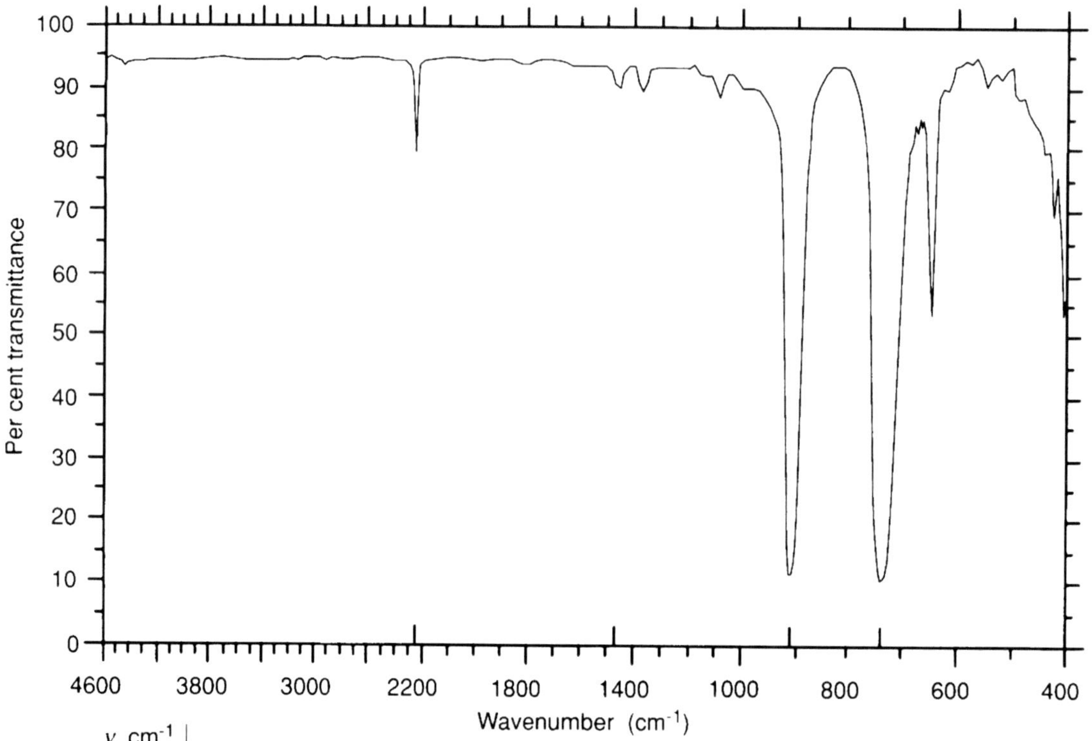

ν cm^{-1}	
2260	C–D stretch
1480	C–Cl stretch overtone
915	C–D bend
740	C–Cl stretch

Figure 10 Infrared spectrum of $CDCl_3$

2500–2000 cm^{-1}

Relatively high frequencies are required to provide the high energies necessary to make the strong triple bonds vibrate – *eg* $CH_3CH_2C{\equiv}N$ (*Fig. 11*).

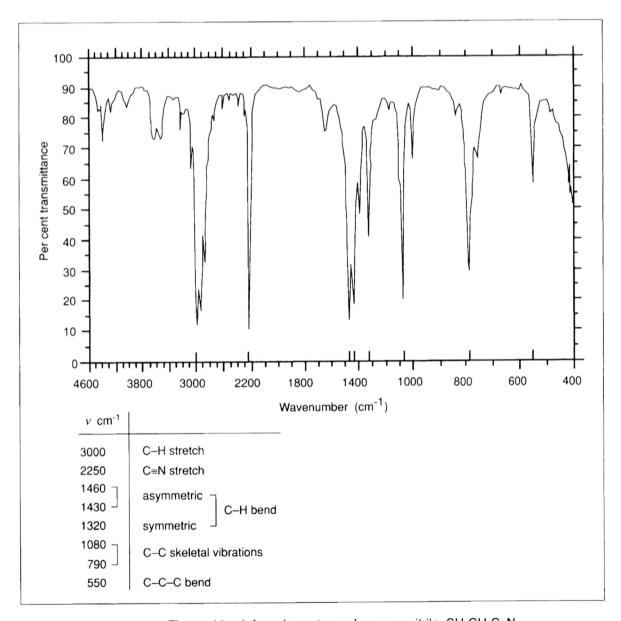

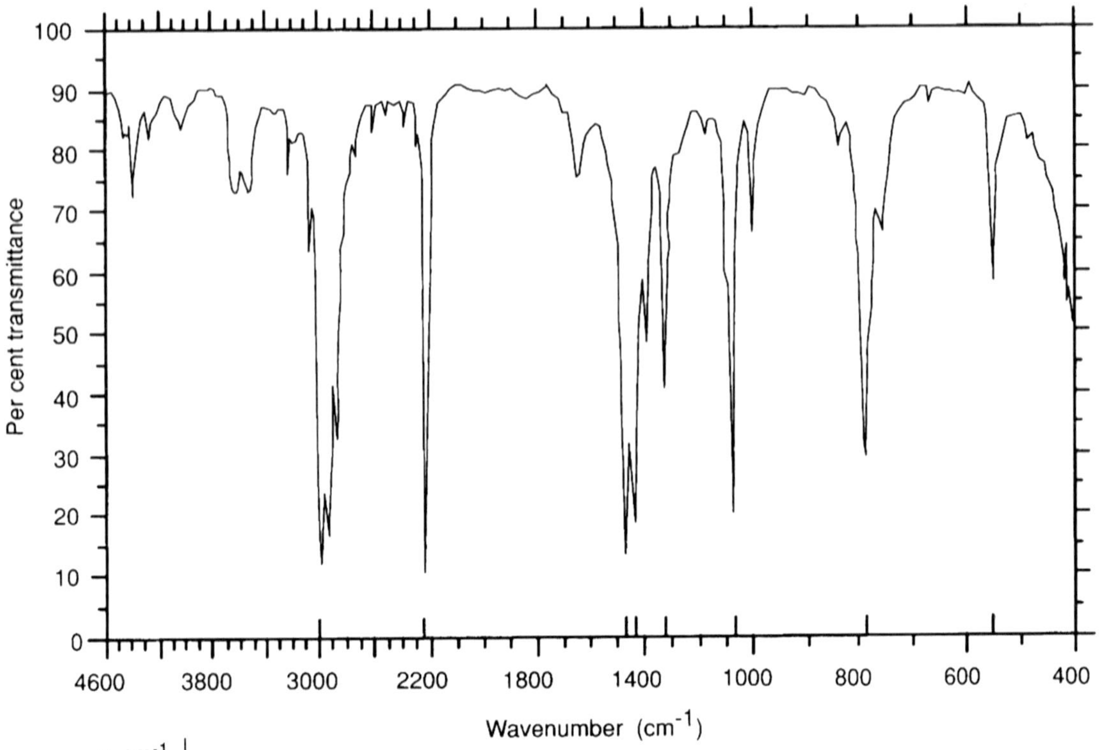

ν cm^{-1}	
3000	C–H stretch
2250	C≡N stretch
1460, 1430	asymmetric C–H bend
1320	symmetric C–H bend
1080, 790	C–C skeletal vibrations
550	C–C–C bend

Figure 11 Infrared spectrum of propanonitrile, $CH_3CH_2C{\equiv}N$

THE ROYAL SOCIETY OF CHEMISTRY

Unilever

2000–1500 cm^{-1}

Double bonds generally absorb in this region – *eg* C=O in propanal (*Fig. 8*). *Figure 12* shows the spectrum of hex-1-ene, with a sharp absorption peak due to the C=C stretch at 1640 cm^{-1}. Other groups absorbing here include C=N and C=O.

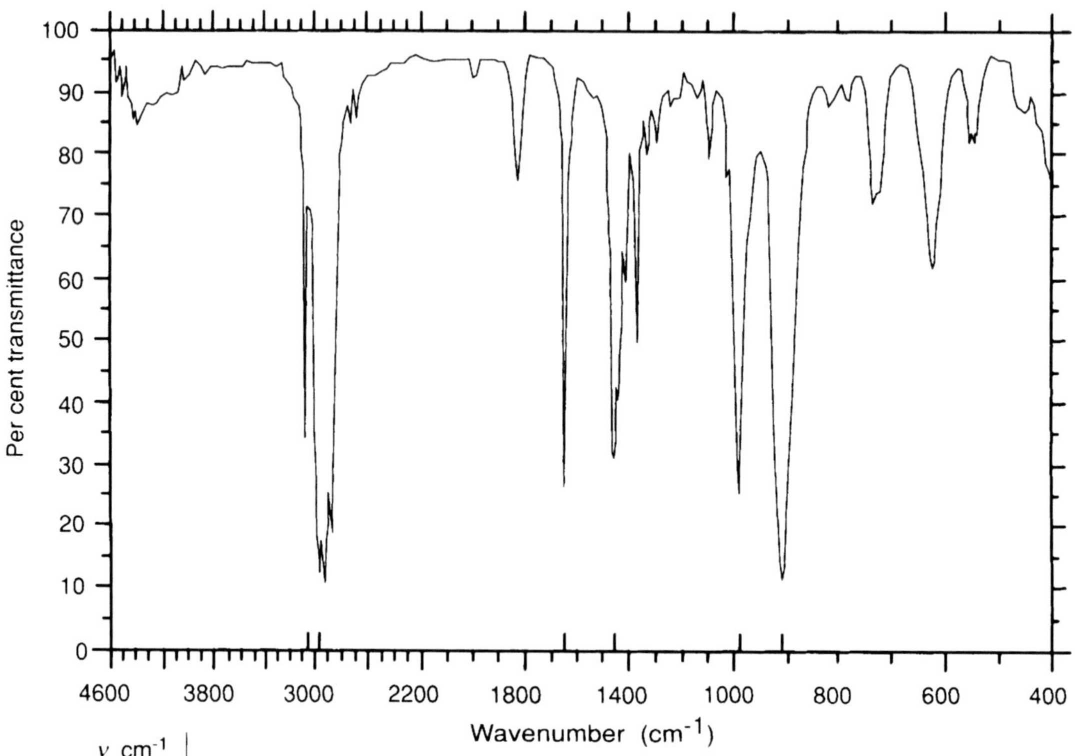

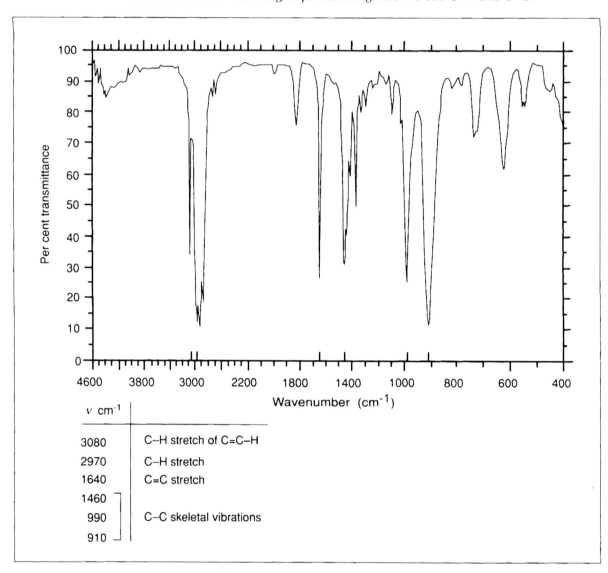

ν cm^{-1}	
3080	C–H stretch of C=C–H
2970	C–H stretch
1640	C=C stretch
1460	C–C skeletal vibrations
990	C–C skeletal vibrations
910	C–C skeletal vibrations

Figure 12 Infrared spectrum of hex-1-ene

Unilever

THE ROYAL SOCIETY OF CHEMISTRY

1500-400 cm^{-1}

Absorption owing to deformations such as rotating, scissoring and some bending depend on the combination of bonds in the molecule. This part of the spectrum is unique to each compound and is often called the 'fingerprint' region. It is rarely used for identifying particular functional groups, but some generalisations can be made – *eg* C–O absorbs at 1300–1020 cm^{-1} (although absorption here is not exclusive to C–O). However, the fingerprint region can help identify particular molecules because no other compound will have the same pattern of absorptions. Many research organisations have reference databases of stored spectra, including the fingerprint region.

Figure 13 shows the infrared spectrum of phenol, which has extensive absorption in the fingerprint region. The other significant absorption at 3600 cm^{-1} is due to the O–H vibration.

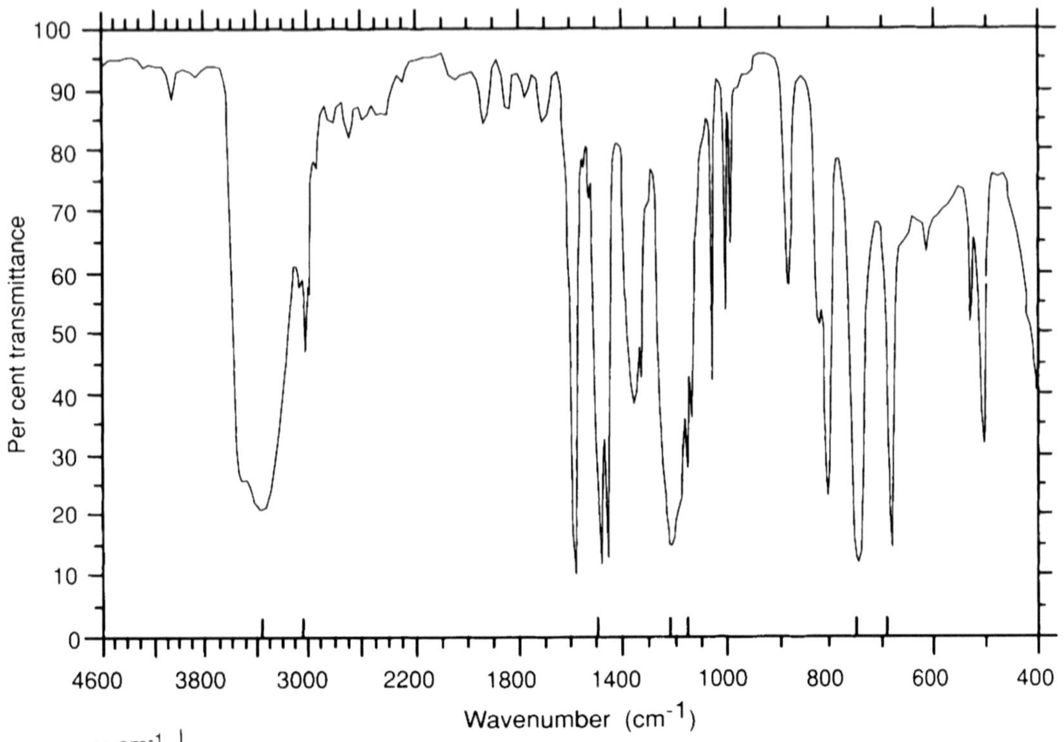

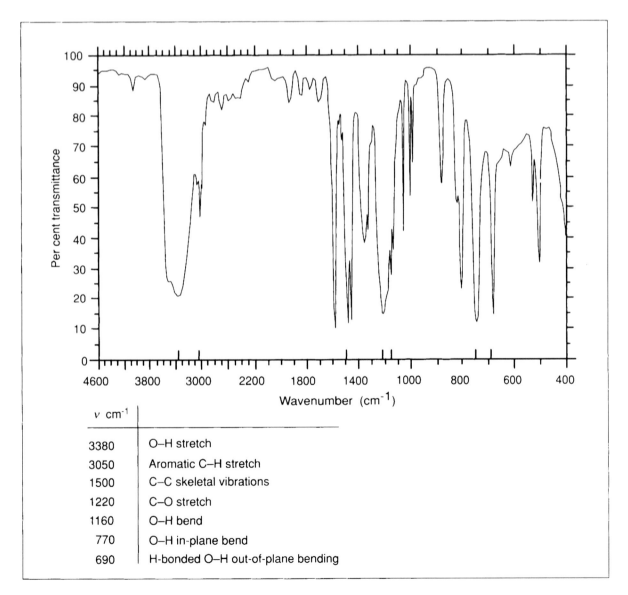

ν cm^{-1}	
3380	O–H stretch
3050	Aromatic C–H stretch
1500	C–C skeletal vibrations
1220	C–O stretch
1160	O–H bend
770	O–H in-plane bend
690	H-bonded O–H out-of-plane bending

Figure 13 Infrared spectrum of phenol

THE ROYAL SOCIETY OF CHEMISTRY

Unilever

Hydrogen bonding

The absorption of the O–H group in alcohols and carboxylic acids does not usually appear as a sharp peak. Instead a broad band is observed because the vibrational mode is complicated by hydrogen bonding. Precisely how broad the band is depends on the degree of hydrogen bonding, and the frequency of an alcoholic O–H stretch depends on whether the alcohol is primary, secondary or tertiary. *Figures 14* and *15* show the spectra of methanol and methanoic acid.

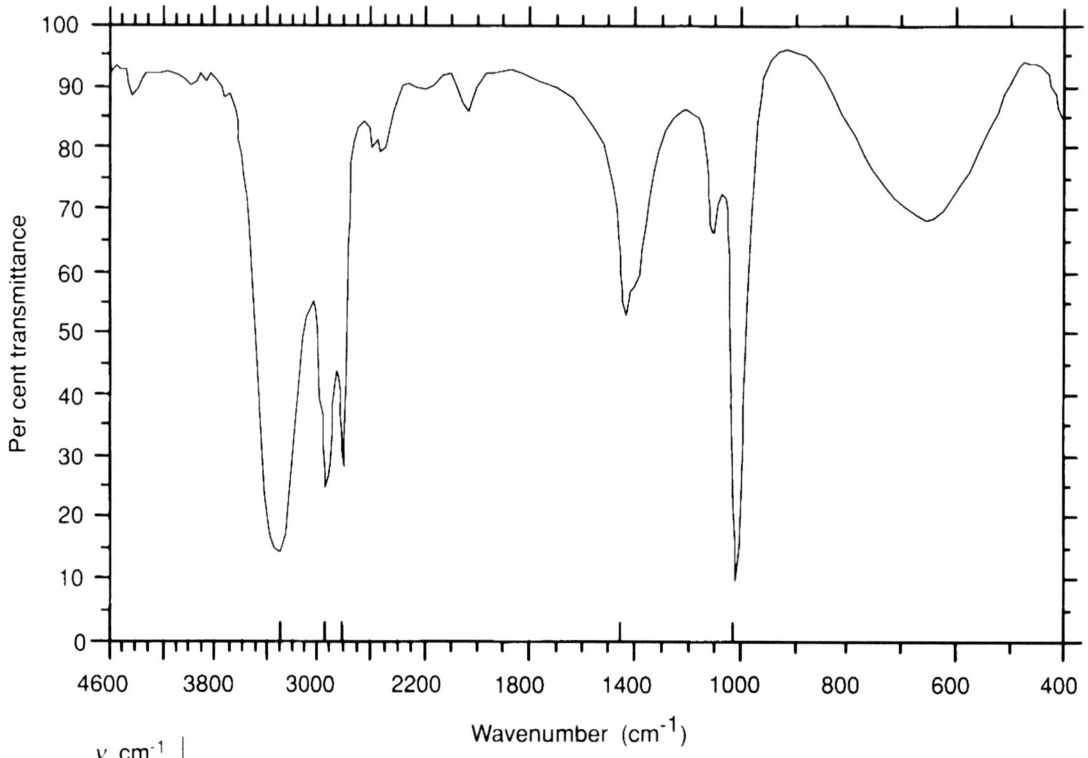

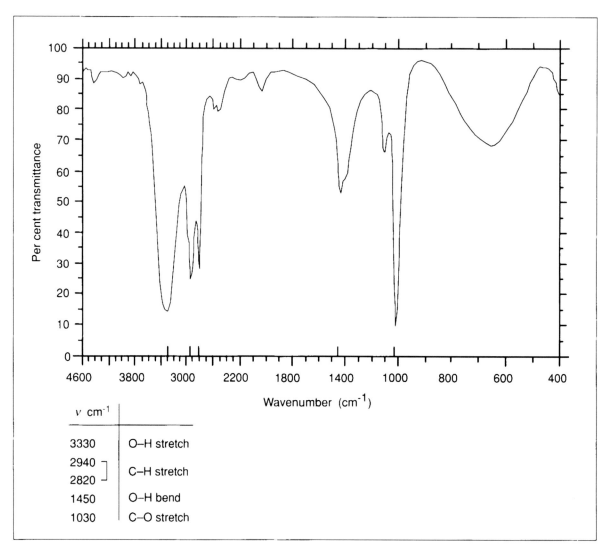

ν cm^{-1}	
3330	O–H stretch
2940, 2820	C–H stretch
1450	O–H bend
1030	C–O stretch

Figure 14 Infrared spectrum of methanol

Unilever

THE ROYAL SOCIETY OF CHEMISTRY

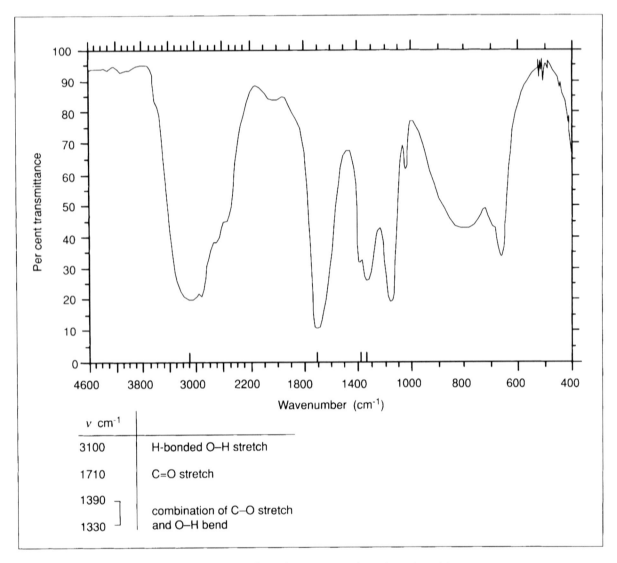

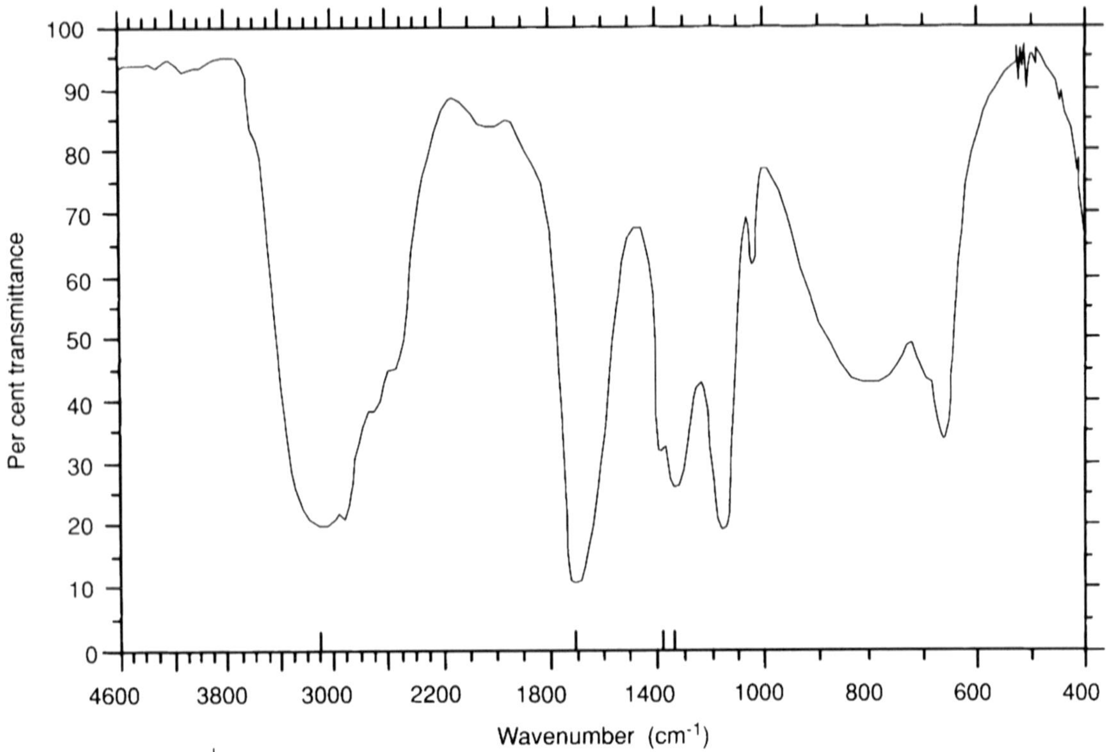

ν cm^{-1}	
3100	H-bonded O–H stretch
1710	C=O stretch
1390 1330	combination of C–O stretch and O–H bend

Figure 15 Infrared spectrum of methanoic acid

Differentiating between intermolecular and intramolecular bonding requires a series of spectra at different dilutions. An absorption band that diminishes as the solution becomes more dilute comes from intermolecular hydrogen bonding (between molecules) – whereas intramolecular bonding shows a much smaller effect.

Applications of infrared absorption

Advances in instrumentation and accessories available now make it possible to obtain spectra of materials previously thought too difficult. In many cases the sample size or concentration was too small, but this problem has been overcome by the construction of specialist cells. These have reflecting surfaces on their insides so that the path length through a liquid, for example, is increased (*Fig. 16*). An increased path length increases absorption according to Beer's Law (page 95).

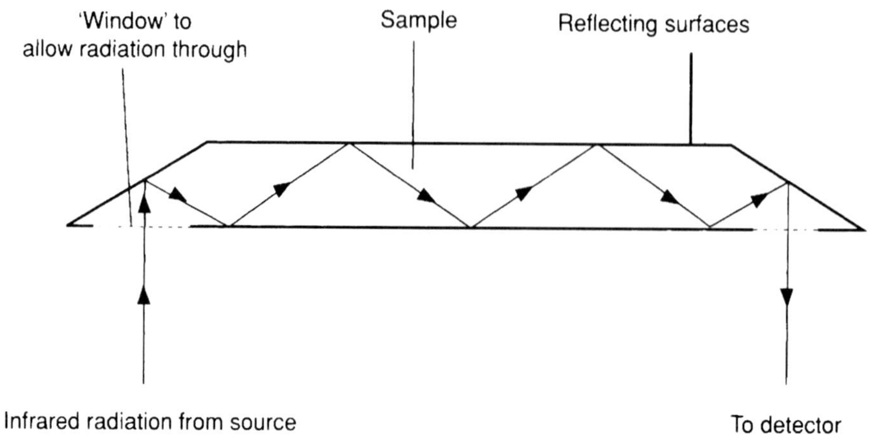

Figure 16 Schematic diagram of a reflectance cell

The infrared spectra of proteins were once rather difficult to observe because of the absorptions due to the solvents used (*ie* water, D_2O and methanol – all of which absorb infrared radiation). By using a reflectance cell and a computer it is now possible to record the spectra of proteins in solution and the pure solvent digitally, and by subtracting one from the other the spectrum of the protein can be obtained.

The secondary structure of proteins changes with pH, solvent, temperature, and exposure to lipids, and it is possible to use the vibrational frequencies of the protein backbone to determine the proportions of α-helix, β-pleated sheet and disordered protein in a sample. Significant regions of the spectrum in this respect are 1670–1620 cm^{-1} and 1305–1200 cm^{-1}.

The trend away from saturated fats in foods has meant that it is now necessary to measure the degree of unsaturation in oils and fats. Oils such as peanut oil can be placed in a reflectance cell to record their spectra and by measuring the areas under the peaks corresponding to unsaturated and saturated environments (*eg* at 3009 cm^{-1} and 2854 cm^{-1} respectively) it is possible to calculate the ratio of unsaturated:saturated bonds.

Microbiological systems can also be identified by using infrared spectroscopy because research has shown that bacteria can be categorised by their unique spectra. The spectra appear as broad complex contours rather than sharply defined peaks, and contain information on the total composition of the bacteria including RNA (ribonucleic acid) and DNA (deoxyribonucleic acid). Studies have extended to the successful identification of viruses, fungi, yeasts and amoebae, by using colonies of fewer than 10^4 cells which measure less than 4×10^{-5} m across.

In the past, some compounds have not yielded useful spectra because of their absorption characteristics – *eg* the spectra of black rubbers, which absorb radiation of all frequencies, have now been recorded. They can be ground to a powder in liquid nitrogen before being made into a potassium bromide pellet. Alternatively, thermogravimetric analysis and FTIR can be combined so that the sample is heated under controlled conditions to provide data on weight loss, and the species lost during heating are analysed by passing them through a cell in an infrared spectrometer. Spectra are collected continuously and are stored on a computer. If the spectra generated suggest that a number of compounds are produced at the same time it is possible to separate them by passing the mixture down a heated gas chromatography column (see Chapter 5 for details of gas chromatography).

The most spectacular combination of techniques is probably the linking of an optical microscope to an FTIR spectrometer. By using this system it is possible to study an object visually on a microscopic scale, and then obtain its infrared spectrum. To cut out stray radiation and to eliminate the risk of recording the spectrum of the background, different sized apertures are used to eliminate the unwanted regions of the visual image. Because some substances reflect radiation from their surfaces it is also possible to obtain two types of infrared spectrum through the microscope:

1. transmissive spectra in which the radiation passing through the sample is measured; and
2. reflective spectra whereby the radiation reflected back towards the radiation source is measured.

Sample sizes down to 2×10^{-5} m can be used, although many scans are necessary and the individual scans have to be added together to obtain the final spectrum.

Examples of other applications of infrared spectroscopy include:

1. the study of particles blocking air filters – the information obtained can be used to discover plant operation difficulties;
2. the study of how fats crystallise in margarines and low fat spreads – this is important because the physical properties of these foodstuffs depend on the crystal structures within them;
3. the categorising or identification of fibres and paint chips in forensic analysis; and
4. the screening of illicit substances – many illicit drugs are diluted or 'cut' by addition of other powdered materials and individual crystals can be identified from their spectra. Many of the powders added to dilute the pure drugs contain carbohydrates.

Drink-driving

A more familiar application (to some!) of infrared absorption is in the analysis of breath samples given by drivers who have been drinking.

Once alcohol has been consumed it is absorbed by the blood through the walls of the stomach and is transported to the liver, where it is removed at a slow rate – approximately one unit per hour. (One unit is the amount of alcohol contained in a glass of sherry, port or wine, a single measure of a spirit, or half a pint of beer.) The consumption rate is usually far greater than this, and the alcohol that cannot be removed by the liver passes through to the heart. From there it is pumped via the blood to the lungs where it is oxygenated prior to being distributed around the rest of the body.

It is while the blood is passing through the lungs that some alcohol is transferred to the breath (*Fig. 17*).

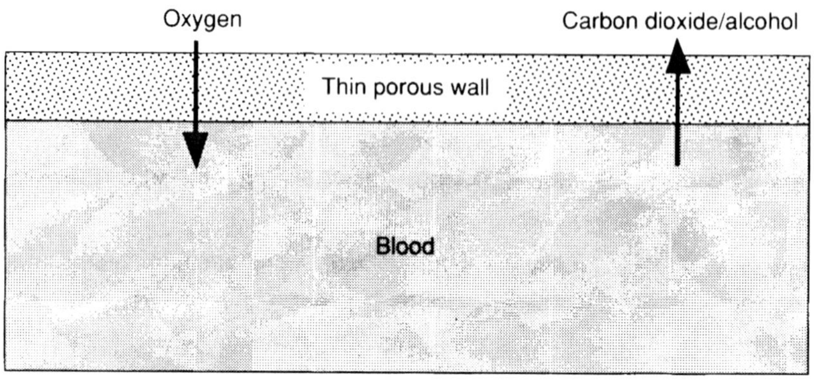

Alcohol and carbon dioxide evaporate from the blood into the breath; oxygen passes in the reverse direction

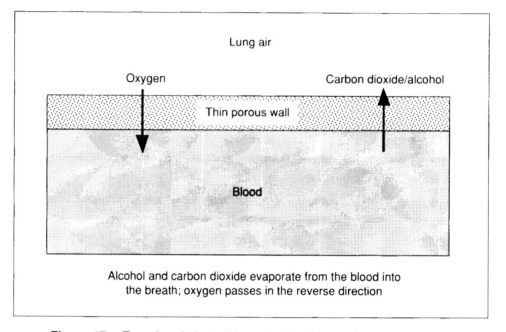

Figure 17 Transfer of alcohol from the blood to the breath

A static equilibrium cannot be established between the blood and breath concentrations of alcohol because the body is not a closed system. However, the relationship between the two concentrations at a given time has been determined and the 'blood/breath ratio' is accepted to be 2300:1 (therefore breath alcohol concentrations in μg per 100 cm^3 can be converted to blood alcohol concentrations in mg per 100 cm^3 by multiplying by 2.3). A small variation does exist from person to person.

Because the concentration of alcohol on the breath is proportional to the concentration in the blood the analysis of exhaled air has become an accepted means of determining whether a person is driving illegally (this is not always the same thing as being fit to drive). The current upper limit to remain within the law (1992) is 35 μg of alcohol per 100 cm^3 of breath, or 80 mg of alcohol per 100 cm^3 of blood.

Although the legal limit is 35 μg per 100 cm^3, the national policy is that prosecution does not follow unless a breath alcohol concentration is at least 40 μg of alcohol per 100 cm^3 of breath. This allows for some variation in the ratio of blood to breath alcohol (the instruments are recalibrated regularly so no compensation need be made for any inaccurate machines). Between 40 and 50 μg of alcohol per 100 cm^3 of breath means that the 'client' is given the option of having a blood or urine test, in which case the sample taken is divided into two. The police have one portion analysed, and the 'client' has the option of having an independent analysis of the other one. If a blood sample, for example, is taken then the figure used to decide whether to take a case to court is the blood alcohol concentration. If the figure is above 50 μg alcohol per 100 cm^3 of breath the 'client' is not given the option of a blood or urine test because the breath alcohol concentration is itself sufficient to convict a driver, and will not disguise a blood alcohol concentration within the legal limit.

Roadside and personal breath test instruments usually contain fuel cells. Typically these consist of a thin permeable film of sintered glass, a ceramic or a polymer, impregnated with a liquid or gel of sodium hydroxide and/or phosphoric acid. The membrane is coated on both sides with a precious metal such as platinum or silver and the metal on one side of the film catalyses the oxidation of the alcohol vapour on the breath of the operator, while the oxygen on the other side of the membrane is reduced. An electromotive force (EMF) is generated, and is converted to a signal which indicates how close the operator is to the legal limit. In the UK these instruments are not used to provide evidence as they are less automated and do not provide a printout.

If a roadside breath test suggests that a driver has too high an alcohol level on their breath, they are taken to a police station where two further breath samples are analysed. (The sample with the lower concentration is used in evidence, the higher reading being disregarded.) British police forces currently use two models, one manufactured by Lion Laboratories and the other by Camic. Both work on the basis of the absorption of infrared radiation by alcohol vapour. (The following description relates to the Lion instrument, the Intoximeter 3000.)

The infrared spectrum of ethanol (*Fig. 18*) shows that the major absorptions are found at 3340 and 2950 cm^{-1}. These correspond to the O–H and C–H vibrations. Alcohol concentrations cannot be determined from the O–H vibration because of the water vapour present in both the atmosphere and the 'client's' breath, therefore the C–H vibration is used.

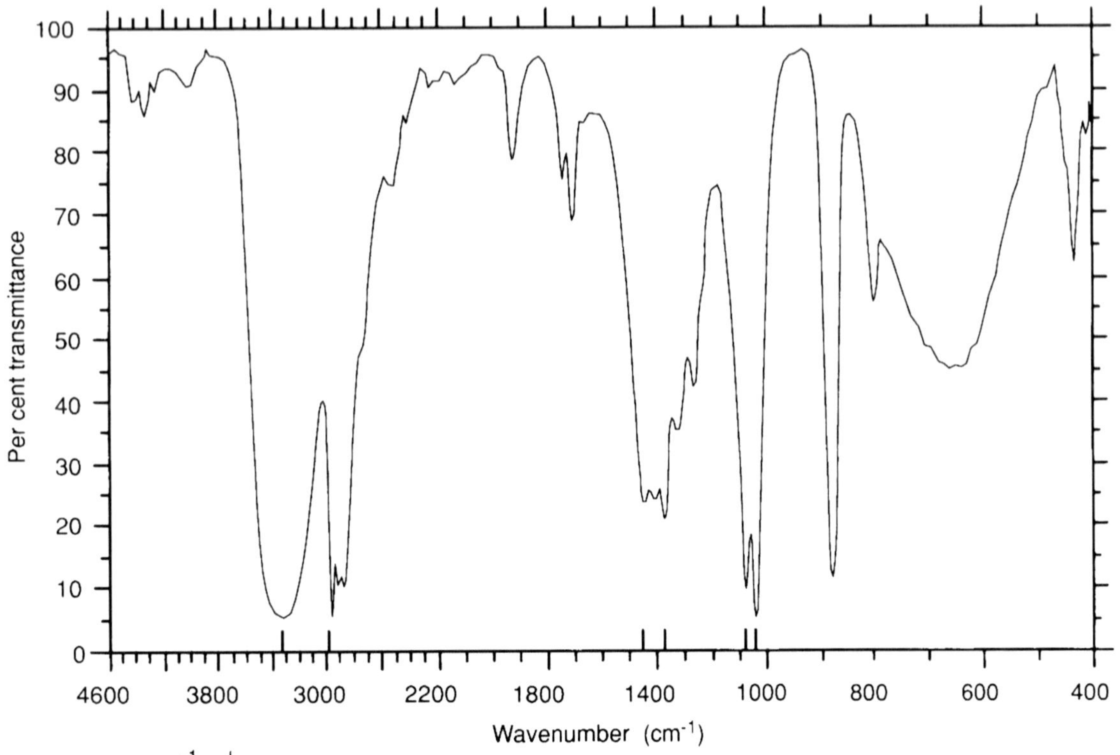

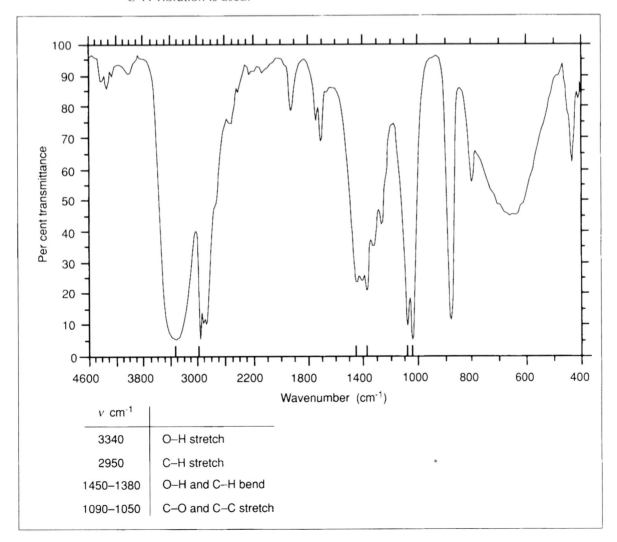

ν cm^{-1}	
3340	O–H stretch
2950	C–H stretch
1450–1380	O–H and C–H bend
1090–1050	C–O and C–C stretch

Figure 18 Infrared absorption spectrum of ethanol

The Intoximeter 3000 uses a nichrome filament operating at 800 °C as its radiation source, and the beam is split by a rotating chopper so that it passes alternately through the sample chamber and an identical reference chamber (*Fig. 19*). The chamber is 280 mm long and has a volume of 70 cm^3, rather larger than the cells used in routine infrared analysis. So that the alcohol does not condense, the chamber is thermostatically heated to 45 °C. The radiation passes through a filter mounted on the solid state photoconducting detector so that only the absorbance at 2950 cm^{-1} is measured. This eliminates interference from other materials on the breath that do not absorb at this frequency.

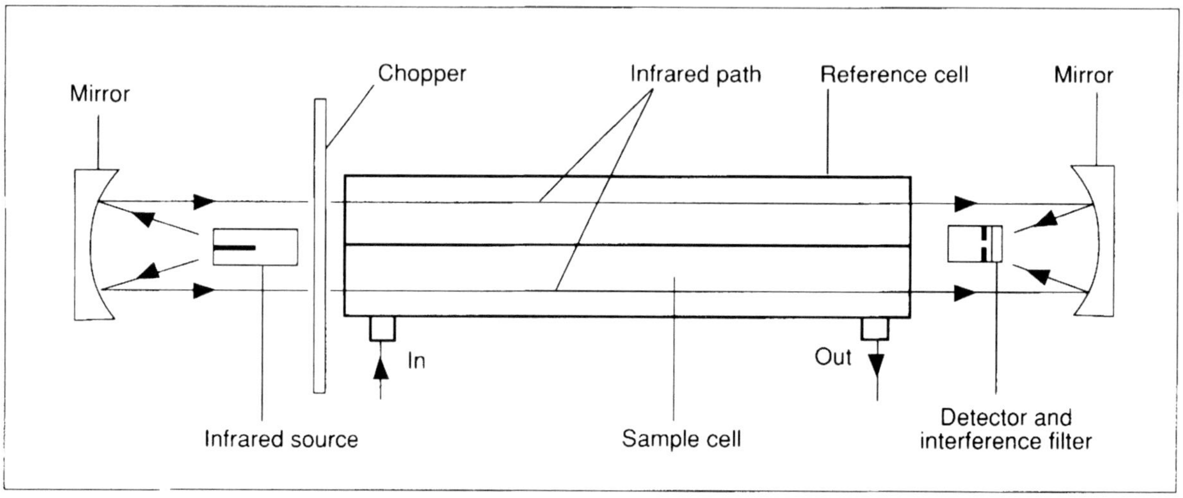

Figure 19 Components of the infrared analyser unit of the Lion Intoximeter 3000

Absorption follows the Beer–Lambert law, and the non-linearity of absorption at higher concentrations is compensated for by suitable software. The instrument is calibrated against a standard, and the absorption is converted to a measure of breath alcohol concentration. This is the figure that is finally given by the machine.

Some other substances also absorb at 2950 cm^{-1}. Of butane, 1,1,1-trichloroethane, and methylbenzene used in solvent abuse only butane registers on the machine. Trials have shown that any signal from butane disappears within the 20 minute period that must elapse between a drink being taken and the test being carried out.

More significantly, propanone absorbs at 2950 cm^{-1} and is found on the breath of some individuals (those with high protein intakes or who are diabetics). A second semiconducting detector is therefore built into the intoximeter that is more sensitive to propanone than to ethanol, while the infrared detector is more sensitive to the alcohol. When the instrument is calibrated the ratio of sensitivity of the two sensors to the two substances is measured and is stored in the computer memory. As a result, if propanone is present on the client's breath the semiconducting sensor will give a higher signal than if ethanol alone is present. The change in ratio of the two signals is then reduced, eliminating the contribution to the infrared absorption by the propanone.

Recent developments

A new technique based on infrared spectroscopy called photoacoustic spectroscopy has been developed over the past decade. In this technique infrared radiation passes through the window of a detector cell and strikes the solid sample. The absorption of the radiation results in sample heating and thermal waves are produced in the sample. These waves are transmitted as pressure waves to the gas around the sample.

In the gas they are acoustic waves with the same frequency as the infrared radiation absorbed, and are detected by a microphone in the cell. The detector cell has a sealed window to allow the infrared radiation through, but not any sounds or vibrations that the microphone might detect.

The waves are generated from within the sample so this method allows investigation of samples at different depths under the surface, depending on the conditions. The technique appears to show most potential in studies of polymers and surface coating properties.

Exercises

It is rare for a structure to be determined on the basis of an infrared spectrum alone. It is more common to use the technique to confirm the presence of functional groups or to support the structures suggested by other techniques. With sufficient other information it is possible to derive a lot of information about the structure or isomer of a compound from its infrared spectrum. Alternatively, the fingerprint region can be used to confirm the presence or otherwise of a compound in a sample.

Exercise 1

The sample appears as a colourless mobile liquid of relative mass 78.5.

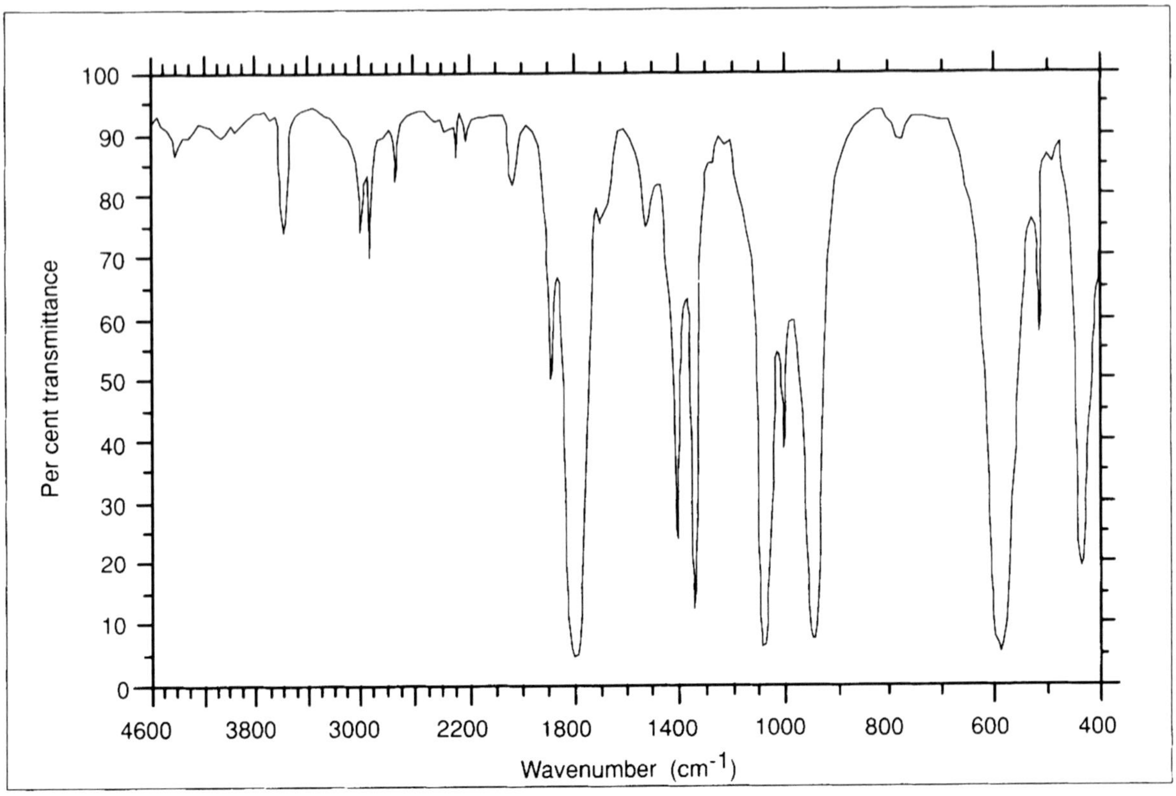

Figure 20

The strong absorption at 1790 cm^{-1} suggests an acid chloride, and the non-integral relative mass suggests that chlorine is present. Subtracting the mass of the COCl group leaves 15 mass units, which corresponds to a methyl group. Therefore the substance is ethanoyl chloride, CH_3COCl.

THE ROYAL SOCIETY OF CHEMISTRY

Unilever

Exercise 2

A solid sample has a composition C 68.9 per cent, H 4.9 per cent, O 26.2 per cent by mass, and a relative mass of 122. From this information and the spectrum alone, it is possible to determine the structure.

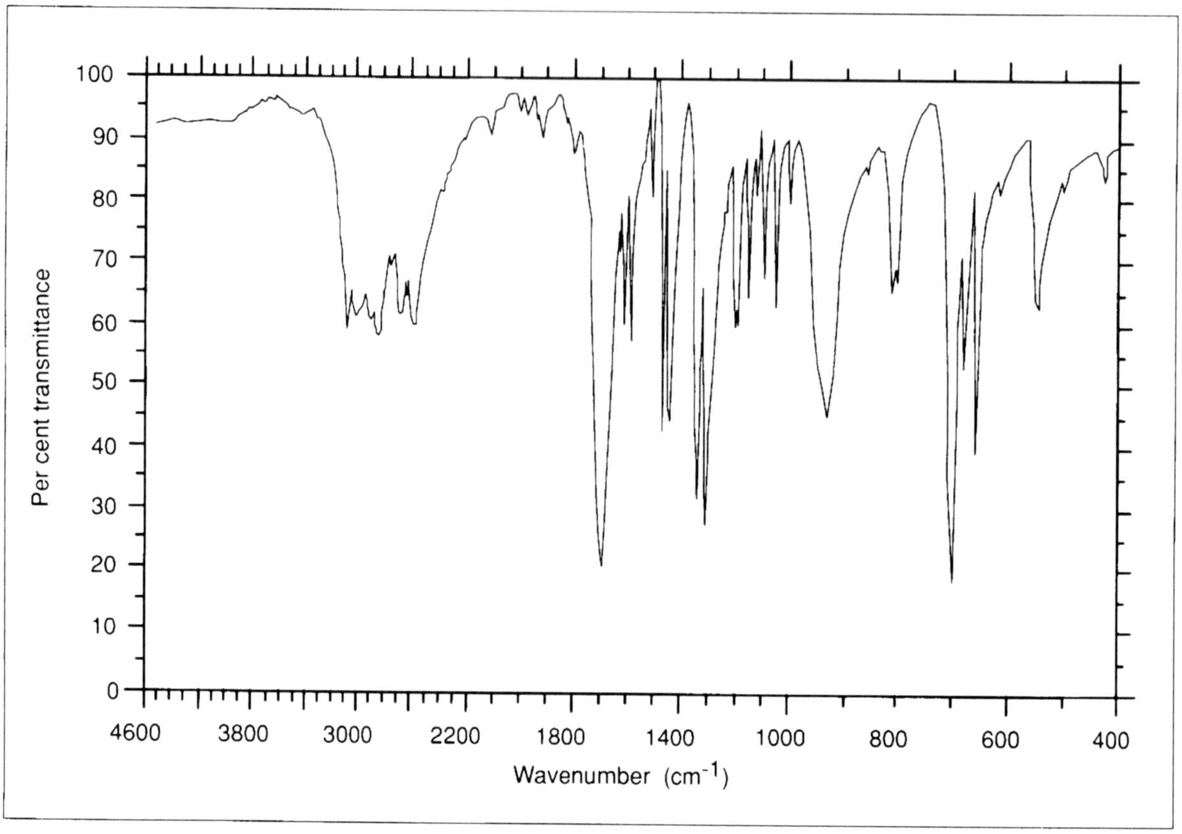

Figure 21

The percentage composition data gives the empirical formula $C_7H_6O_2$. This has relative mass 122. From the infrared spectrum, the strong absorption at 1680 cm^{-1} suggests a carbonyl group, and the peak at 2930 cm^{-1} is consistent with a carboxylic O–H stretch. The sharp peak at 1480 cm^{-1} indicates the C–C stretch of an aromatic ring. This should be accompanied by a sharp spectral line at 2950-3050 cm^{-1} as the aromatic ring hydrogen atoms vibrate. This is observed at the side of the carboxylic O–H absorption peak. The structure is effectively solved now, because an aromatic ring bonded to a carboxylic acid group gives a relative mass of 77+45=122. Thus we can conclude that the sample is benzoic acid, C_6H_5COOH.

Unilever

THE ROYAL SOCIETY OF CHEMISTRY

Exercise 3

The spectra below are known to be of ethanoic acid and ethanoic anhydride. Determine which is which and explain the difference between the spectra.

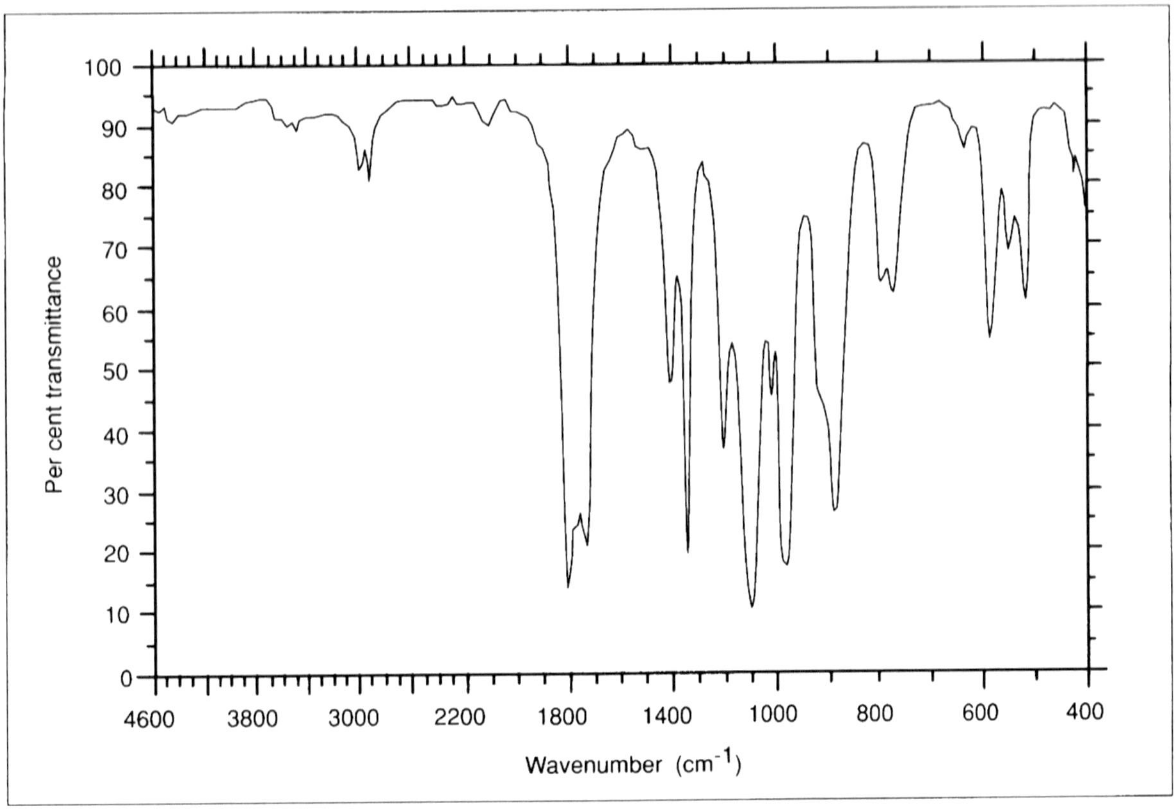

Figure 22

THE ROYAL SOCIETY OF CHEMISTRY

Unilever

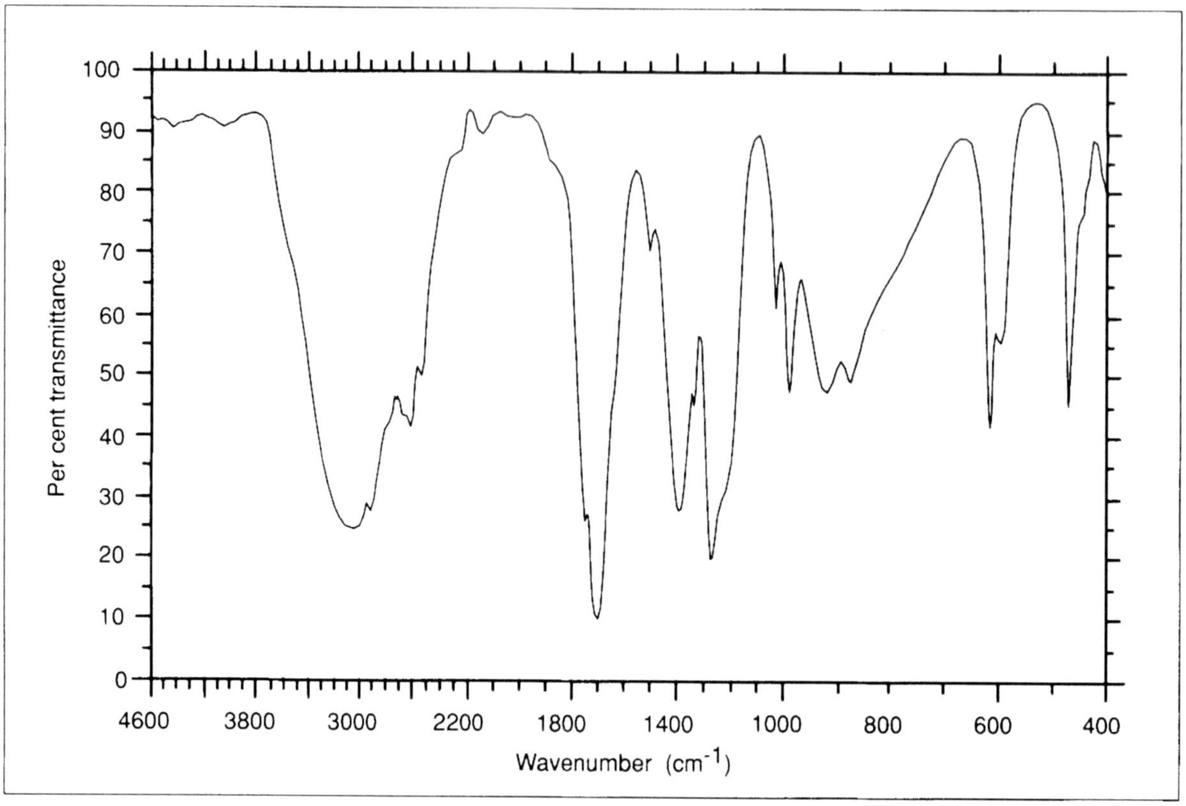

Figure 23

Both spectra contain absorptions in the 1600–1800 cm^{-1} region due to the carbonyl stretch. However, only *Fig. 23* has an absorption corresponding to the hydrogen bonded O–H bond found in the acid. This is at 2600–3400 cm^{-1}. Thus, *Fig. 22* is the spectrum of ethanoic anhydride, and *Fig. 23* is the spectrum of ethanoic acid.

Unilever

THE ROYAL SOCIETY OF CHEMISTRY

Exercise 4

The spectrum below (*Fig. 24*) is of a C_6 straight chain hydrocarbon. One bond is not a single bond. What is the compound?

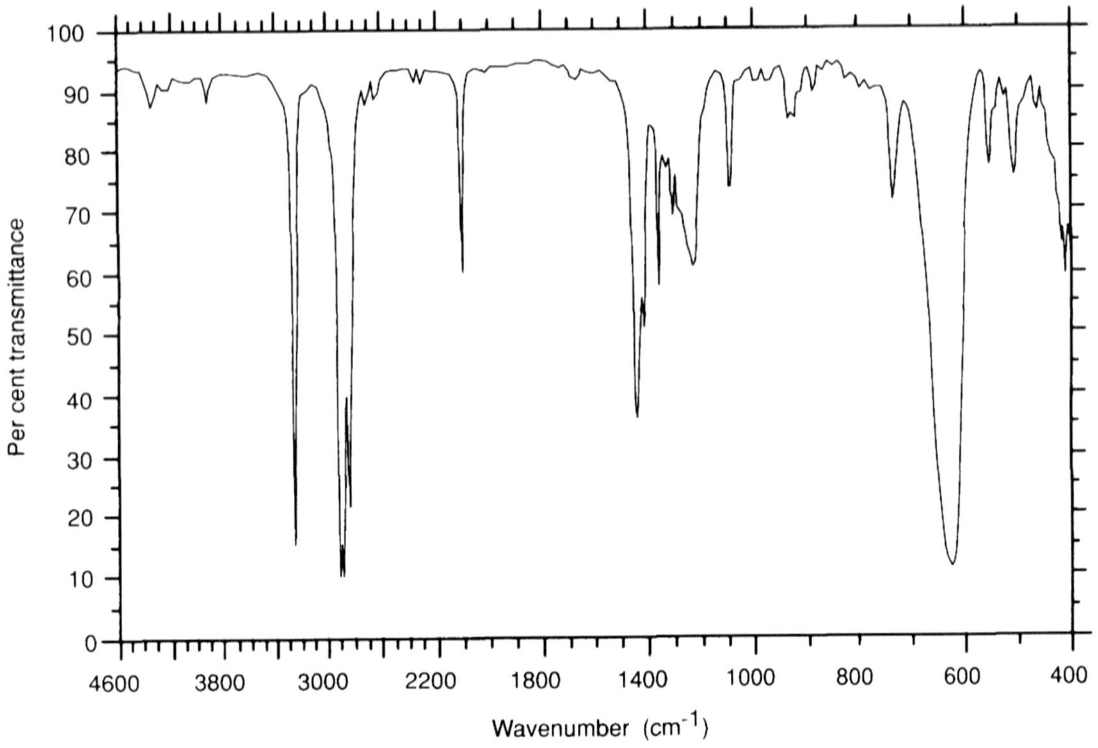

Figure 24

Figure 24 shows an absorption at 2120 cm^{-1}, indicating the presence of a triple bond, C≡C. Therefore, the compound is a hexyne. The position of the triple bond can be determined from the peak at 3350 cm^{-1}, because this suggests an ethynic hydrogen vibration, C≡C–H. Therefore, the triple bond is at the end of the chain, and *Fig. 24* is the spectrum of hex-1-yne. This is supported by the C–H bending frequency of the C≡C–H group centred on 615 cm^{-1}.

THE ROYAL SOCIETY OF CHEMISTRY

Unilever

Exercise 5

Figures 25 and 26 are the infrared spectra of the isomers ethoxyethane, $(C_2H_5)_2O$ and butan-1-ol, $CH_3CH_2CH_2CH_2OH$. From the spectra decide which is which.

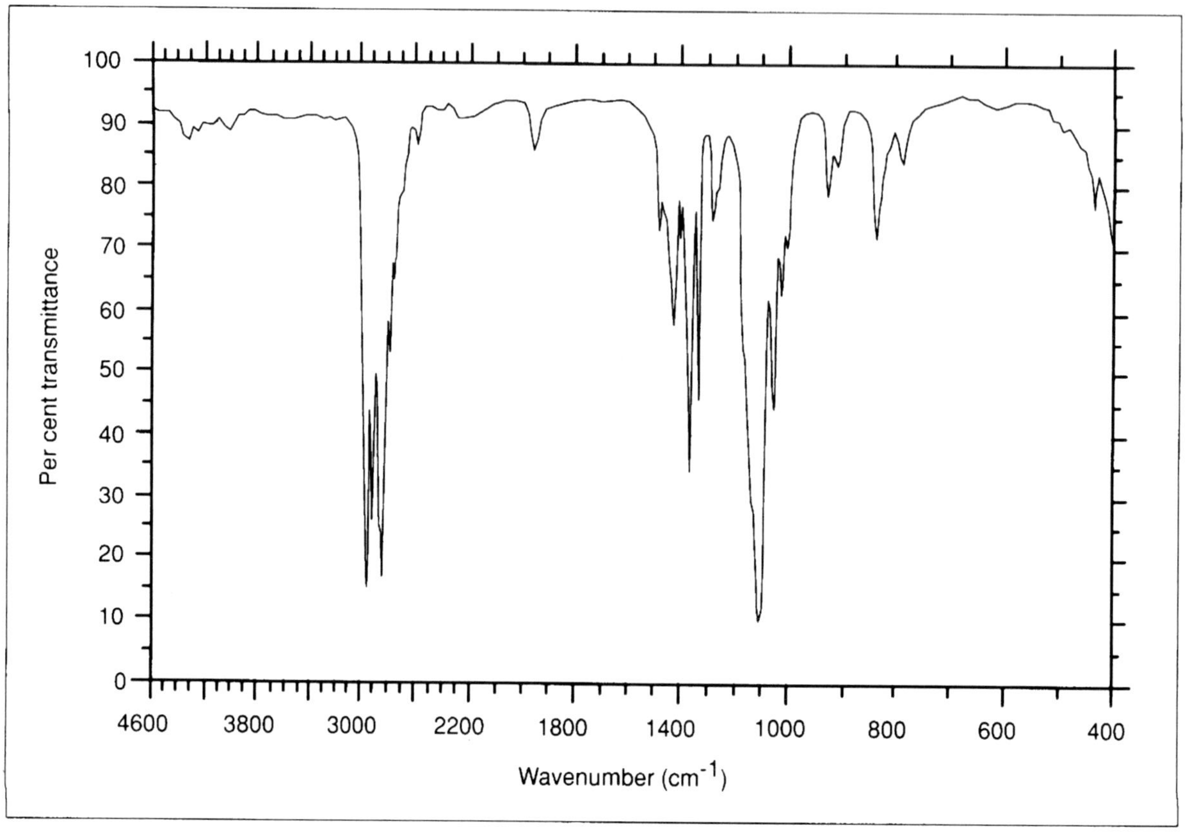

Figure 25

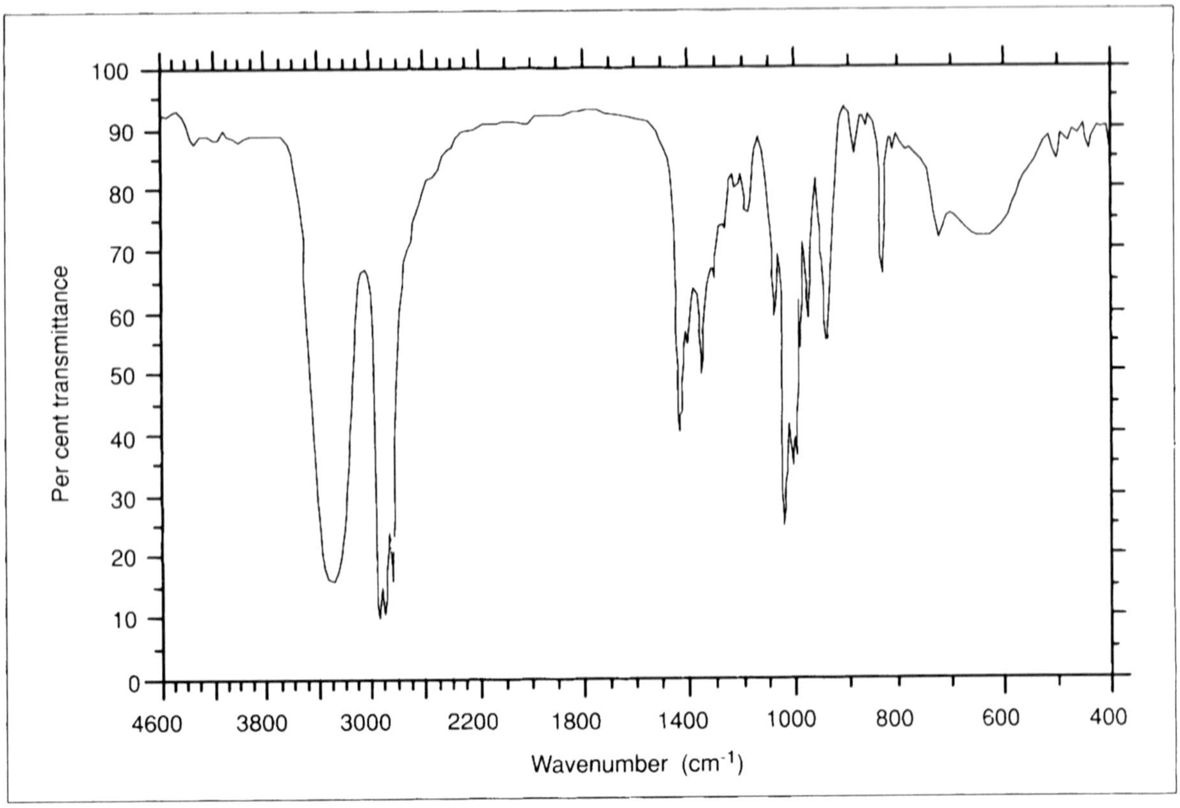

Figure 26

Both spectra show absorptions at *ca* 3000 cm^{-1}, consistent with the vibration of the C–H bond in saturated compounds. *Figure 26* also shows an absorption at 3300 cm^{-1} due to the hydrogen bonded O–H stretch of an alcohol, so *Fig. 26* represents the spectrum of butan-1-ol. There is a very strong absorption at 1150-1300 cm^{-1} in *Fig. 25*. Although this is in the fingerprint region, we know that we are looking for data consistent with an ether, and it is so strong and correlates so well with the C–O stretching frequency that there is little doubt that it is the ether, ethoxyethane. It could be argued that in cases like this, once one isomer has been conclusively identified the other one needs no further analysis, but real life situations are rarely that simple!

THE ROYAL SOCIETY OF CHEMISTRY

Unilever

4. Ultraviolet/visible spectroscopy

Visible light absorption is known to all of us, because this is what causes objects to be coloured. For example, a blue dye appears blue because the light at the red end of the spectrum is absorbed, leaving the blue light to be transmitted (*Fig. 1*).

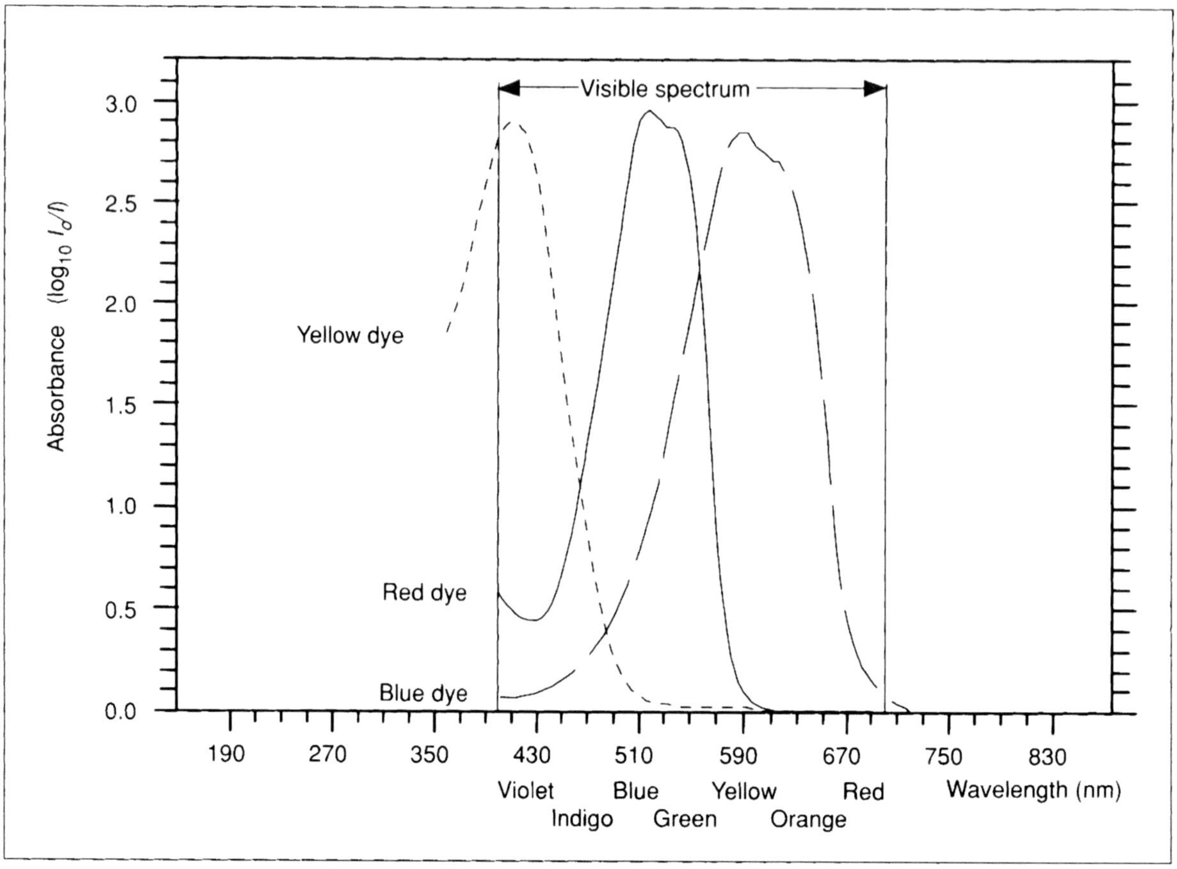

Figure 1 Absorption of light by dyes

The theory

Visible light lies in the wavelength range 4.0 – 7.0 x 10^{-7} m (*Fig. 2*). To keep the numbers more manageable it is usually quoted in nanometres (10^{-9} m) so that the range becomes 400–700 nm. When light is absorbed by a material, valence (outer) electrons are promoted from their normal (ground) states to higher energy (excited) states.

Unilever

THE ROYAL SOCIETY OF CHEMISTRY

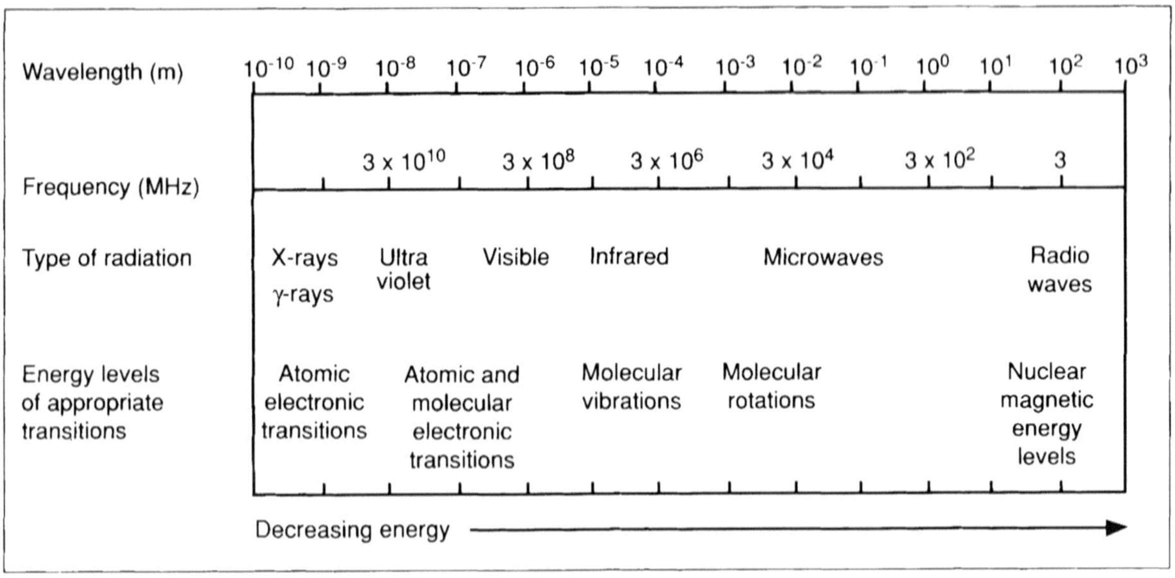

Figure 2 Regions of the electromagnetic spectrum

The energy of visible light depends on its frequency, and is approximately equivalent to 170 kJ mol^{-1} (mole of photons) for red light and 300 kJ mol^{-1} for blue light. The promotion of electrons to different energy levels is not restricted to electromagnetic radiation in the visible part of the spectrum; it can also occur in the ultraviolet region. To encompass the majority of electron transitions the spectrum between 190 and 900 nm is usually considered (Table 1).

Table 1 Frequency, wavelength and energy of radiation in the part of the spectrum used for ultraviolet/visible spectroscopy

Frequency (ν) (Hz)	Wavelength (λ) (m)	(nm)	Energy (kJ mol^{-1})	
3.33×10^{14}	9.0×10^{-7}	900	137.5	(infrared)
4.29×10^{14}	7.0×10^{-7}	700	171.2	(red light)
7.50×10^{14}	4.0×10^{-7}	400	299.3	(blue light)
1.58×10^{15}	1.9×10^{-7}	190	630.5	(ultraviolet)

Frequency, wavelength and energy are interrelated:

$$c = \nu\lambda$$

where c = velocity of light (3.00×10^{8} ms^{-1})
ν = frequency in Hz
λ = wavelength in m

and

$$E = h\nu \quad \text{or} \quad E = h\nu L \text{ for one mole of photons}$$

where E = energy of one mole of radiation
h = Planck's constant (6.63×10^{-34} Js)
L = Avogadro constant (6.02×10^{23} mol^{-1})

The origin of the absorptions

Valence electrons can generally be found in one of three types of electron orbital:

1. single, or σ, bonding orbitals;
2. double or triple bonds (π bonding orbitals); and
3. non-bonding orbitals (lone pair electrons).

Sigma bonding orbitals tend to be lower in energy than π bonding orbitals, which in turn are lower in energy than non-bonding orbitals. When electromagnetic radiation of the correct frequency is absorbed, a transition occurs from one of these orbitals to an empty orbital, usually an antibonding orbital, σ^* or π^* (*Fig. 3*).

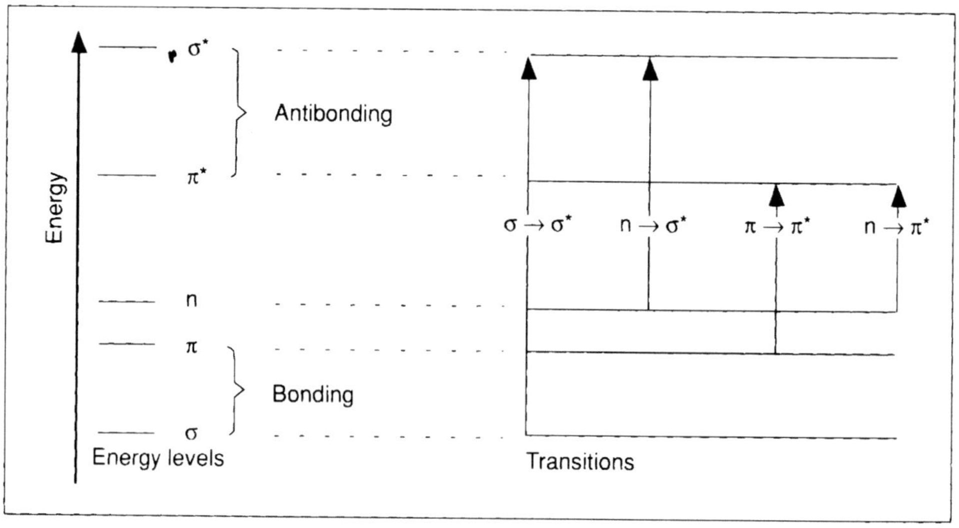

Figure 3 Electron transitions in ultraviolet/visible spectroscopy

The exact energy differences between the orbitals depends on the atoms present and the nature of the bonding system.

Most of the transitions from bonding orbitals are of too high a frequency (too short a wavelength) to measure easily, so most of the absorptions observed involve only $\pi \rightarrow \pi^*$, $n \rightarrow \sigma^*$ and $n \rightarrow \pi^*$ transitions. A common exception to this is the d→d transition of d-block element complexes, but in most cases a knowledge of the precise origin of transitions is not required.

The spectrometer

Because only small numbers of absorbing molecules are required, it is convenient to have the sample in solution (ideally the solvent should not absorb in the ultraviolet/visible range however, this is rarely the case). In conventional spectrometers electromagnetic radiation is passed through the sample which is held in a small square-section cell (usually 1 cm wide internally). Radiation across the whole of the ultraviolet/visible range is scanned over a period of approximately 30 s, and radiation of the same frequency and intensity is simultaneously passed through a reference cell containing only the solvent. Photocells then detect the radiation transmitted and the spectrometer records the absorption by comparing the difference between the intensity of the radiation passing through the sample and the reference cells (*Fig. 4*). In the latest spectrometers radiation across the whole range is monitored simultaneously, see page 111.

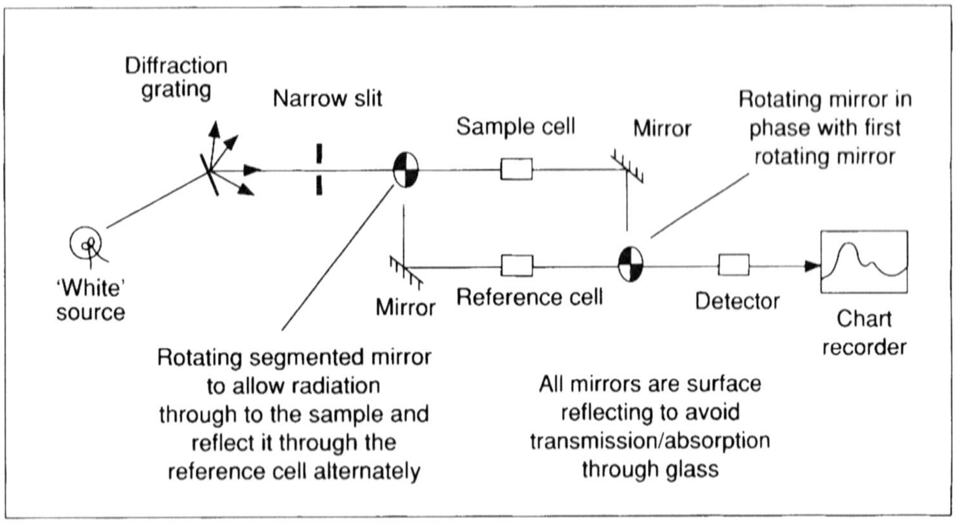

Figure 4 Diagram showing how the ultraviolet/visible spectrometer works

No single lamp provides radiation across the whole of the range required, so two are used. A hydrogen or deuterium discharge lamp covers the ultraviolet range, and a tungsten filament (usually a tungsten/halogen lamp) covers the visible range. The radiation is separated according to its frequency/wavelength by a diffraction grating followed by a narrow slit. The slit ensures that the radiation is of a very narrow waveband – *ie* it is monochromatic.

The cells in the spectrometer must be made of pure silica for ultraviolet spectra because soda glass absorbs below 365 nm, and pyrex glass below 320 nm.

Detection of the radiation passing through the sample or reference cell can be achieved by either a photomultiplier or a photodiode, that converts photons of radiation into tiny electrical currents; or a semiconducting cell (that emits electrons when radiation is incident on it) followed by an electron multiplier similar to those used in mass spectrometers (see page 8). The spectrum is produced by comparing the currents generated by the sample and the reference beams.

Modern instruments are self-calibrating, though the accuracy of the calibration can be checked if necessary. Wavelength checks are made by passing the sample beam through glass samples (containing holmium oxide) that have precise absorption peaks, and the absorption is calibrated by passing the sample beam through either a series of filters, each with a specific and known absorption, or a series of standard solutions.

Absorption laws

Beer's law tells us that absorption is proportional to the number of absorbing molecules – *ie* to the concentration of absorbing molecules (this is only true for dilute solutions) – and Lambert's law tells us that the fraction of radiation absorbed is independent of the intensity of the radiation. Combining these two laws, we can derive the Beer-Lambert Law:

$$\log_{10} \frac{I_o}{I} = \varepsilon l c$$

where I_o = the intensity of the incident radiation
I = the intensity of the transmitted radiation

ε = a constant for each absorbing material, known as the molar absorption coefficient (called the molar extinction coefficient in older texts) and having the units $mol^{-1}\ dm^3\ cm^{-1}$, but by convention the units are not quoted
l = the path length of the absorbing solution in cm
c = the concentration of the absorbing species in $mol\ dm^{-3}$

The value of $\log_{10}(I_o/I)$ is known as the absorbance of the solution (in older texts it is referred to as the optical density), and can be read directly from the spectrum, often as 'absorbance units'. A useful constant is the molar absorption coefficient, ε, because it is independent of concentration and path length, whereas absorption depends upon both. The other useful piece of information is the wavelength at which maximum absorption occurs. This is given the symbol λ_{max} (*Fig. 5*). These two pieces of information alone are frequently sufficient to identify a substance, although identification is not the most common use of this technique. Conversely, if the values of ε and λ_{max} are known, the concentration of its solution can be calculated – this is the more common application. The values of both ε and λ_{max} are strongly influenced by the nature of the solvent, and for organic compounds, by the degree of substitution and conjugation.

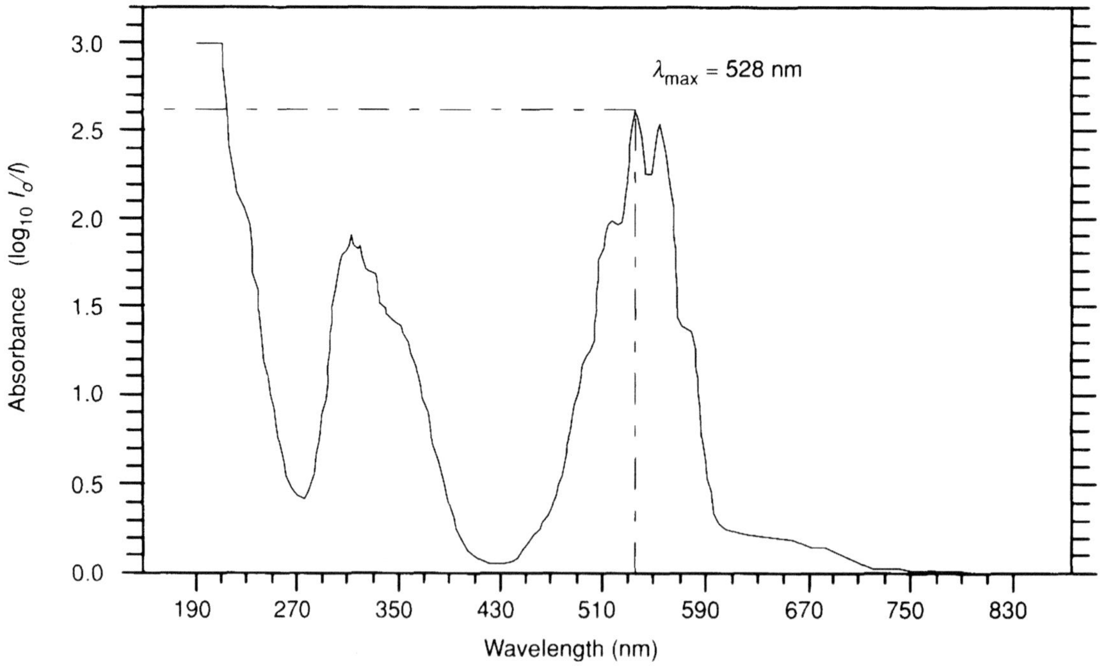

The solution used to obtain this spectrum contained 1.227 mg of potassium manganate(VII) in 5.0 cm³ of solution. From the spectrum ε is found to be 1.68×10^3

Figure 5 Ultraviolet/visible spectrum of potassium manganate (VII), $KMnO_4$, showing λ_{max}

Absorption curves

The energies of the orbitals involved in electronic transitions have fixed values, and as energy is quantised, it would be expected that absorption peaks in ultraviolet/visible spectroscopy should be sharp peaks. However this is rarely, if ever, actually observed. Instead, broad absorption peaks are seen. This is because a number of vibrational energy levels are available at each electronic energy level, and transitions can occur to and from the different vibrational levels (*Fig. 6*). This results in peak broadening. The situation is further complicated by the fact that different rotational energy levels are also available to absorbing materials (Table 2). Only in a few cases, often in the vapour phase or in non-polar solvents, can fine structure be observed – *eg* the vibrational fine structure of the 260 nm band of benzene (*Fig. 7*).

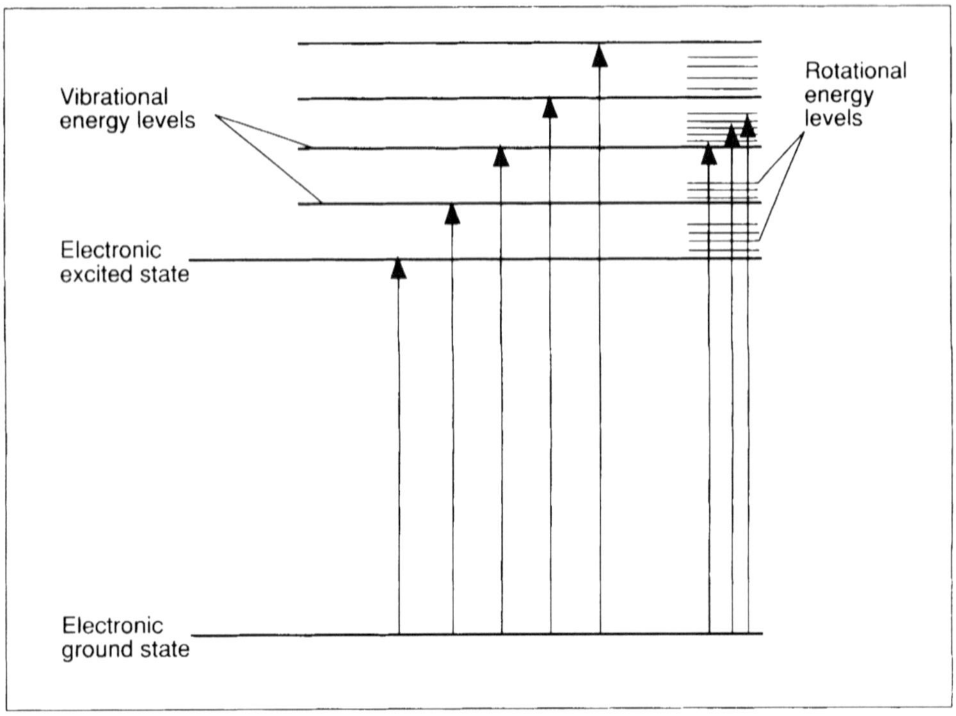

Figure 6 Electronic and vibrational levels

Table 2 Energy level differences

between electronic levels	≈ 100	kJ mol^{-1}
between vibrational levels	≈ 1	kJ mol^{-1}
between rotational levels	≈ 0.01	kJ mol^{-1}

THE ROYAL SOCIETY OF CHEMISTRY

Unilever

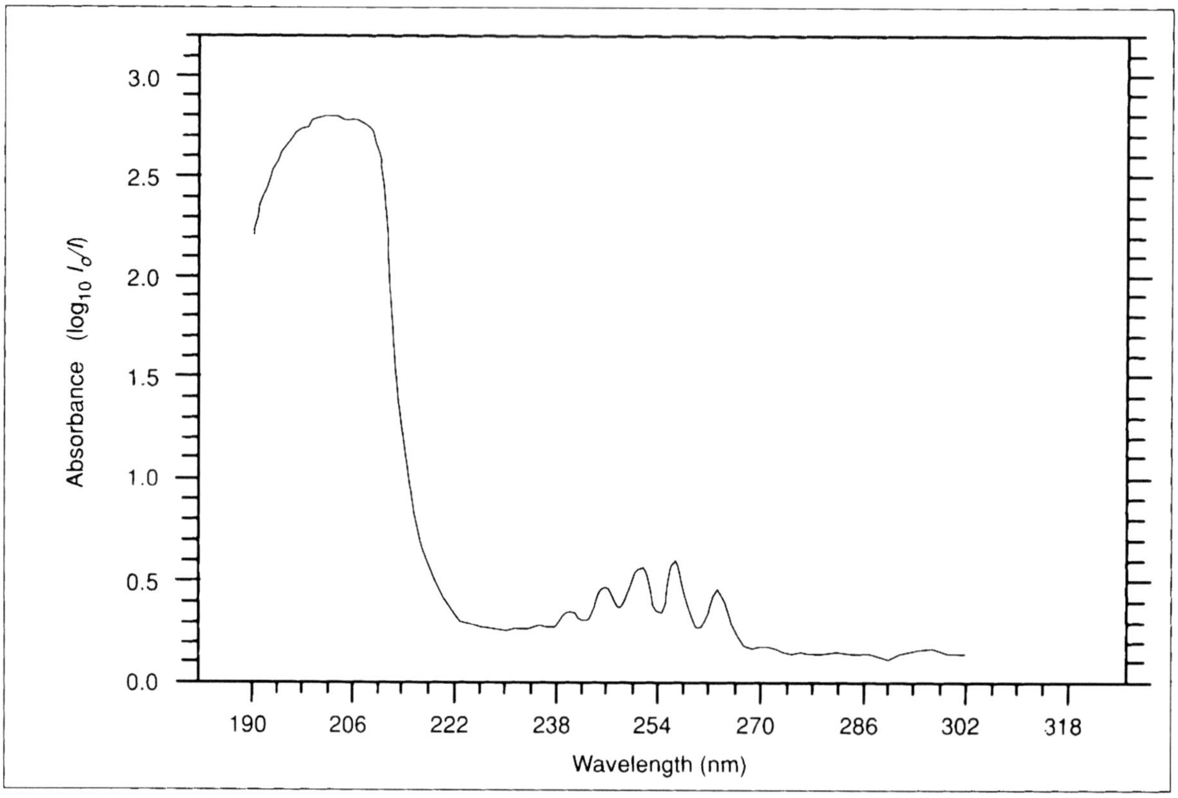

Figure 7 Ultraviolet/visible spectrum of benzene showing vibrational fine structure

Copper complexes

Uncomplexed copper(II) ions appear white or colourless because the transition between the highest occupied orbital and the lowest unoccupied orbital is at too short a wavelength to see. In hydrated copper (II) ions the energy levels of the 3d orbitals are split (*Fig. 8*), and when visible light is absorbed a transition is possible. For the hydrated copper (II) ion the absorption occurs at the red end of the spectrum (*Fig. 9*) hence the complex appears blue. The amount of splitting of the energy levels depends on the ligand, so if ammonia replaces water and ΔE increases, the colour of the complex becomes blue/violet *ie* absorption occurs at the middle of the visible spectrum (*Fig. 10*). Note that ε is larger (both spectra contain the same concentration of copper ions).

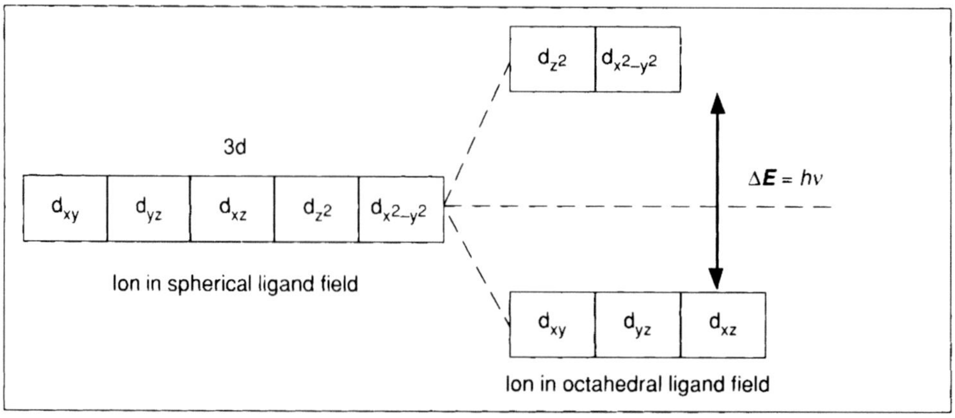

Figure 8 Splitting of d orbital energy levels in $Cu(H_2O)_6^{2+}$

Unilever

THE ROYAL SOCIETY OF CHEMISTRY

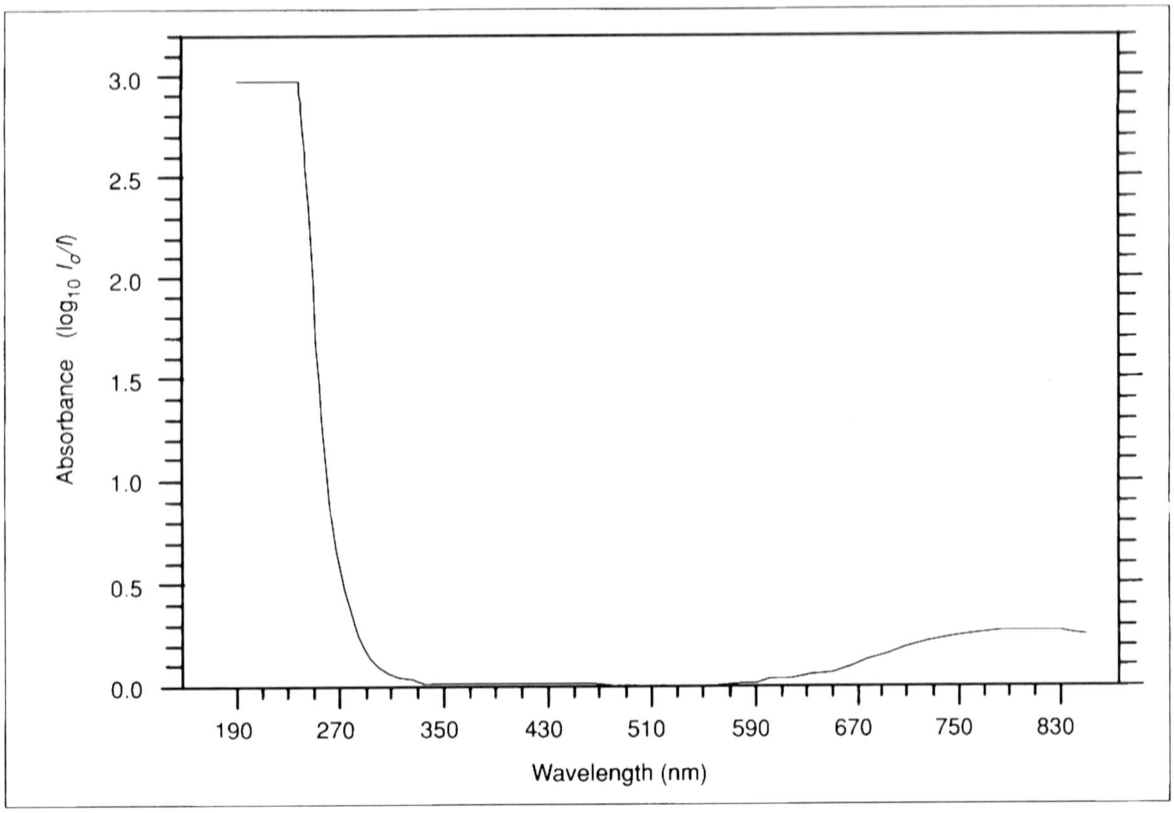

Figure 9 Ultraviolet/visible spectrum of $Cu(H_2O)_6^{2+}$

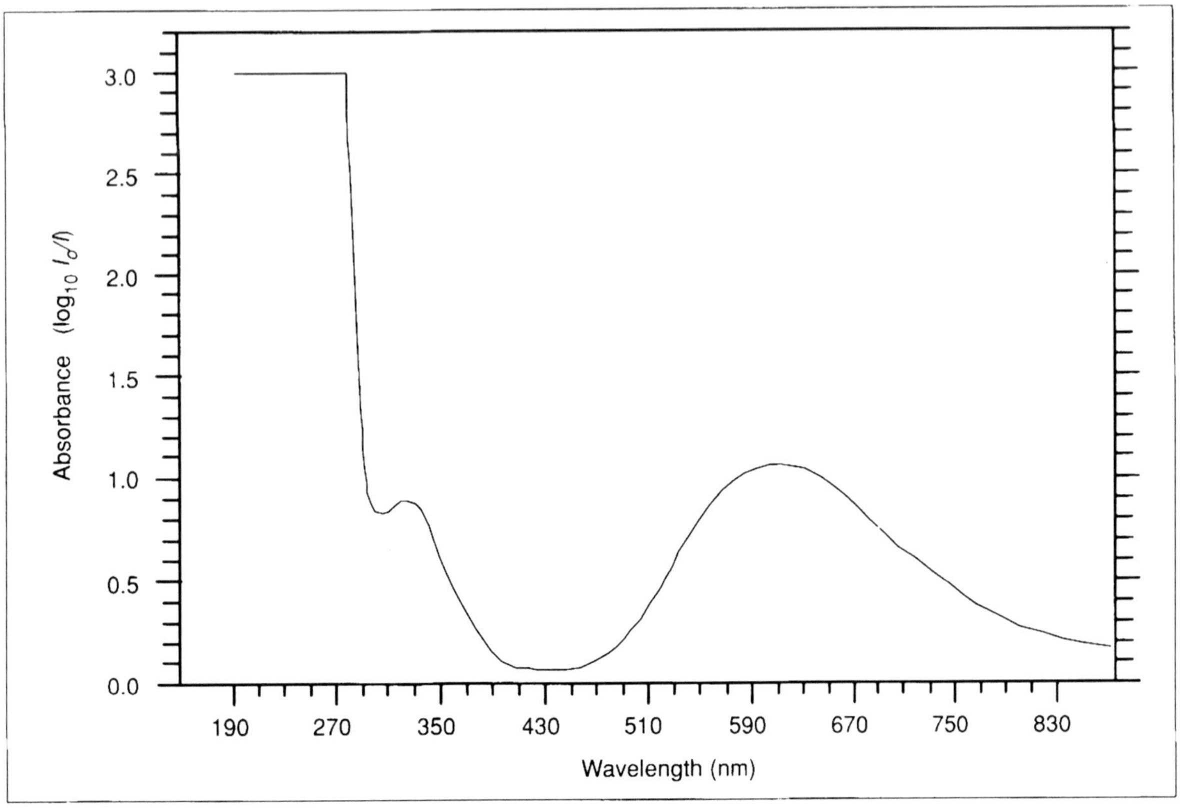

Figure 10 Ultraviolet/visible spectrum of $Cu(NH_3)_4(H_2O)_2^{2+}$

THE ROYAL SOCIETY OF CHEMISTRY

Unilever

Changing the ligands around a transition metal ion can increase the stability of the complex formed because the difference between the energy levels of the d orbitals increases (*Fig. 11*). (Stability in this context is taken to be the extent to which the complex will form from, or dissociate into, its constituents at equilibrium.)

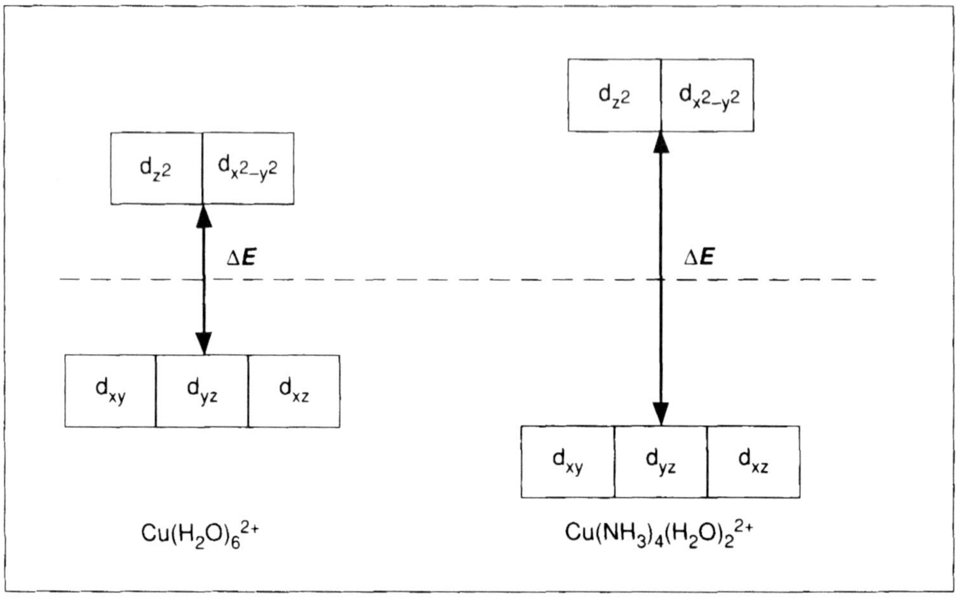

Figure 11 Increased splitting of d orbital energy levels

Some ligands giving increasing separation of energy levels (and therefore increasing stability to the ions) are:

I^-	smallest separation of energy levels
Br^-	
Cl^-	
OH^-	
F^-	
H_2O	
NH_3	
$NH_2CH_2CH_2NH_2$	
CN^-	largest separation of energy levels

This list forms part of the spectrochemical series and the order can vary depending on the metal ion in the complex.

Stability of complex ions

It is possible to measure the relative stabilities of complexes through equilibrium constants called stability constants. If the successive replacement of water molecules by ammonia molecules in the $Cu(H_2O)_6^{2+}$ ion is used as an example, the first substitution can be described by the equation

$$Cu(H_2O)_6^{2+} + NH_3 \rightleftharpoons Cu(H_2O)_5NH_3^{2+} + H_2O$$

The equilibrium constant for this is

$$K_1 = \frac{[Cu(H_2O)_5NH_3^{2+}]}{[Cu(H_2O)_6^{2+}][NH_3]}$$

Similarly for the substitution of the second water molecule:

$$Cu(H_2O)_5NH_3^{2+} + NH_3 \rightleftharpoons Cu(H_2O)_4(NH_3)_2^{2+} + H_2O$$

there is an equilibrium constant

$$K_2 = \frac{[Cu(H_2O)_4(NH_3)_2^{2+}]}{[Cu(H_2O)_5NH_3^{2+}][NH_3]}$$

Likewise replacement of the third and fourth water molecules give rise to the constants K_3 and K_4:

$$Cu(H_2O)_4(NH_3)_2^{2+} + NH_3 \rightleftharpoons Cu(H_2O)_3(NH_3)_3^{2+} + H_2O$$

$$K_3 = \frac{[Cu(H_2O)_3(NH_3)_3^{2+}]}{[Cu(H_2O)_4(NH_3)_2^{2+}][NH_3]}$$

$$Cu(H_2O)_3(NH_3)_3^{2+} + NH_3 \rightleftharpoons Cu(H_2O)_2(NH_3)_4^{2+} + H_2O$$

$$K_4 = \frac{[Cu(H_2O)_2(NH_3)_4^{2+}]}{[Cu(H_2O)_3(NH_3)_3^{2+})][NH_3]}$$

It is also possible to quote a constant for the replacement of four water molecules together.

$$K = \frac{[Cu(H_2O)_2(NH_3)_4^{2+}]}{[Cu(H_2O)_6^{2+}][NH_3]^4}$$

This is in fact the product of the four step-wise constants *ie*

$$K = K_1 \times K_2 \times K_3 \times K_4$$

The values of the constants quoted are:

$K_1 = 1.78 \times 10^4$ $K_2 = 4.07 \times 10^3$ $K_3 = 9.55 \times 10^2$
$K_4 = 1.74 \times 10^2$ $K = 1.20 \times 10^{13}$

Complexes with large stability constants are more stable than those with small constants. For example, using chloride ions to replace the water ligands surrounding copper ions gives a less stable complex. This is demonstrated simply by adding aqueous ammonia to a solution containing the $CuCl_4^{2-}$ ion – the solution turns from yellow to deep blue. The reduced stability is also evident in the smaller stability constants:

$K_1 = 631$ $K_2 = 398$ $K_3 = 3.09$ $K_4 = 5.37$ $K = 4.17 \times 10^5$

THE ROYAL SOCIETY OF CHEMISTRY

Unilever

Chromophores

Many organic molecules absorb ultraviolet/visible radiation and this is usually because of the presence of a particular functional group. The groups that actually absorb the radiation are called chromophores.

Mathematical treatments of the energy levels of orbital systems suggest that some electronic transitions are statistically probable (said to be 'allowed', and these absorptions are strong, and tend to have ε values in excess of 10 000). Other transitions have a probability of zero – they are not expected to occur at all – and are said to be 'forbidden' but they frequently do occur, to give weak bands with ε values that rarely exceed 1 000. Some particularly useful forbidden transitions are: d→d absorptions of transition metals; the $n \rightarrow \pi^*$ absorption of carbonyl groups at *ca* 280 nm; and the $\pi \rightarrow \pi^*$ absorption of aromatic compounds at *ca* 230–330 nm, depending on the substituents on the benzene ring.

Factors affecting absorption

The solvent

The excited states of most $\pi \rightarrow \pi^*$ transitions are more polar than their ground states because a greater charge separation is observed in the excited state. If a polar solvent is used the dipole–dipole interaction reduces the energy of the excited state more than the ground state, hence the absorption in a polar solvent such as ethanol will be at a longer wavelength (lower energy, hence lower frequency) than in a non-polar solvent such as hexane (*Figs. 12a* and *12b*).

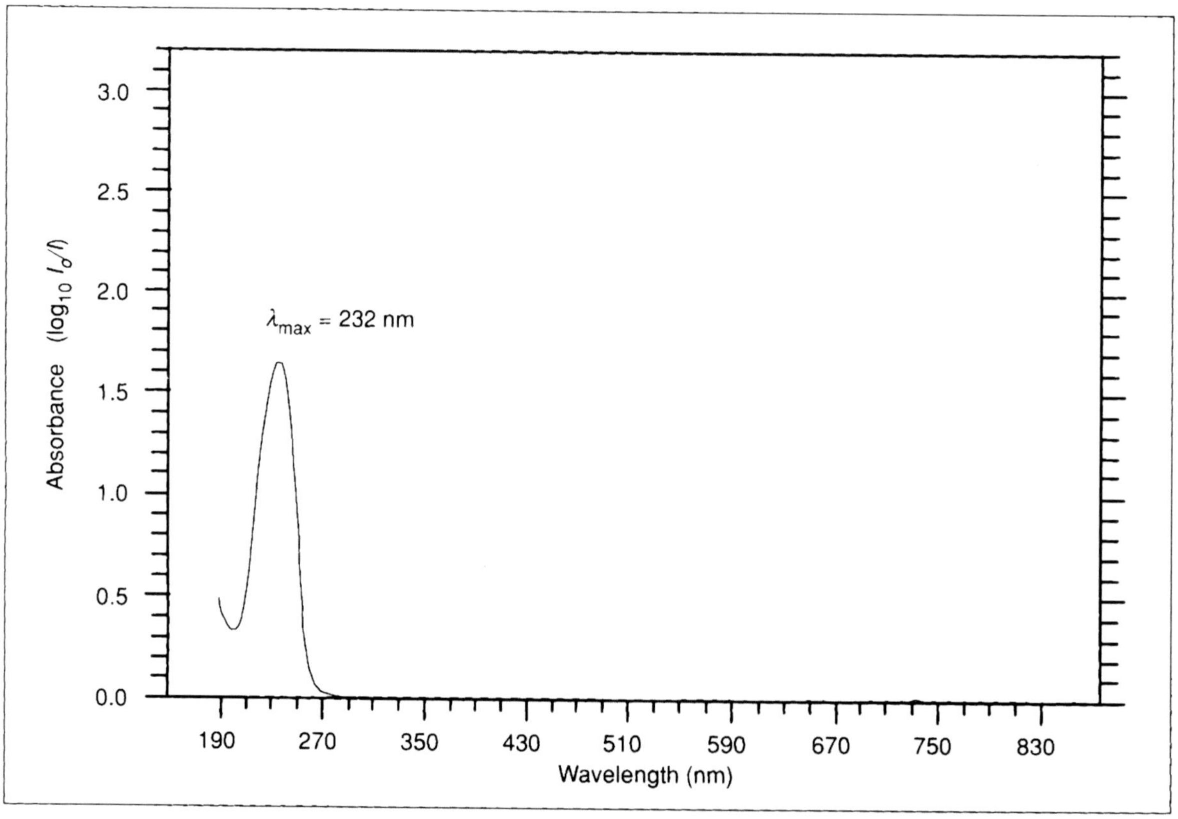

Figure 12a Ultraviolet/visible spectrum of 4-methyl-3-penten-2-one (mesityl oxide) in hexane

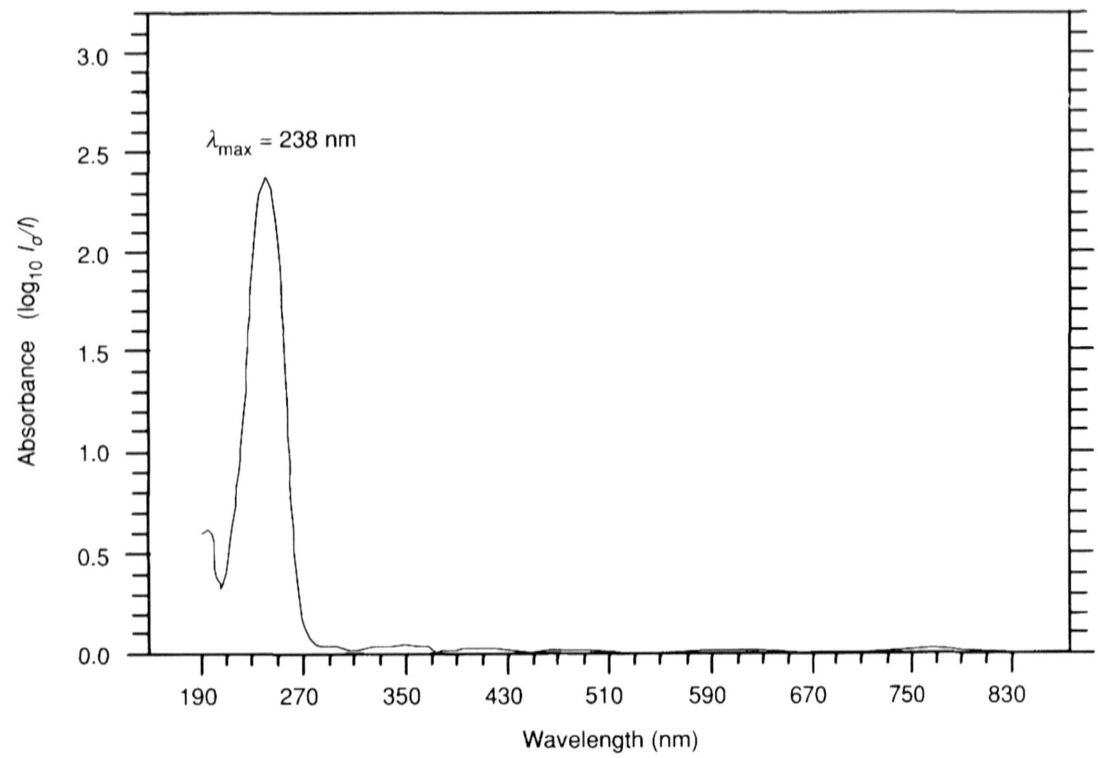

Figure 12b Ultraviolet/visible spectrum of 4-methyl-3-penten-2-one (mesityl oxide) in ethanol

The reverse is also observed if the excited state reduces the degree of hydrogen bonding. Here the transitions are $n \rightarrow \pi^*$ and the shift of wavelength is due to the lesser extent that the solvent can hydrogen bond to the excited state. Carbonyl groups in particular hydrogen bond to their solvent. For example changing from hexane to water as the solvent for propanone, the absorption maximum moves from 280 to 257 nm (*Figs. 13a, 13b* and Table 3).

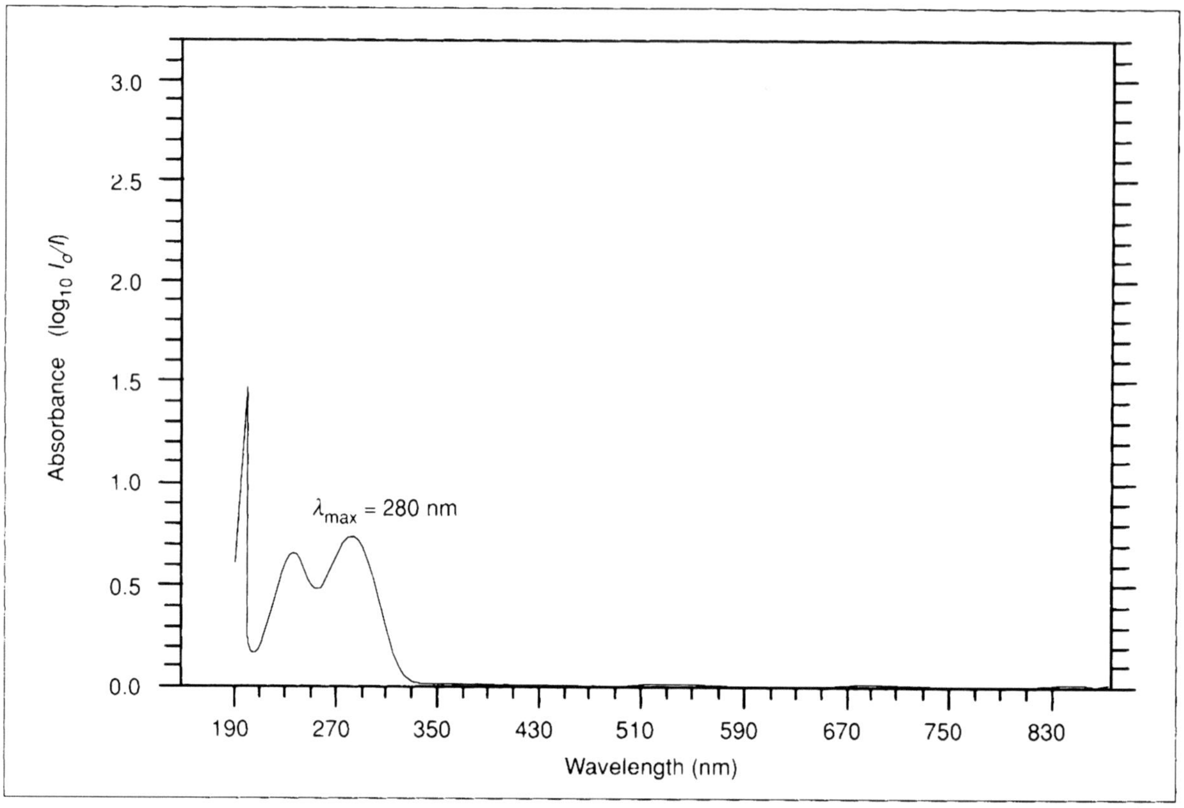

Figure 13a Ultraviolet/visible spectrum of propanone in hexane

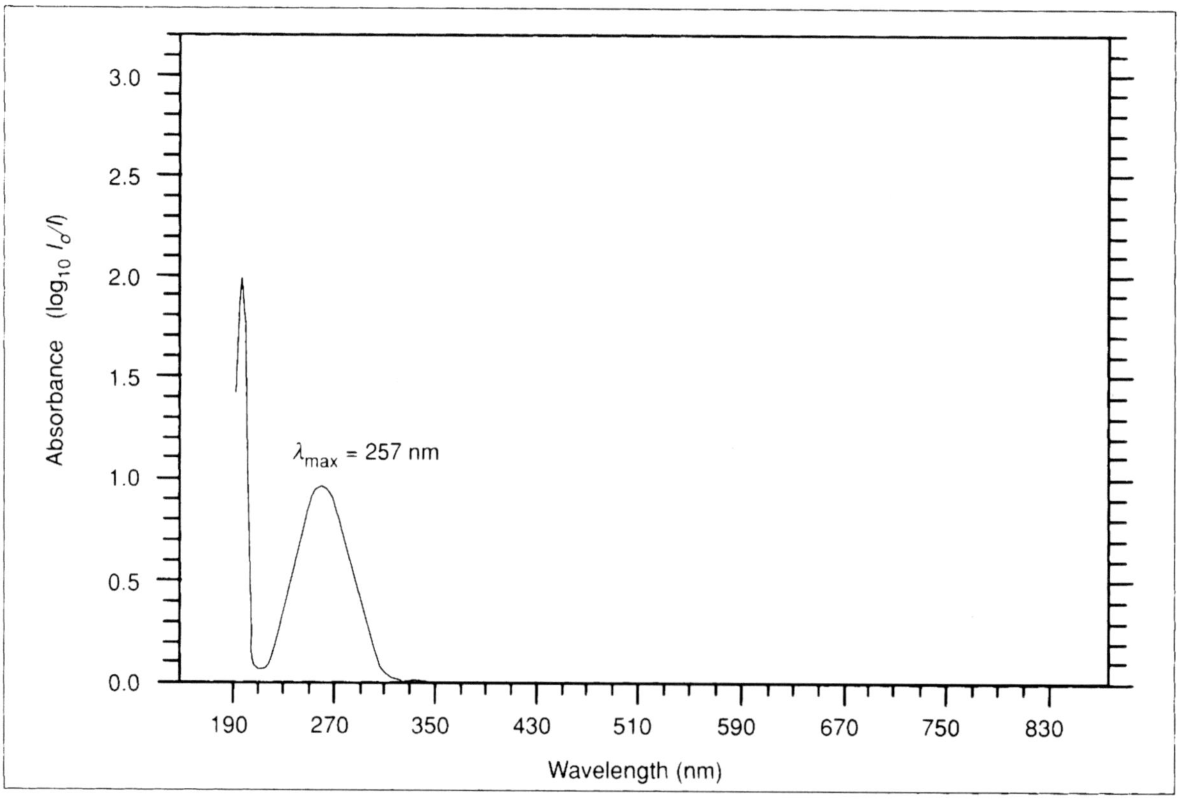

Figure 13b Ultraviolet/visible spectrum of propanone in water

Unilever

THE ROYAL SOCIETY OF CHEMISTRY

Table 3 The effect of the solvent on the absorption maximum of propanone

Solvent	λ_{max}/nm	ε at λ_{max}
Hexane	280	14.8
Trichloromethane	277	17.0
Ethanol	271	15.2
Water	257	17.4

Care must be taken when choosing a solvent, because many solvents absorb in the ultraviolet region. The minimum wavelengths at which some solvents are useful are given in Table 4.

Table 4 Minimum wavelength at which different solvents are useful

Solvent	Minimum wavelength (nm)
Ethanonitrile	190
Water	191
Cyclohexane	195
Hexane	201
Methanol	203
Ethanol	204
Ethoxyethane	215
Dichloromethane	220
Trichloromethane	237
Tetrachloromethane	257

Degree of conjugation

Ethene, containing only one double bond, has an absorption maximum at 185 nm (ε = 10 000). If the carbon chain length is increased this peak shifts to a slightly longer wavelength because the σ bonded electrons of the alkyl group interact with the π bond electrons in the double bond (*ie* the energy of the excited state is reduced).

The shift in wavelength is small compared with the effect of increasing the number of double bonds, especially if the electrons in the π systems (the double bonds) can interact with each other. The simplest example is buta-1,3-diene, $CH_2=CH-CH=CH_2$ (*Fig. 14*). Buta-1,3-diene has an absorption maximum at 220 nm, with an absorption coefficient of 20 000 – *ie* both the wavelength and the intensity of the absorption have increased. This difference arises because instead of the double bonds absorbing in isolation of each other the π system extends over the length of the carbon chain – *ie* the system is conjugated (or delocalised) – and lowers the energy of the excited state.

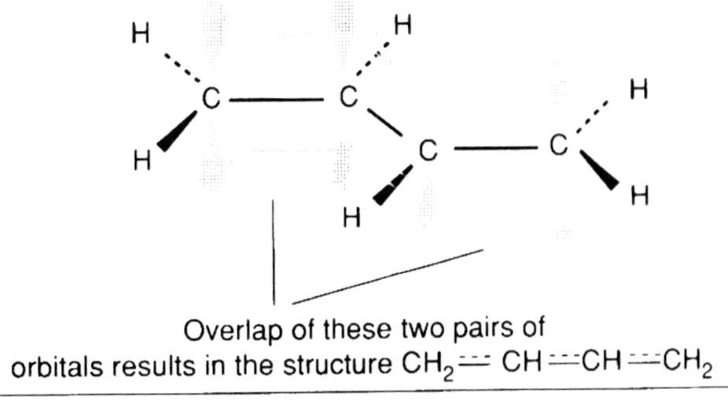

Figure 14 Conjugation in buta-1,3-diene

The longer the conjugated carbon chain in the absorbing system, the greater the intensity of the absorption. This is shown by the spectra of the polyenes CH_3-$(CH=CH)_n$-CH_3, where n=3,4 and 5 (*Fig. 15*).

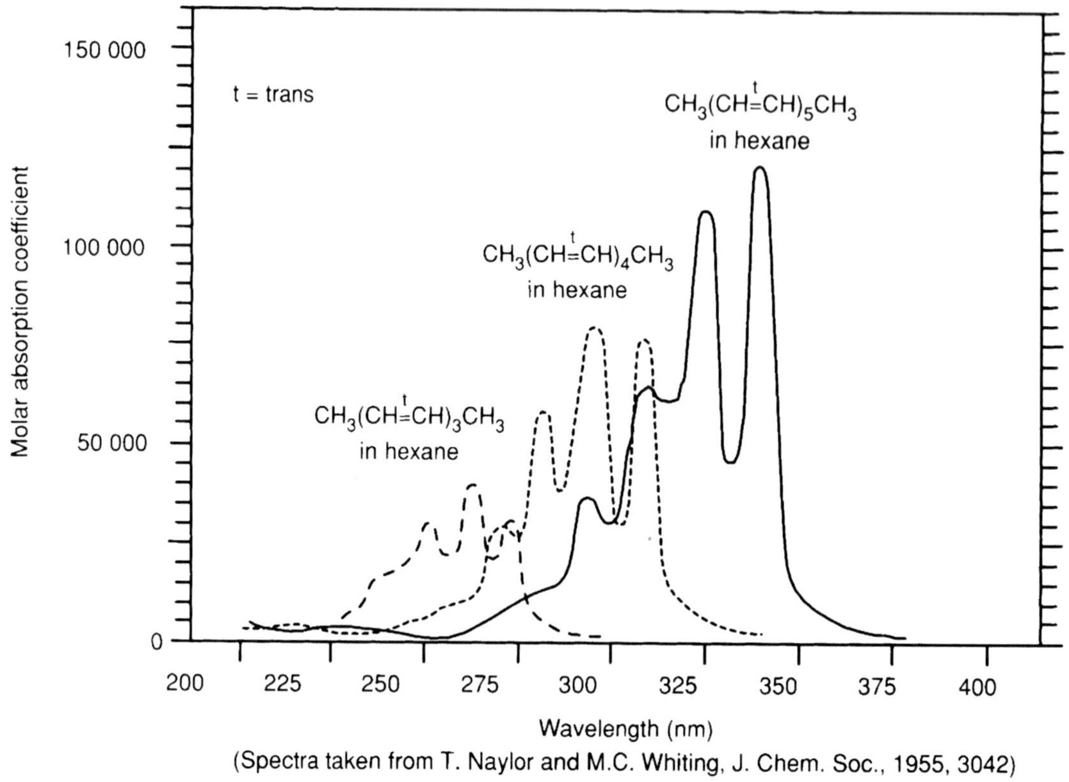

Figure 15 Ultraviolet/visible spectra of the polyenes $CH_3(CH=CH)_nCH_3$, where n=3,4 and 5

β-carotene, a vitamin found in carrots, and used in food colouring, has eleven conjugated double bonds (*Fig. 16*) and its absorption maximum has shifted out of the ultraviolet and into the blue region of the visible spectrum, hence it appears bright orange (*Fig. 17*).

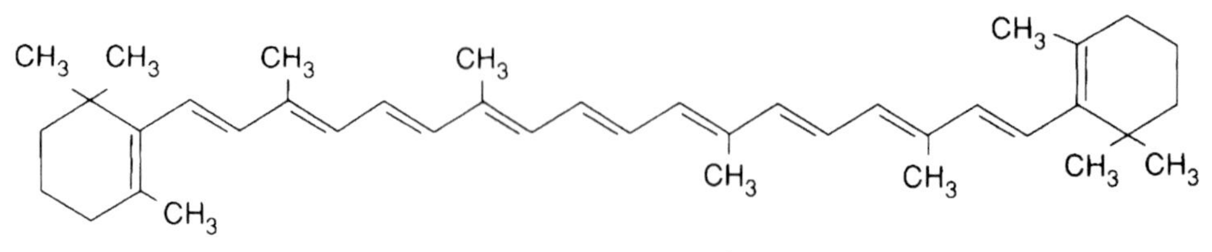

Figure 16 β-carotene

Unilever

THE ROYAL SOCIETY OF CHEMISTRY

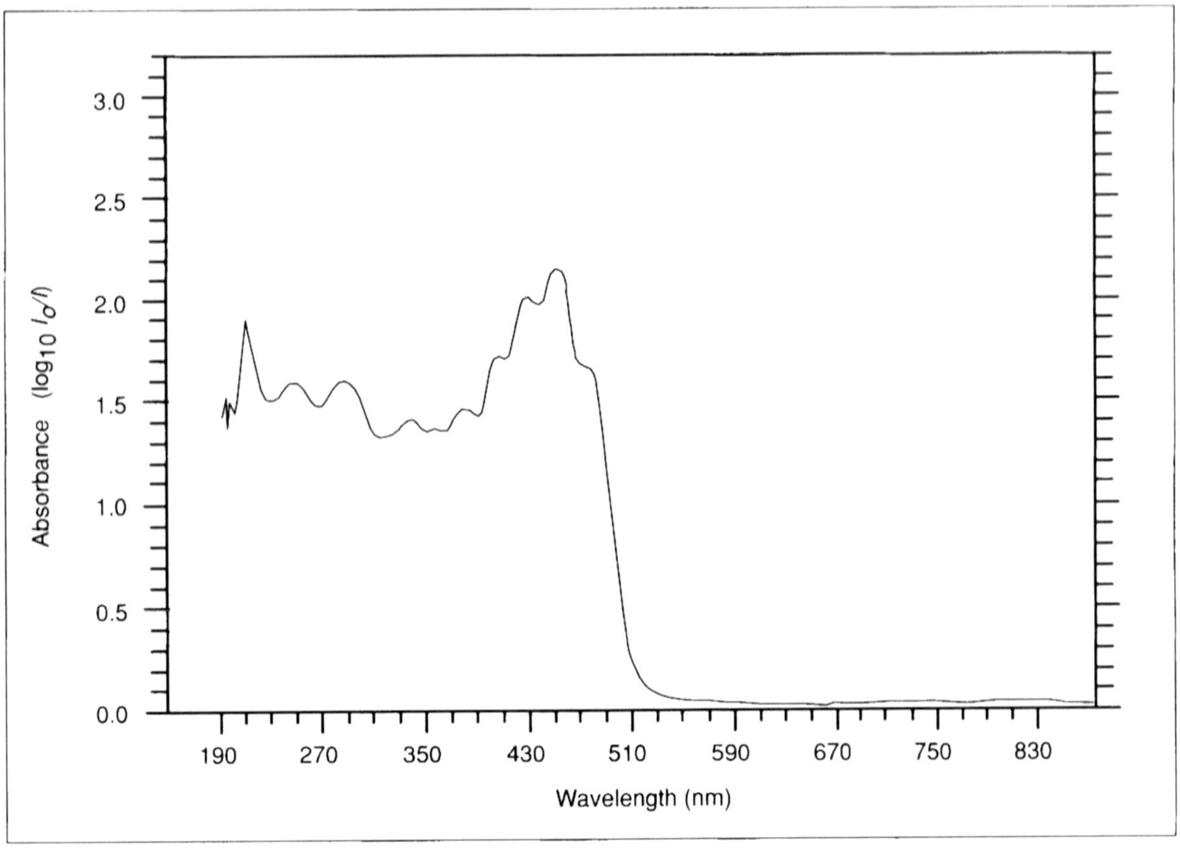

Figure 17 Ultraviolet/visible spectrum of β-carotene

THE ROYAL SOCIETY OF CHEMISTRY

Unilever

Conjugation does not have to be restricted to atoms of the same type. For example, the unsaturated carbonyl compound 3-butene-2-one, CH_2=CH–CO–CH_3 (methyl vinyl ketone), has an absorption maximum at 212 nm with $\varepsilon = 7.125 \times 10^5$ (*Fig. 18*). Neither the C=C double bond nor the carbonyl on their own have intense maxima above 200 nm.

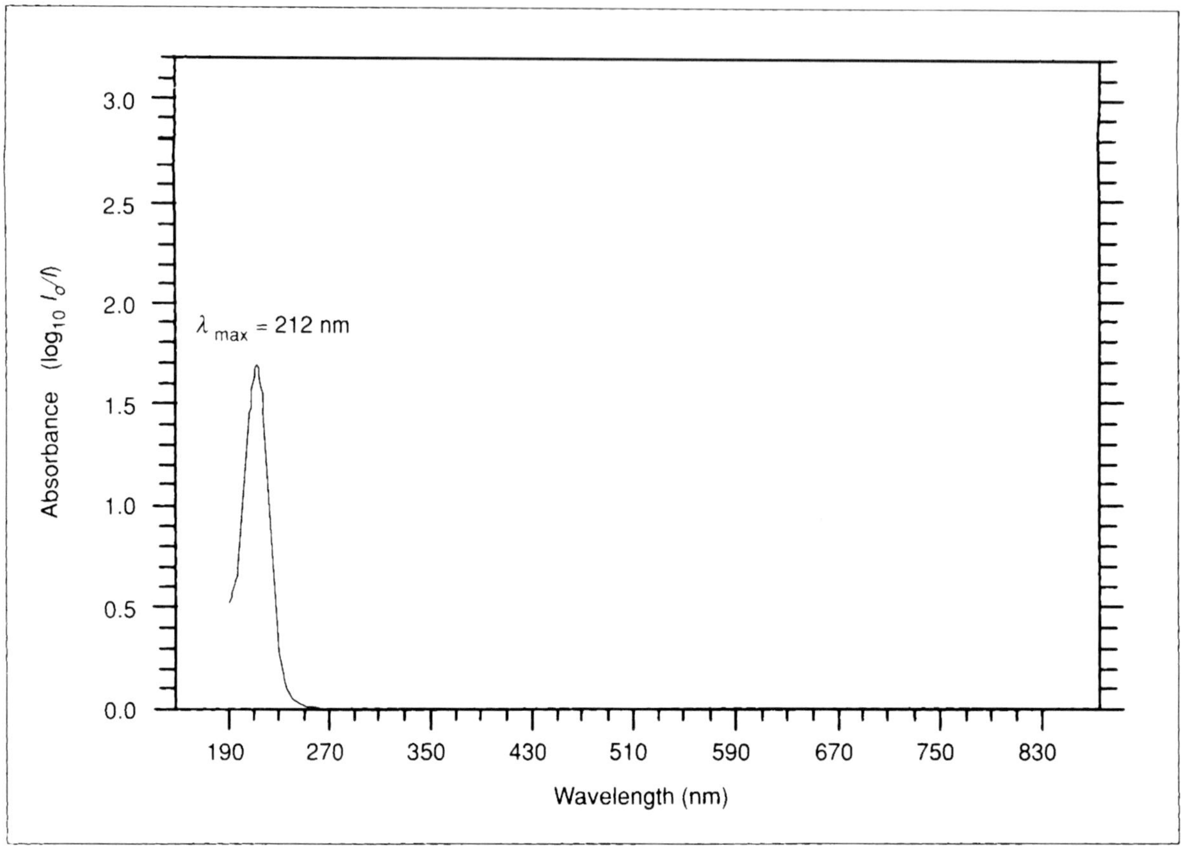

Figure 18 Ultraviolet/visible spectrum of CH_2=CH-CO-CH_3

Acid-base indicators

Absorption is advantageous in the acid-base indicators. A small change in the chemical structure of the indicator molecule can cause a change in the chromophore and it will absorb in different parts of the visible spectrum. The spectra of phenolphthalein and litmus (*Figs. 19a, 19b, 20a* and *20b*) illustrate this point.

Unilever

THE ROYAL SOCIETY OF CHEMISTRY

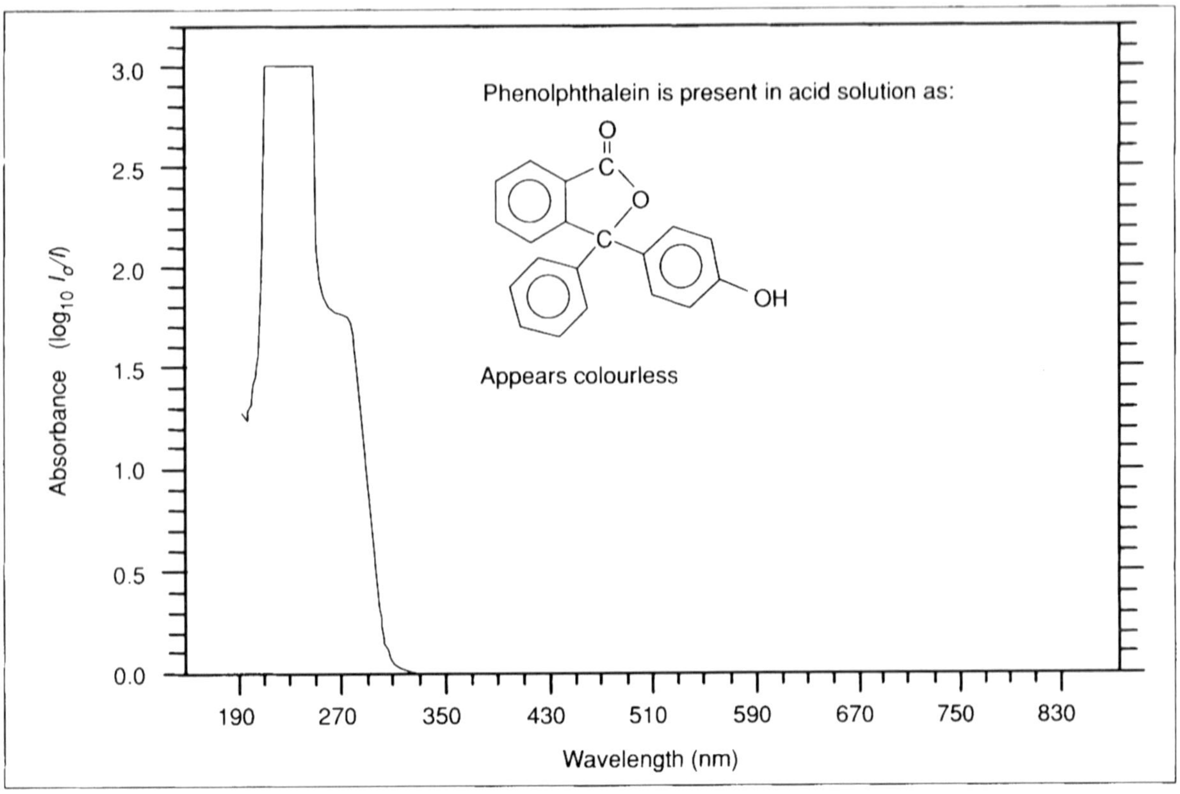

Figure 19a Ultraviolet/visible spectrum of phenolphthalein (pH = 1)

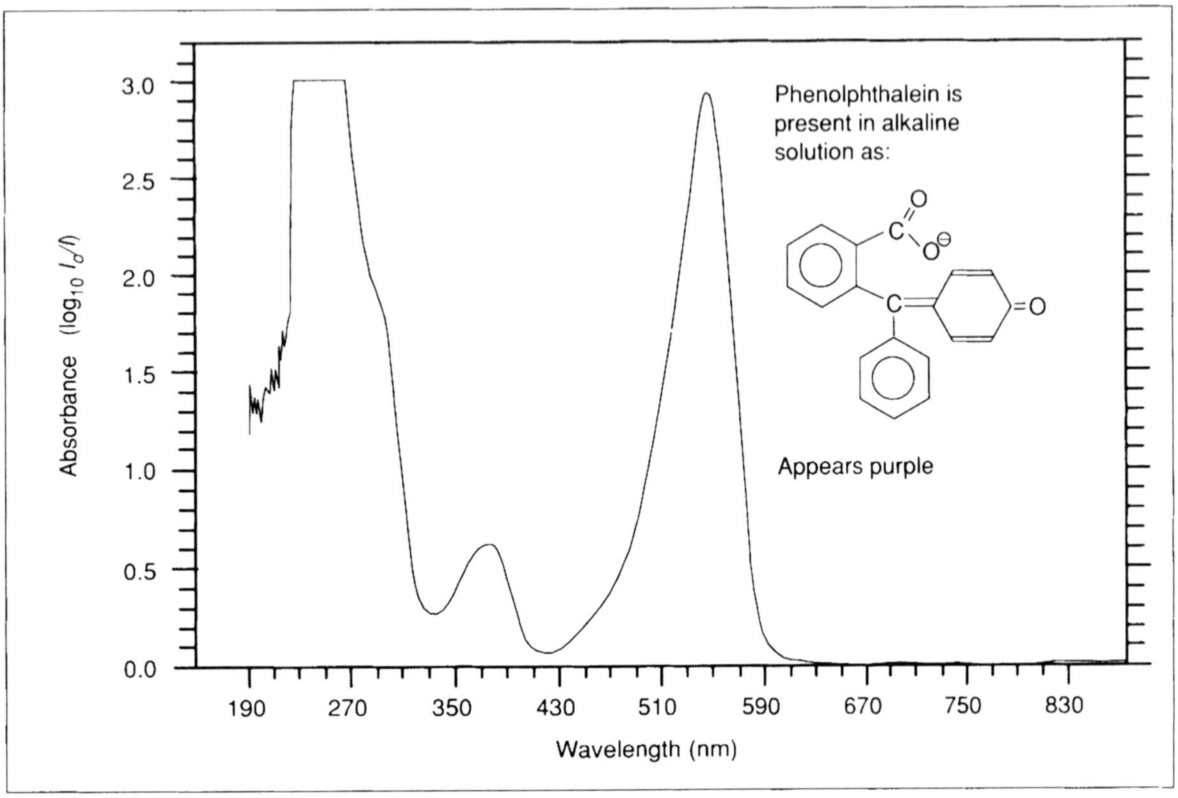

Figure 19b Ultraviolet/visible spectrum of phenolphthalein (pH = 13)

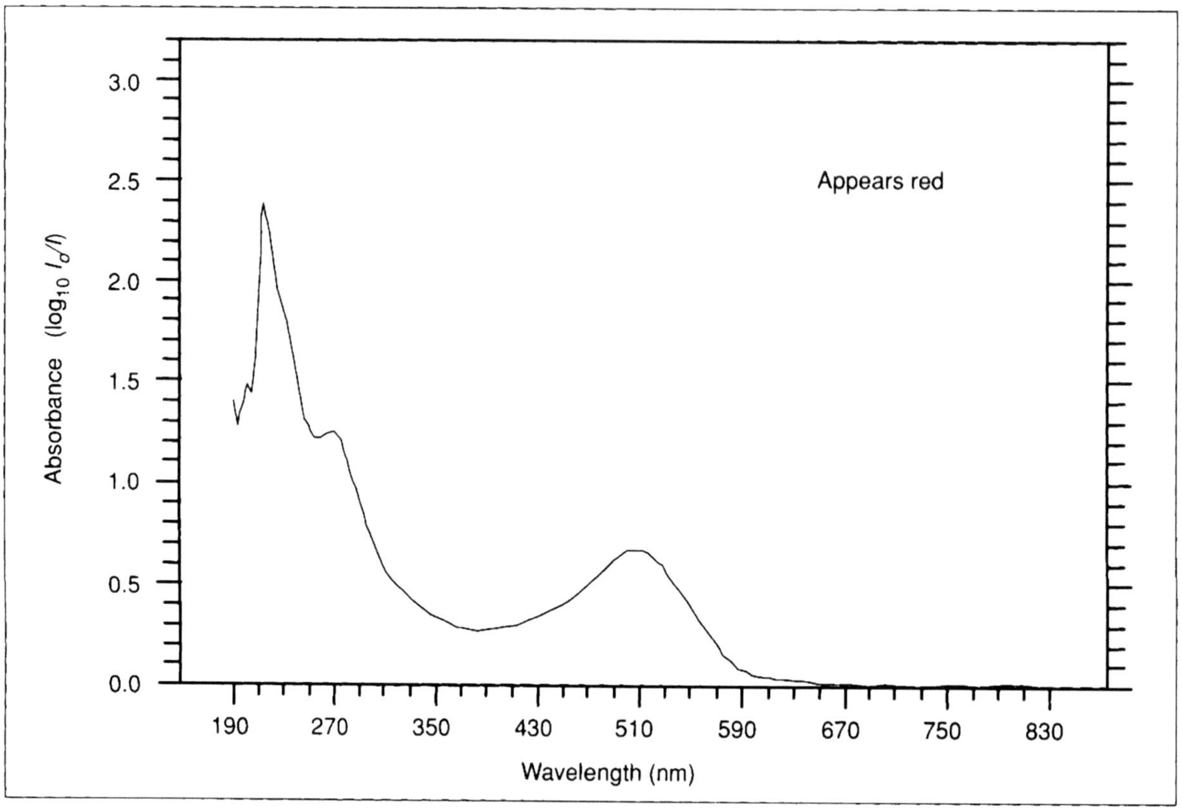

Figure 20a Ultraviolet/visible spectrum of litmus (pH = 1)

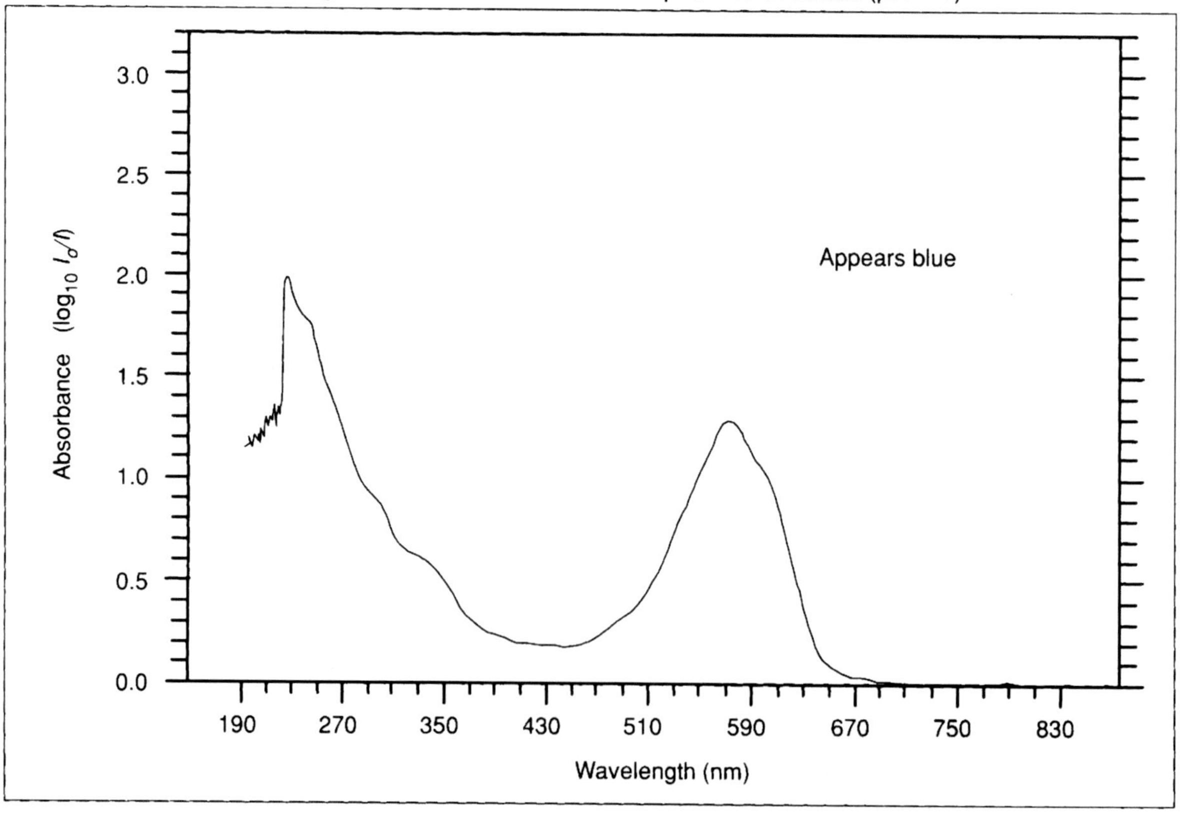

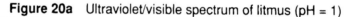

Figure 20b Ultraviolet/visible spectrum of litmus (pH = 13)

Unilever

THE ROYAL SOCIETY OF CHEMISTRY

Recent trends

Single beam, as well as double beam instruments are now on the market. These have the advantage that they are capable of measuring a spectrum very quickly. The principles of the single beam instrument are the same as for the double beam, but data on the reference are taken first, followed by the sample. The complete spectrum can be obtained very quickly. The diode array detector can be used to achieve this (see page 155). A computer can then interpret the two sets of data and plot the spectrum on a chart. This type of system can be in joint application with another technique *eg* the outflow from a chromatography column is passed through a small volume cell (often less than 10^{-2} cm^3) so that its ultraviolet/visible spectrum can be obtained as it flows through.

This has several advantages:

1 the different solutes do not have to be separated and collected in individual tubes so that their spectra may be obtained subsequently;

2 each spectrum can be determined in a fraction of a second; and

3 the spectra are stored by a computer so the spectrum of each solute can be compared with a library of known compounds.

One application is in the dope testing of race horses (page 150).

An alternative method of monitoring for particular substances is to pass the outflow from a chromatography column through a cell in a variable wavelength spectrometer. If the radiation is set at a wavelength known to be absorbed by the substance its presence will be shown by an absorption peak on the chart output of the spectrometer. An example of this is given in the section on caffeine/theobromine analysis (page 138).

Applications of ultraviolet/visible spectroscopy

In research, ultraviolet/visible spectroscopy is used more extensively in assaying than in identification. The trace metal content of an alloy, such as manganese in steel, can be determined by firstly reacting the sample to get the metal into solution as an ion. The ion is then complexed or made to react so that it is in a form that can be measured – *eg* manganese as the manganate(VII) ion. When the spectrum is recorded, the most useful piece of information is the absorbance because if the absorption coefficient of the chromophore is known the concentration of the solution can be calculated, and hence the mass of the metal in the sample.

The same principle can be applied to drug metabolites. Samples are taken from various sites around the body and their solutions are analysed to determine the amount of drug reaching those parts of the body. A useful feature of this type of analysis is the ability to calculate very small concentrations (of the order 0.0001 mol dm^{-3}) with extreme accuracy. It is important that the absorbance of the solution remains below two for quantitative measurements because of limitations of the instrument and solute-solute interactions that can cause deviations from the Beer-Lambert law.

The absorption of ultraviolet light is a feature of optical whiteners put into washing powders. The whitener absorbs radiation in the near ultraviolet and re-emits it in the visible range (*Fig. 21*).

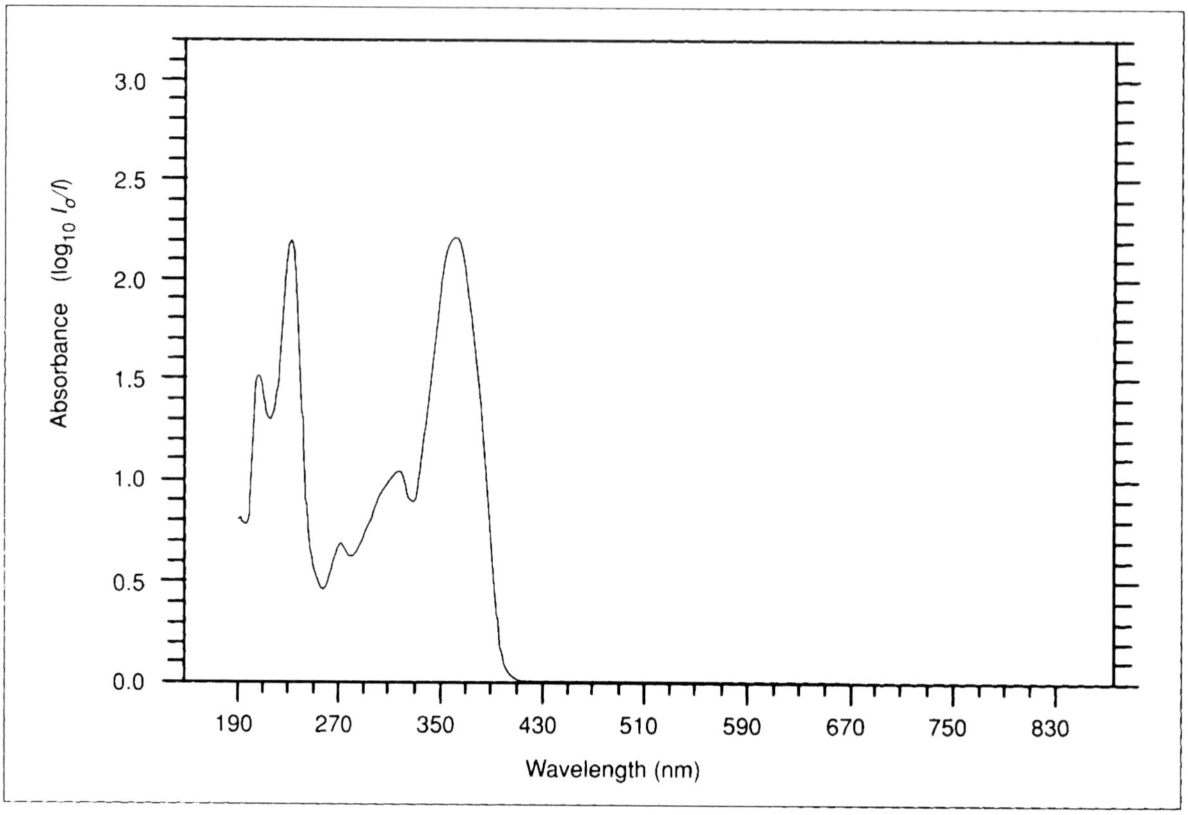

Figure 21 Ultraviolet/visible spectrum of an optical whitener used in a washing powder

Optical whiteners are also added to many toothpastes and detergent powders. They are often evident in discos that have ultraviolet lighting. White shirts and blouses frequently appear purple/blue, and people with false teeth should not smile!

Other applications include adding ultraviolet absorbing inks to water marks on paper so that they show up under an ultraviolet lamp; postcoding of household valuables with ultraviolet sensitive ink; and using invisible (but ultraviolet fluorescent) inks for signatures in building society savings books.

Unilever

THE ROYAL SOCIETY OF CHEMISTRY

Exercise

Ultraviolet/visible spectroscopy can also be used to study reaction rates. If a reagent or a product of the reaction absorbs radiation at a particular frequency the spectrometer can be set to measure the absorption at that frequency as a function of time.

The rate of hydrolysis of an ester (4-nitrophenylethanoate)

4-Nitrophenylethanoate hydrolyses in alkaline solution to give 4-nitrophenoxide ions and ethanoate ions:

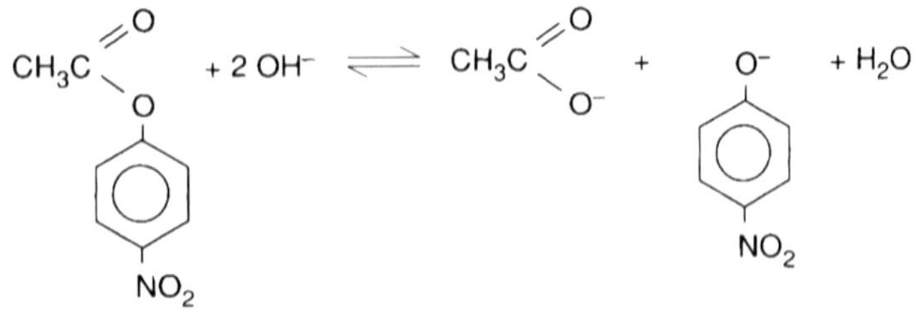

In the experiment the concentration of hydroxide ions is kept constant by a buffer solution and the progress of the reaction is followed by the absorption of light of wavelength 400 nm. The solution turns yellow as the 4-nitrophenoxide ion is liberated.

25.0×10^{-6} dm^3 (25.0 μl) of a solution of 0.01 mol dm^{-3} 4-nitrophenylethanoate in methanol is injected into 2.50 cm^3 of a buffer solution in a 1 cm cell. The buffer, at pH 10.9, consists of 0.01 mol dm^{-3} sodium carbonate neutralised with 1.0 mol dm^{-3} hydrochloric acid to the required pH. The reaction is followed by monitoring the appearance of the 4-nitrophenoxide chromophore at 400 nm with time. The absorption curve at 400 nm is shown in *Fig. 22.*

From the absorption curve determine:

1. the order of the reaction with respect to 4-nitrophenylethanoate;
2. the rate constant, k ; and
3. the absorption coefficient of the 4-nitrophenoxide ion.

Answers

Absorbance, on the y-axis, is directly proportional to the concentration of the 4-nitrophenoxide ion since

$$\text{Absorbance} = \log_{10} \frac{I_o}{I} = \varepsilon l c$$

The concentration of the 4-nitrophenoxide ion at time t, [4-nitrophenoxide]$_t$, is therefore given by

$$[\text{4-nitrophenoxide}]_t \ \alpha \ \text{absorbance}$$

From the equation for the reaction it is clear that for each mole of 4-nitrophenoxide ion formed one mole of the ester is hydrolysed. Thus, the concentration of the remaining ester is equal to its initial concentration less the concentration of 4-nitrophenoxide ion at that time, *ie*

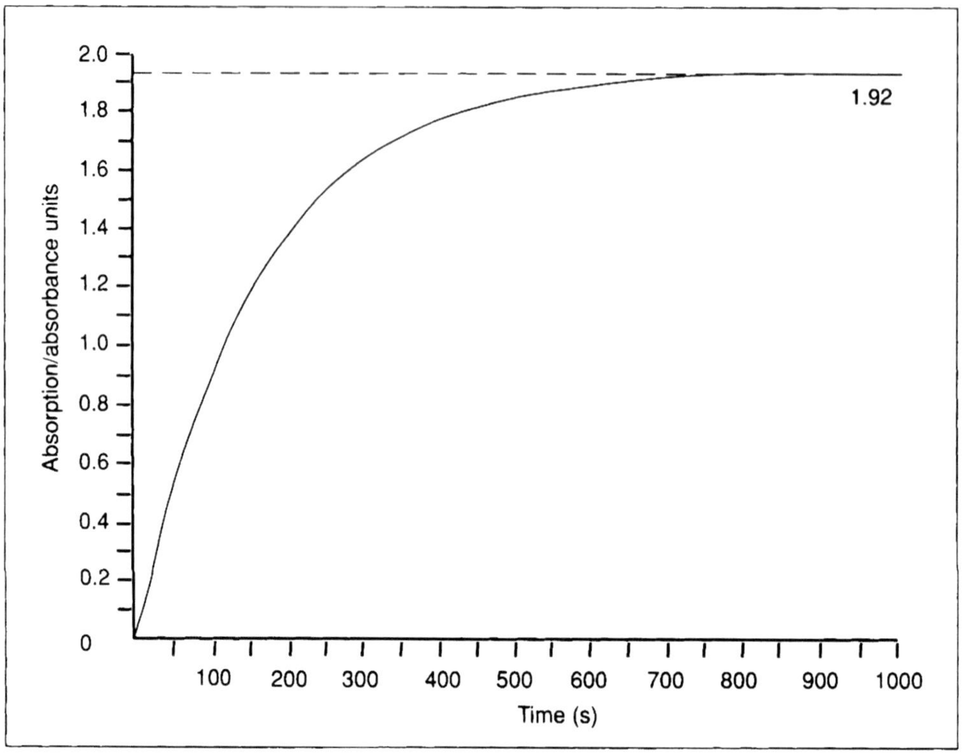

Figure 22 Absorption at 400 nm in a hydrolysis experiment

$[\text{ester}]_t = [\text{ester}]_o - [\text{4-nitrophenoxide}]_t$

where $[\text{ester}]_t$ = concentration of ester at time t
$[\text{ester}]_o$ = original concentration of ester
$[\text{4-nitrophenoxide}]_t$ = concentration of 4-nitrophenoxide ion at time t

The concentration of the ester is therefore given by

$[\text{ester}]_t \; \alpha \; \text{absorbance}_\infty - \text{absorbance}_t$

From *Fig. 22* the absorption at $t = \infty$ is 1.92, so

$[\text{ester}]_t \; \alpha \; 1.92 - \text{absorbance}_t$

Once the concentration of the ester at different times has been calculated, graphs can be plotted as follows to determine the order of the reaction with respect to the ester:

zero order:	[ester] versus time will be a straight line
first order:	ln [ester] versus time will be a straight line
second order:	1/[ester] versus time will be a straight line

From *Fig. 22*, the following data can be deduced:

Time (s)	**Absorbance units [4-nitrophenoxide]**	**$[ester]_t$**	**$\ln [ester]_t$**	**$1/[ester]_t$**
0	0.00	1.92	0.65	0.52
100	0.89	1.03	0.03	0.97
200	1.37	0.55	–0.60	1.82
300	1.63	0.29	–1.24	3.45
400	1.76	0.16	–1.83	6.25
500	1.84	0.08	–2.53	12.50
600	1.88	0.04	–3.22	25.00

The only graph that produces a straight line is ln [ester] versus time, so the reaction must be first order with respect to the ester.

The gradient of this plot is equal to $-k$ where k is the rate constant. From the graph,

$$\text{Gradient} = -k = \frac{-3.22 - 0.03}{500} = \frac{-3.19}{500} = -0.0064\ \text{s}^{-1}$$

Thus $\underline{k = 0.0064\ \text{s}^{-1}}$

From this it is possible to calculate the half-life of the reaction quite simply:

$$t_{1/2} = \frac{\ln 2}{k} = \frac{0.6931}{0.0064} = 108.3\ \text{s}$$

The absorption coefficient of the 4-nitrophenoxide ion can be calculated as follows:

From the Beer–Lambert law

$$\text{absorbance} = \log_{10}\frac{I_o}{I} = \varepsilon l c$$

The absorbance can be measured directly from *Fig. 22*, and the path length of the cell is 1 cm (*ie* l= 1). The 0.01 mol dm^{-3} ester solution was diluted from 0.025 cm^3 to 2.525 cm^3 (0.025 cm^3 +2.5 cm^3), *ie* by a factor of 101. Its final concentration was therefore

$$\frac{0.01}{101} = 0.000099\ \text{mol dm}^{-3}$$

Substituting these values,

$$1.92 = \varepsilon \times 1 \times 0.000099$$

$$\varepsilon = \frac{1.92}{0.000099} = 19\,392\ \text{mol}^{-1}\ \text{dm}^3\ \text{cm}^{-1}$$

By convention the units are not usually quoted, and absorption coefficients are commonly given to fewer significant figures.

Thus, $\underline{\varepsilon = 19\,400}$.

THE ROYAL
SOCIETY OF
CHEMISTRY

Unilever

5. Chromatography

Chromatography is usually introduced as a technique for separating and/or identifying the components in a mixture. The basic principle is that components in a mixture have different tendencies to adsorb onto a surface or dissolve in a solvent. It is a powerful method in industry, where it is used on a large scale to separate and purify the intermediates and products in various syntheses.

The theory

There are several different types of chromatography currently in use – *ie* paper chromatography; thin layer chromatography (TLC); gas chromatography (GC); liquid chromatography (LC); high performance liquid chromatography (HPLC); ion exchange chromatography; and gel permeation or gel filtration chromatography.

Basic principles

All chromatographic methods require one static part (the stationary phase) and one moving part (the mobile phase). The techniques rely on one of the following phenomena: adsorption; partition; ion exchange; or molecular exclusion.

Adsorption

Adsorption chromatography was developed first. It has a solid stationary phase and a liquid or gaseous mobile phase. (Plant pigments were separated at the turn of the 20th century by using a calcium carbonate stationary phase and a liquid hydrocarbon mobile phase. The different solutes travelled different distances through the solid, carried along by the solvent.) Each solute has its own equilibrium between adsorption onto the surface of the solid and solubility in the solvent, the least soluble or best adsorbed ones travel more slowly. The result is a separation into bands containing different solutes. Liquid chromatography using a column containing silica gel or alumina is an example of adsorption chromatography (*Fig. 1*).

The solvent that is put into a column is called the eluent, and the liquid that flows out of the end of the column is called the eluate.

Partition

In partition chromatography the stationary phase is a non-volatile liquid which is held as a thin layer (or film) on the surface of an inert solid. The mixture to be separated is carried by a gas or a liquid as the mobile phase. The solutes distribute themselves between the moving and the stationary phases, with the more soluble component in the mobile phase reaching the end of the chromatography column first (*Fig. 2*). Paper chromatography is an example of partition chromatography.

Unilever

THE ROYAL SOCIETY OF CHEMISTRY

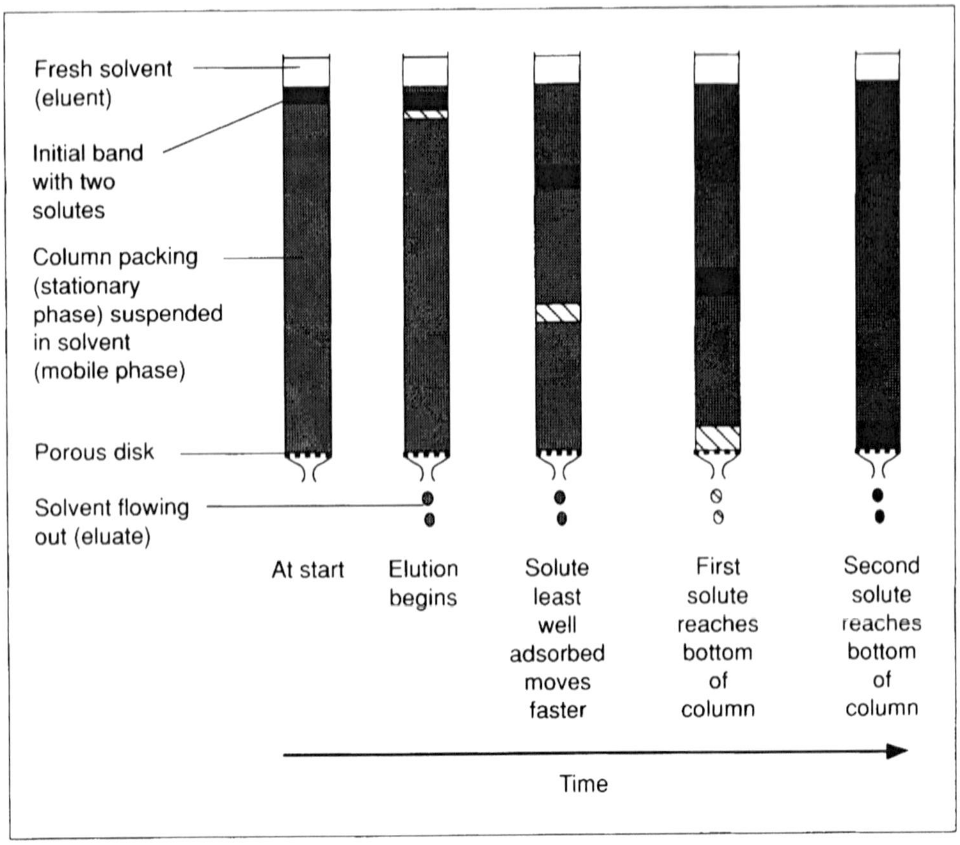

Figure 1 Adsorption chromatography using a column

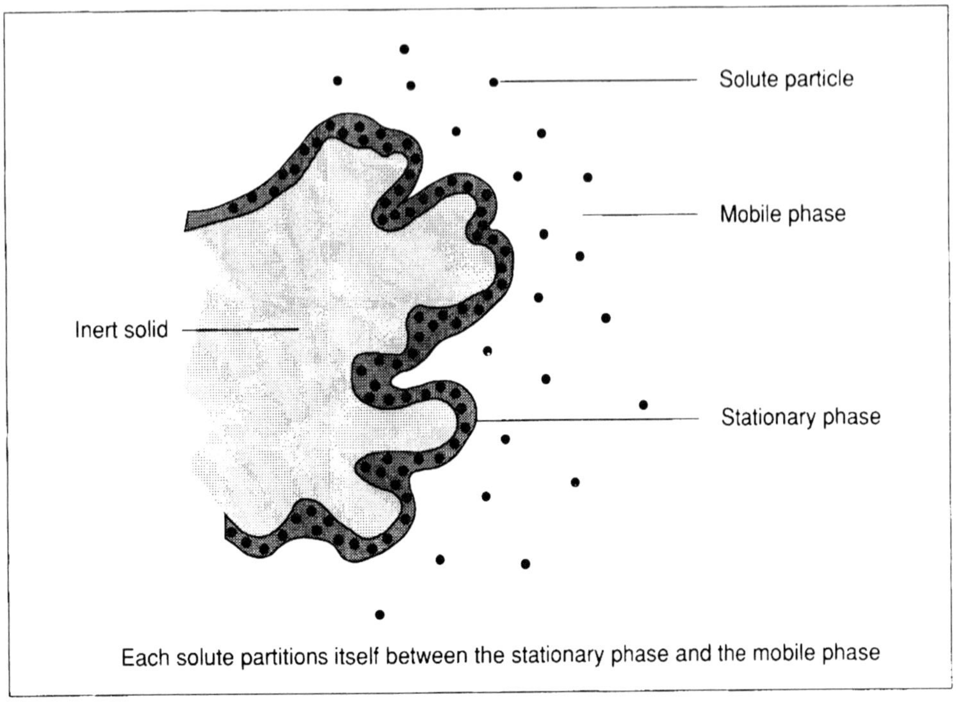

Figure 2 Partition chromatography

Ion exchange

Ion exchange chromatography is similar to partition chromatography in that it has a coated solid as the stationary phase. The coating is referred to as a resin, and has ions (either cations or anions, depending on the resin) covalently bonded to it and ions of the opposite charge are electrostatically bound to the surface. When the mobile phase (always a liquid) is eluted through the resin the electrostatically bound ions are released as other ions are bonded preferentially (*Fig. 3*). Domestic water softeners work on this principle.

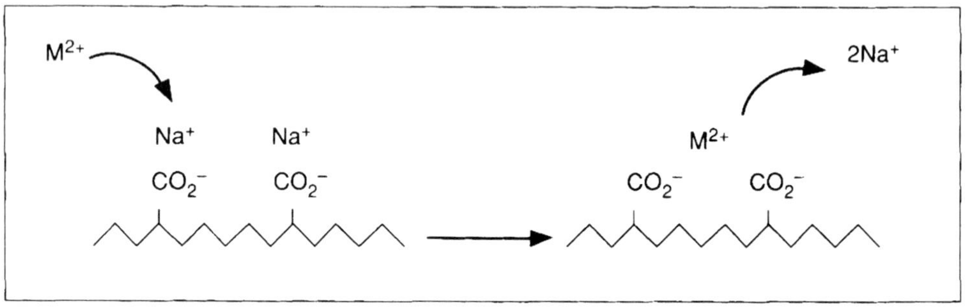

Figure 3 Ion exchange chromatography

Molecular exclusion

Molecular exclusion differs from other types of chromatography in that no equilibrium state is established between the solute and the stationary phase. Instead, the mixture passes as a gas or a liquid through a porous gel. The pore size is designed to allow the large solute particles to pass through uninhibited. The small particles, however, permeate the gel and are slowed down so the smaller the particles, the longer it takes for them to get through the column. Thus separation is according to particle size (*Fig. 4*).

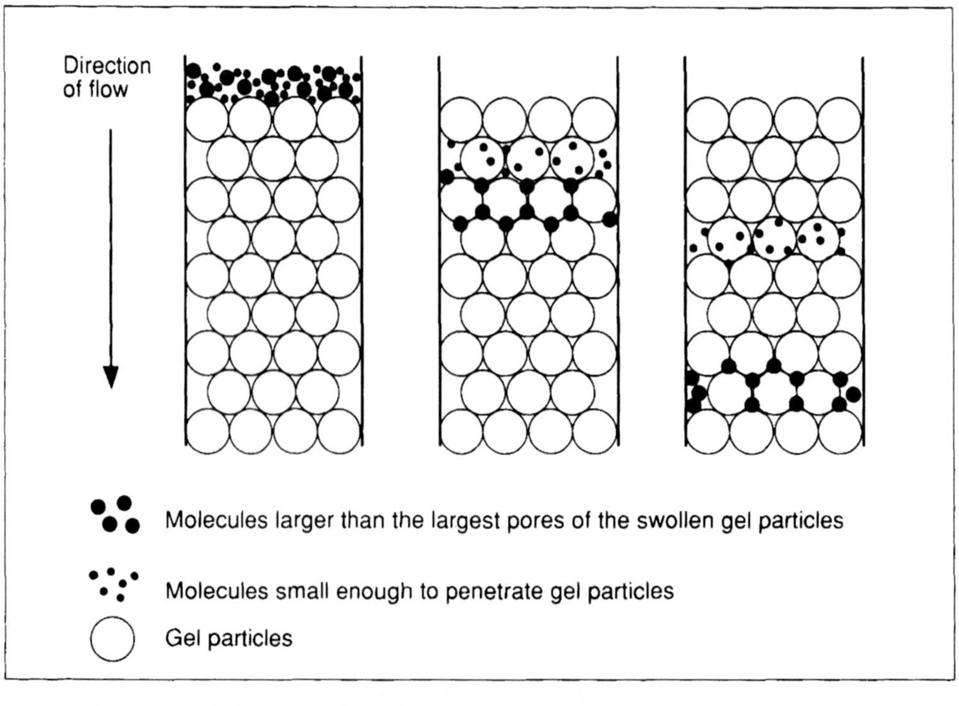

Figure 4 Gel permeation chromatography

Unilever

THE ROYAL SOCIETY OF CHEMISTRY

Chromatographic techniques

Paper chromatography

This is probably the first, and the simplest, type of chromatography that people meet. A drop of a solution of a mixture of dyes or inks is placed on a piece of chromatography paper and allowed to dry. The mixture separates as the solvent front advances past the mixture. Filter paper and blotting paper are frequently substituted for chromatography paper if precision is not required. Separation is most efficient if the atmosphere is saturated in the solvent vapour (*Fig. 5*).

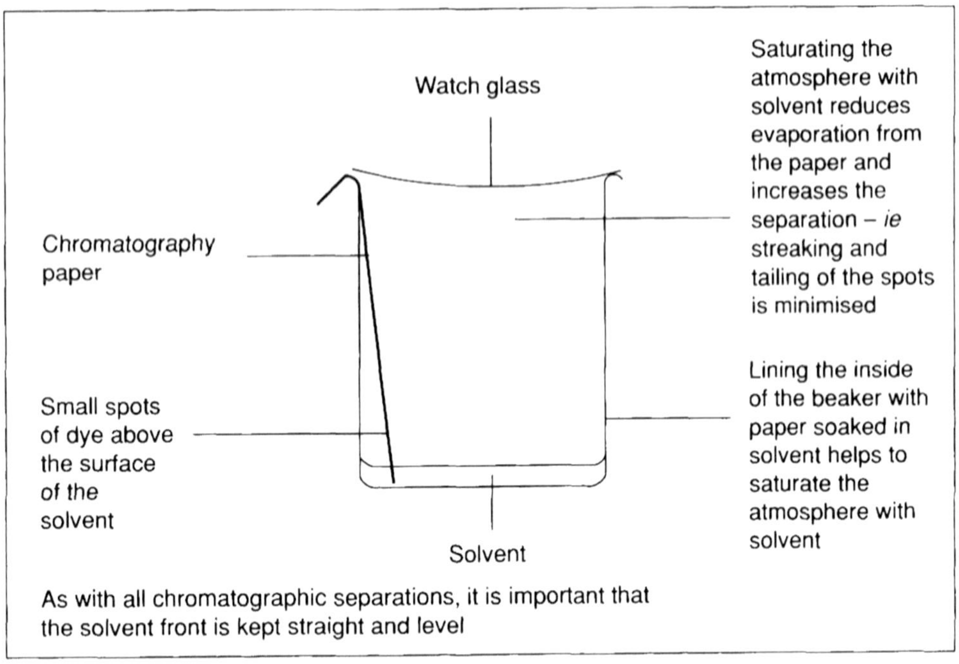
Watch glass
Saturating the atmosphere with solvent reduces evaporation from the paper and increases the separation – *ie* streaking and tailing of the spots is minimised
Chromatography paper
Small spots of dye above the surface of the solvent
Lining the inside of the beaker with paper soaked in solvent helps to saturate the atmosphere with solvent
Solvent
As with all chromatographic separations, it is important that the solvent front is kept straight and level

Figure 5 Paper chromatography

Some simple materials that can be separated by using this method are inks from fountain and fibre-tipped pens, food colourings and dyes. The components can be regenerated by dissolving them out of the cut up paper.

The efficiency of the separation can be optimised by trying different solvents, and this remains the way that the best solvents for industrial separations are discovered (some experience and knowledge of different solvent systems is advantageous).

Paper chromatography works by the partition of solutes between water in the paper fibres (stationary phase) and the solvent (mobile phase). Common solvents that are used include pentane, propanone and ethanol. Mixtures of solvents are also used, including aqueous solutions, and solvent systems with a range of polarities can be made. A mixture useful for separating the dyes on Smarties is a 3:1:1 mixture (by volume) of butan-1-ol:ethanol:0.880 ammonia solution.

As each solute distributes itself (equilibrates) between the stationary and the mobile phase, the distance a solute moves is always the same fraction of the distance moved by the solvent. This fraction is variously called the retardation factor or the retention ratio, and is given the symbol R or R_f:

THE ROYAL SOCIETY OF CHEMISTRY

Unilever

$$\text{Retention ratio} = \frac{\text{distance moved by solute}}{\text{distance moved by solvent}} = R_f$$

So as long as the correct solvent and type of chromatography paper are used, a component can be identified from its retention ratio (*Fig. 6*).

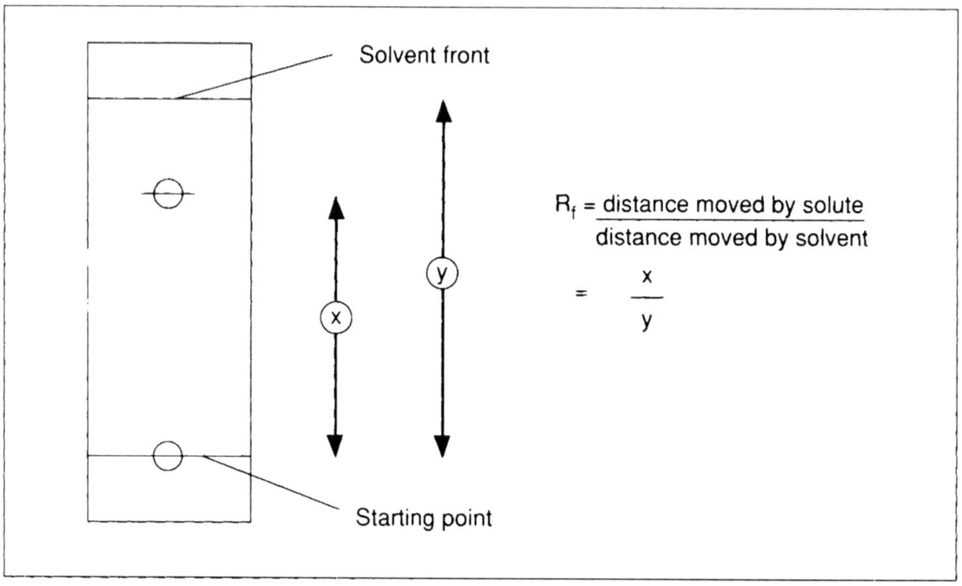

Figure 6 The retention ratio, R_f

It is possible that two solutes have the same R_f values using one solvent, but different values using another solvent (*eg* this occurs with some amino acids). This means that if a multi component system is not efficiently separated by one solvent the chromatogram can be dried, turned through 90^0, and run again using a second solvent, (*Fig. 7*).

Thin layer chromatography (TLC)

Thin layer chromatography is similar to paper chromatography, but the stationary phase is a thin layer of a solid such as alumina or silica supported on an inert base such as glass, aluminum foil or insoluble plastic. The mixture is 'spotted' at the bottom of the TLC plate and allowed to dry. The plate is placed in a closed vessel containing solvent (the mobile phase) so that the liquid level is below the spot.

TLC has advantages over paper chromatography in that its results are more reproducible, and that separations are very efficient because of the much smaller particle size of the stationary phase.

The solvent ascends the plate by capillary action, the liquid filling the spaces between the solid particles. This technique is usually done in a closed vessel to ensure that the atmosphere is saturated with solvent vapour and that evaporation from the plate is minimised before the run is complete. The plate is removed when the solvent front approaches the top of the plate and the position of the solvent front recorded before it is dried (this allows the R_f value to be calculated).

TLC has applications in industry in determining the progress of a reaction by studying the components present; and in separating reaction intermediates. In the latter case a line of the reaction mixture is 'painted' across the TLC plate instead of a single spot, and the line of product after separation is cut out of the plate and dissolved in an appropriate solvent.

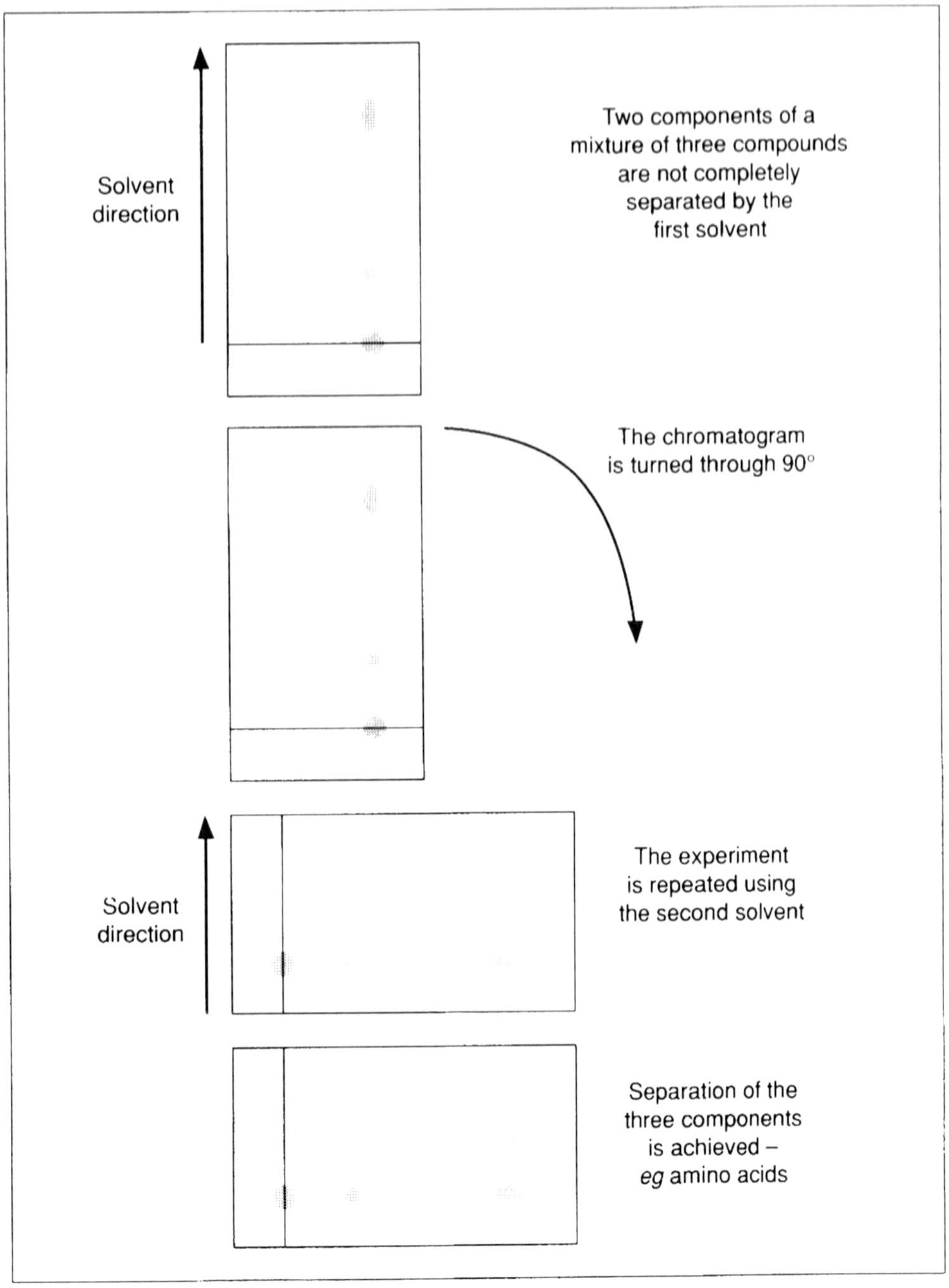

Figure 7 Using two solvents to separate a multi component mixture

Many spots are not visible without the plates being 'developed'. This usually involves spraying with a solution that is reversibly adsorbed or reacts in some way with the solutes. Two examples of developing solutions are iodine in petroleum ether (useful for identifying aromatic compounds, especially those with electron donating groups – *eg* $C_6H_5NH_2$) and ninhydrin (useful for identifying amino acids). Iodine vapour is also used to develop plates in some cases. Alternatively, specially prepared plates can be used that fluoresce in ultraviolet light. The plates are used in the normal manner, but once dried they are placed under an ultraviolet lamp. Solute spots mask fluorescence on the surface of the plate – *ie* a dark spot is observed. Some compounds have their own fluorescence which can be used for identification, or

THE ROYAL SOCIETY OF CHEMISTRY

Unilever

retardation factors can be used to identify known solutes.

Radioactive solutes can be identified on TLC plates by passing the plates under a Geiger counter with a narrow window. A chart recorder plots the count rate as the plate passes under the counter (*Fig. 8*). Accurate quantitative data can be derived by integrating the peaks (this is not shown in the diagram).

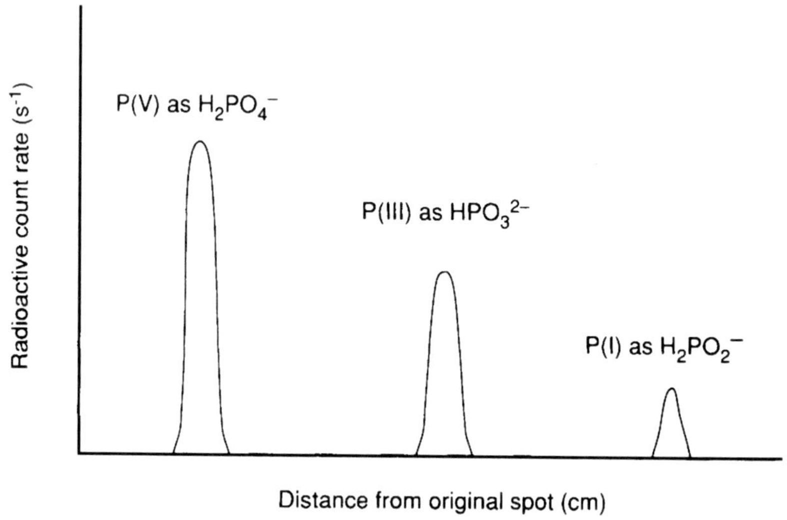

This plate showed radioactive ^{32}P in oxidation states (I), (III) and (V) as a result of the nuclear reaction $^{35}_{17}Cl + ^{1}_{0}n \rightarrow ^{32}_{15}P + ^{4}_{2}\alpha$

Potassium chloride crystals were irradiated with neutrons, and the ^{35}Cl nuclei absorbed a neutron and emitted an α particle to leave ^{32}P nuclei, which decay by β emission: $^{32}_{15}P \rightarrow ^{32}_{16}S + ^{0}_{-1}\beta$

The oxidation state distribution of radiophosphorus can be seen once the irradiated crystals are dissolved in solution and studied using the solvent system methanol:2–aminopropane:dichloroethanoic acid:ethanoic anhydride (100:30:5:3 by volume). The areas under the peaks show that the majority of the radiophosphorus was in oxidation state (V).

Figure 8 Count rate on a TLC plate

A relatively new method of detecting components on TLC plates is to scan the lane along which a mixture has travelled with a beam of fixed wavelength light. The reflected light from the lane (A) is measured relative to the radiation reflected from outside the lane (B):

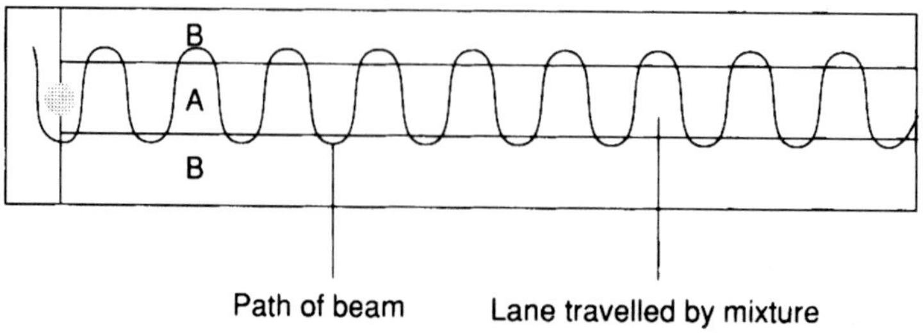

The plates can be scanned for ultraviolet/visible absorption; or natural fluorescence. Ultraviolet sensitive plates or already developed plates can also be read.

Gas chromatography (GC)

This technique uses a gas as the mobile phase, and the stationary phase can either be a solid or a non-volatile liquid (in which case small inert particles such as diatomaceous earth are coated with the liquid so that a large surface area exists for the solute to equilibrate with). If a solid stationary phase is used the technique is described as gas-solid adsorption chromatography, and if the stationary phase is liquid it is called gas-liquid partition chromatography. The latter is more commonly used, but in both cases the stationary phase is held in a narrow column in an oven and the stationary phase particles are coated onto the inside of the column.

> Diatomaceous earth is made from the skeletons of a single-celled non-flowering plant. The skeletons are made of hydrated silica, and are ground to a fine powder of the required particle size.
>
> The advantage of diatomaceous earth is that it has fewer silanol (Si–OH) groups than silica and is less prone to electrostatic attractions (*eg* hydrogen bonds). The polarity of the OH groups can be reduced or eliminated by esterifying or silanising them.
>
> Esterification with ethanoic acid:
>
> $$\text{–Si–O–H} + CH_3COOH \rightarrow \text{–Si–OCOCH}_3 + H_2O$$
>
> Silanising:
>
> ```
> O–H     O–H                                  (CH3)3Si–O         O–Si(CH3)3
>   \     /                                              \       /
> –Si–O–Si–  +  (CH3)3SiNHSi(CH3)3  →                    –Si–O–Si–   +  NH3
>               (hexamethyldisilazane)
> ```
>
> Diatomaceous earth is also known as kieselguhr – the clay Nobel used as the inert base for dynamite.

Practical details

For separation or identification the sample must be either a gas or have an appreciable vapour pressure at the temperature of the column – it does not have to be room temperature. The sample is injected through a self sealing disc (a rubber septum) into a small heated chamber where it is vaporised if necessary (*Fig. 9*). Although the sample must all go into the column as a gas, once it is there the temperature can be below the boiling point of the fractions as long as they have appreciable vapour pressures inside the column. This ensures that all the solutes pass through the column over a reasonable time span. The injector oven is usually 50–100 °C hotter than the start of the column.

The sample is then taken through the column by an inert gas (known as the carrier gas) such as helium or nitrogen which must be dry to avoid interference from water molecules. It can be dried by passing it through anhydrous copper(II) sulphate or self-indicating silica (silica impregnated with cobalt(II) chloride). Unwanted organic solvent vapours can be removed by passing the gas through activated charcoal. The column is coiled so that it will fit into the thermostatically controlled oven.

The temperature of the oven is kept constant for a straightforward separation, but if there are a large number of solutes, or they have similar affinities for the stationary phase relative to the mobile phase, then it is common for the temperature of the

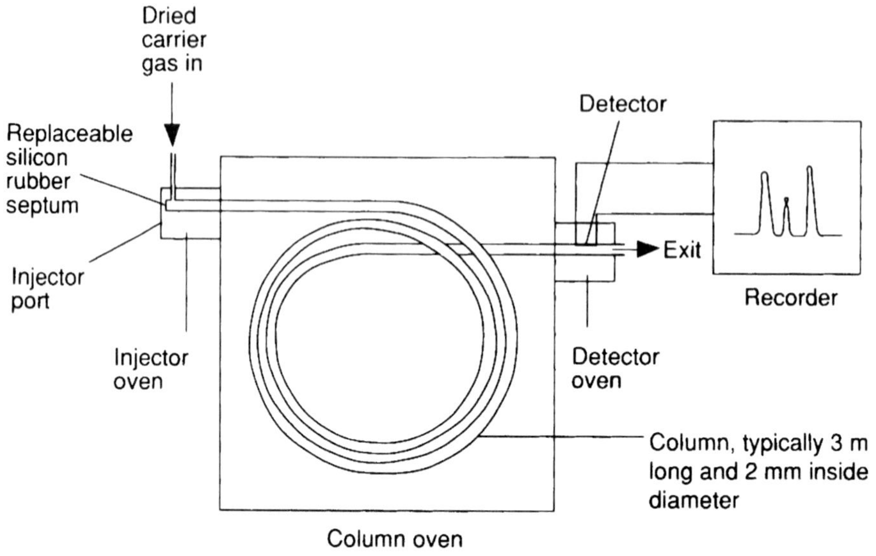

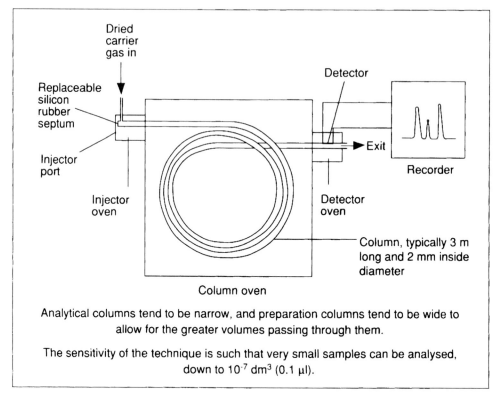

Analytical columns tend to be narrow, and preparation columns tend to be wide to allow for the greater volumes passing through them.

The sensitivity of the technique is such that very small samples can be analysed, down to 10^{-7} dm^3 (0.1 μl).

Figure 9 The gas chromatograph

column to be increased gradually over a required range. This is done by using computer control, and gives a better separation if solute boiling points are close, and a faster separation if some components are relatively involatile.

The solutes progress to the end of the column, to a detector. Two types of detector are commonly used: thermal conductivity detectors and flame ionisation detectors. Thermal conductivity detectors respond to changes in the thermal conductivity of the gas leaving the column. A hot tungsten–rhenium filament is kept in an oven set at a given temperature so that all solutes are in the gaseous phase (*Fig. 10*). When the carrier gas – helium, for example, which will have been warmed to the temperature of the detector block – leaves the column it cools the hot filament. However, if a solute emerges with the helium it will cool the filament less (unless the solute is hydrogen, because only hydrogen has a thermal conductivity greater than helium) and the temperature of the filament will rise. Its resistance will then increase, and that can be measured.

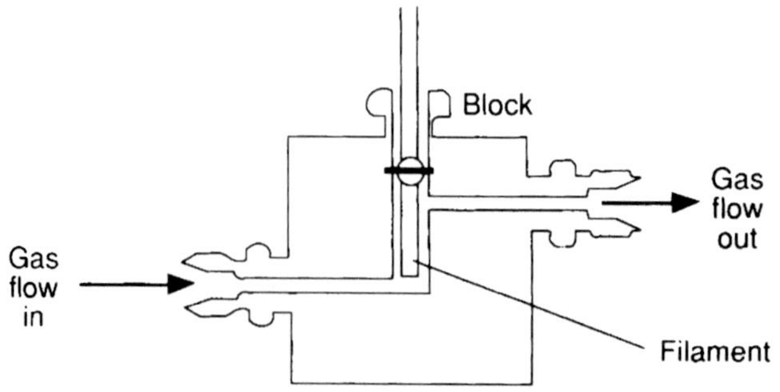

Figure 10 A simple thermal conductivity detector

So that the change in resistance can be monitored directly a second circuit measures the resistance from the pure carrier gas (*Fig. 11*).

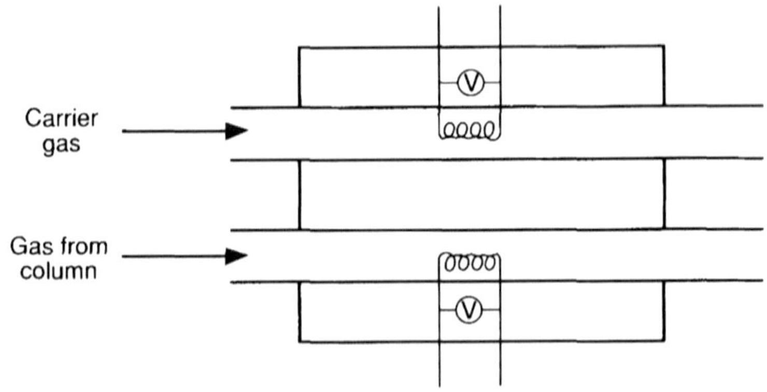

Figure 11 Comparative thermal conductivity detector

The performance of a drug *in vivo* is frequently studied by using GC–MS (see page 22). After giving volunteers a dose of the drug, blood plasma samples are taken at varying intervals, perhaps up to 24 h, and the parent drug and its metabolites are separated and identified. GC–MS has the advantage over other techniques that it is particularly sensitive – *eg* it can differentiate between the drug and its metabolites when their ultraviolet spectra are very similar.

In reality, most thermal conductivity detectors have four filaments in a Wheatstone bridge arrangement, two filaments in the exit gas from the column and two in the reference gas stream (*Fig. 12*).

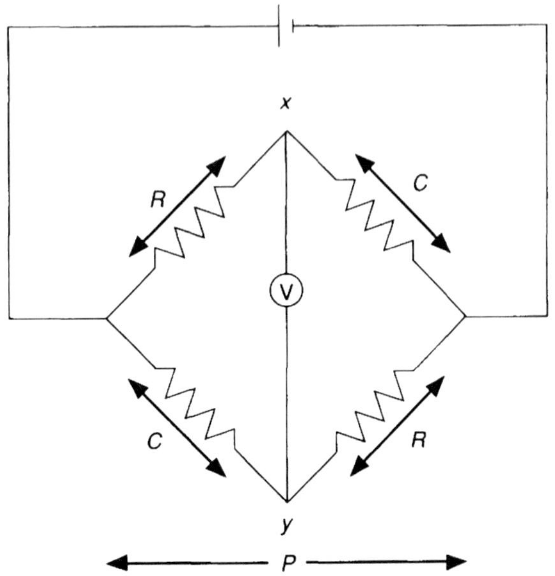

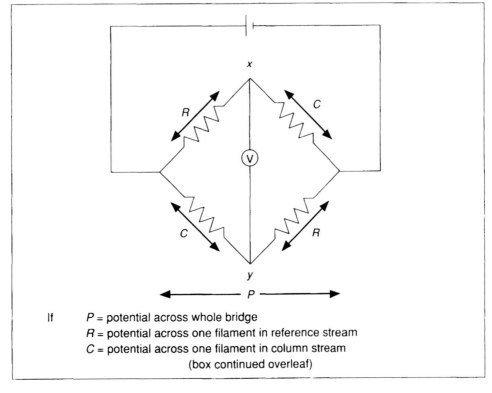

If *P* = potential across whole bridge
R = potential across one filament in reference stream
C = potential across one filament in column stream

(box continued overleaf)

Figure 12 Wheatstone bridge arrangement of filaments

the potential at $x = \frac{R}{C+R} \times P$ potential at $y = \frac{C}{C+R} \times P$

therefore potential between x and y, measured on voltmeter, is

$$x\text{-}y = \frac{R-C}{C+R} \times P$$

If the gases in the two sides of the detector are the same, their conductivities will be the same so they will cool the filaments equally and $R-C= 0$. The potential between x and y will then be zero. However, if the gases have different compositions then $R-C$ will not be zero and a value will be recorded.

Figure 12 continued

Flame ionisation detection (FID) is particularly useful for detecting organic compounds and this technique is by far and away the most common GC detection system. The gas from the column is mixed with hydrogen and air, and is then burned. Some CH• radicals, which are formed on combustion, are then oxidised to CHO^+ ions and these ions allow a current to be transmitted via a cathode, the current is then converted to a signal on a chart recorder (*Fig. 13*).

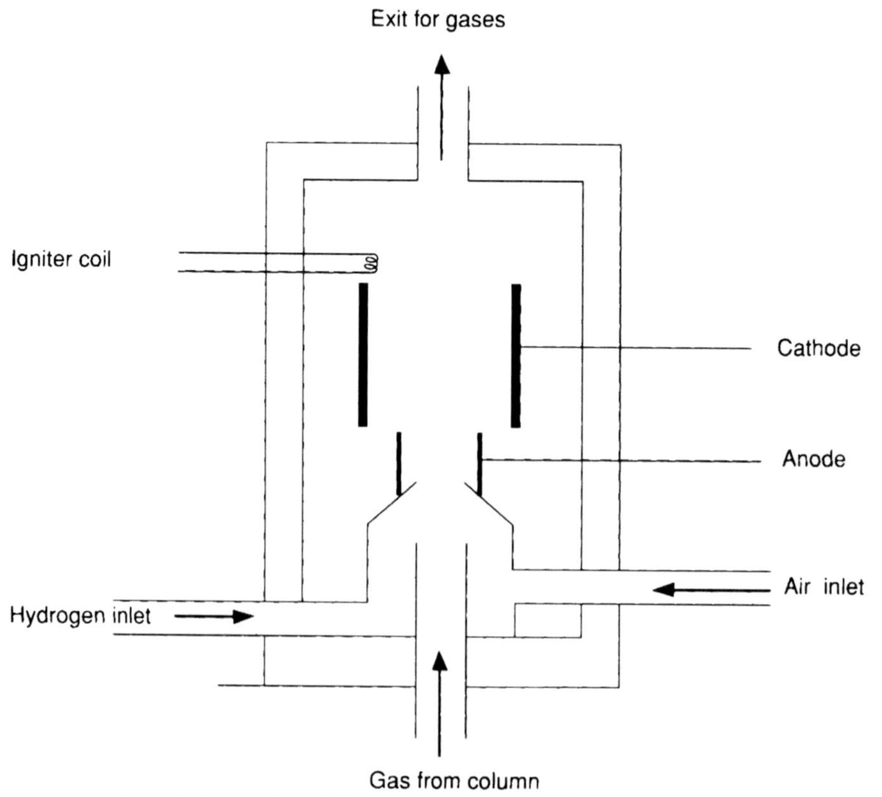

Figure 13 Flame ionisation detector

Although the number of ions produced is small (perhaps one in 10^5 carbon atoms produce an ion) the proportion is constant. The current produced is also proportional to the number of ions produced, so the total signal on the chart peak is proportional to the amount of that solute in the mixture.

A flame ionisation detector is approximately 1000 times as sensitive as the thermal conductivity detector for organic materials, but is of no use if the solutes do not burn or produce ions. Detectors are available that detect nitrogen and phosphorus as their ions, but these rely on excited rubidium atoms rather than a flame, to produce ions (see Box).

The nitrogen–phosphorus detector (NPD)

The NPD is fundamentally similar to the flame ionisation detector. They both work by forming ions and subsequently detecting them as a minute electrical current. However, a major difference arises in the way the ions are formed.

The eluate (exit gases) of the GC is forced through a jet in the presence of air and hydrogen gas. The mixture passes over the surface of a heated rubidium salt in the form of a bead. The excited rubidium atoms (Rb^*) selectively ionise nitrogen and phosphorus. The ions formed allow a small electric current to flow between two charged surfaces which, under different operating conditions, gives a response to either nitrogen or phosphorus containing compounds, or both (*Fig. 14*). To differentiate between nitrogen and phosphorus containing compounds the retention time of each solute in the column is used.

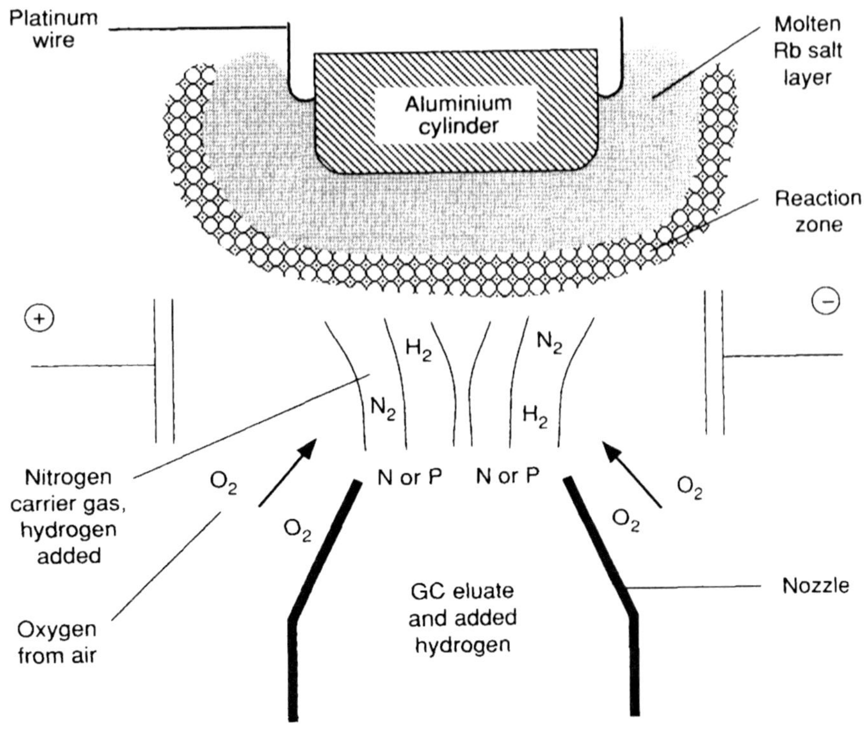

Diagram adapted by courtesy of Hewlett Packard

Figure 14 The nitrogen–phosphorus detector

The rubidium bead is mounted on a small aluminium cylinder, and a current is supplied through a platinum wire (the current is a few pico amps). This current heats the bead and excites the rubidium atoms so that they can ionise nitrogen and phosphorus.

The hydrogen is used to maintain the temperature of the bead and consequently only a small flow rate is necessary (Table 1).

Table 1 Typical flow rates in the flame ionisation detector and the nitrogen–phosphorus detector /$cm^3\ min^{-1}$

	Air	Hydrogen
FID	300	30
NPD	100	4

The NPD can be set for phosphorus detection to determine very low concentrations of pesticides such as malathion, the active ingredient of Prioderm, a treatment used for killing head lice. The structure of malathion is:

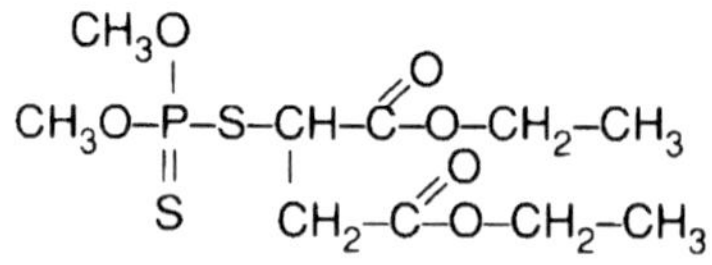

Once a mixture has been separated by GC its components need to be identified. For known substances this can be done from a knowledge of the time it takes for solutes to reach the detector once they have been injected into the column. These are known as retention times and will vary depending on each of the following:

1 the flow rate of the carrier gas;

2 the temperature of the column;

3 the length and diameter of the column;

4 the nature of and interactions between the solute and the stationary and mobile phases; and

5 the volatility of the solute.

Each material to be identified by GC is run through the column so that its retention time (the time for the components to pass through the column) can be determined. For compounds of completely unknown structure or composition the solutes must be collected individually and then analysed by using another method – *eg* mass spectrometry.

Liquid chromatography (LC)

Liquid chromatography is similar to gas chromatography but uses a liquid instead of a gaseous mobile phase. The stationary phase is usually an inert solid such as silica gel ($SiO_2.xH_2O$), alumina ($Al_2O_3.xH_2O$) or cellulose supported in a glass column (*Fig. 15*).

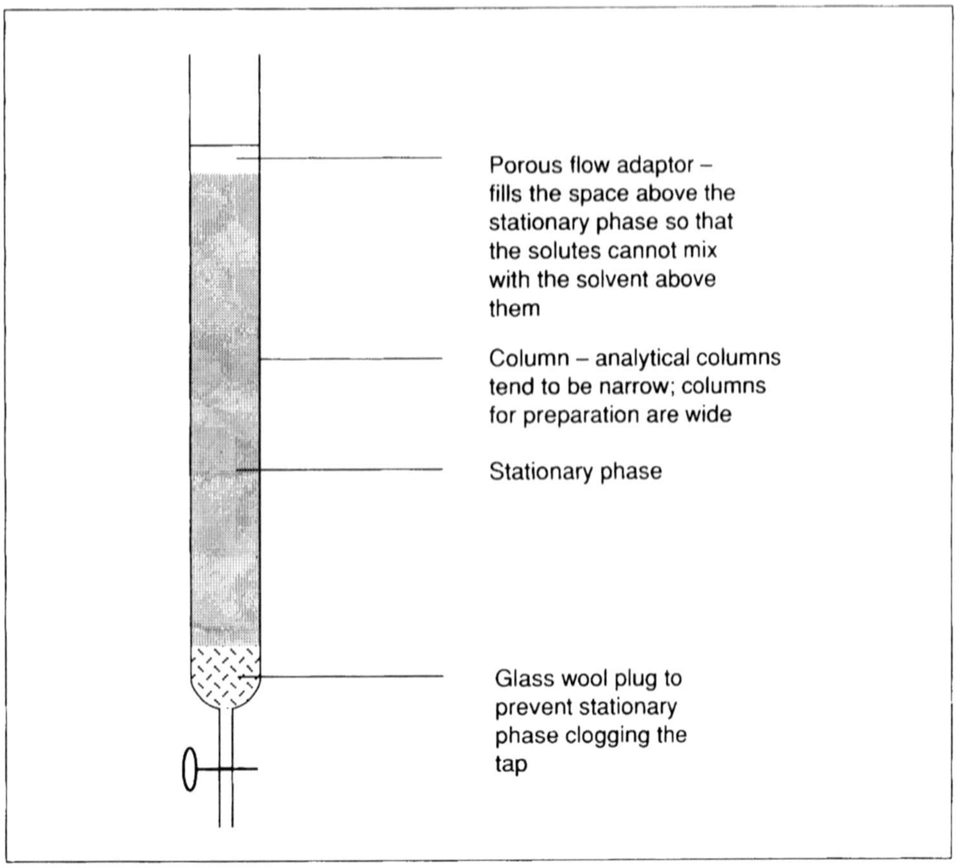

Figure 15 Liquid chromatography column

The adsorbing properties of silica and alumina are reduced if they absorb water, but the reduction is reversed by heating to 200–400 °C. Silica is slightly acidic, and readily adsorbs basic solutes. On the other hand, alumina is slightly basic and strongly adsorbs acidic solutes. Other stationary phases that can be used include magnesia, $MgO.xH_2O$ (good for separating unsaturated organic compounds); and dextran (a polymer of glucose) cross-linked with propan-1,2,3-triol (glycerol, $CH_2OHCHOHCH_2OH$), which is sold as Sephadex and can separate compounds such as purines. Sephadex has the structure shown in *Fig. 16*.

Purines are a class of nitrogen containing bicyclic organic compounds that include adenine and guanine, the bases found in DNA.

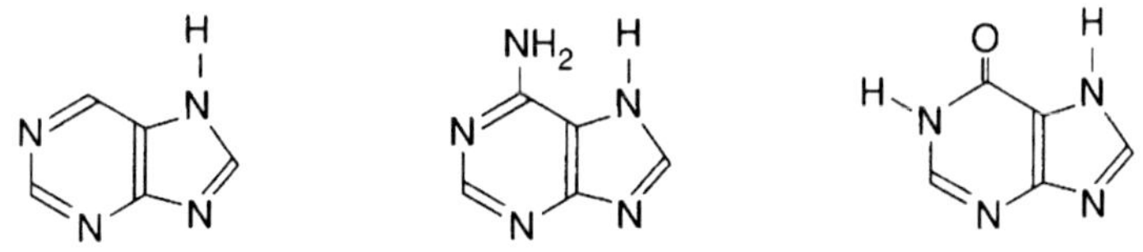

Purine Adenine Guanine

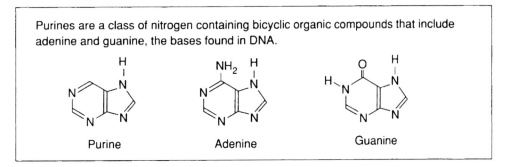

A wide range of solvents are used in this technique, including hydrocarbons, aromatic compounds, alcohols, ketones, halogenocompounds and esters. A mixture of solvents can also be used. The optimum solvent is chosen by running experiments on a small scale using TLC plates.

THE ROYAL SOCIETY OF CHEMISTRY

Unilever

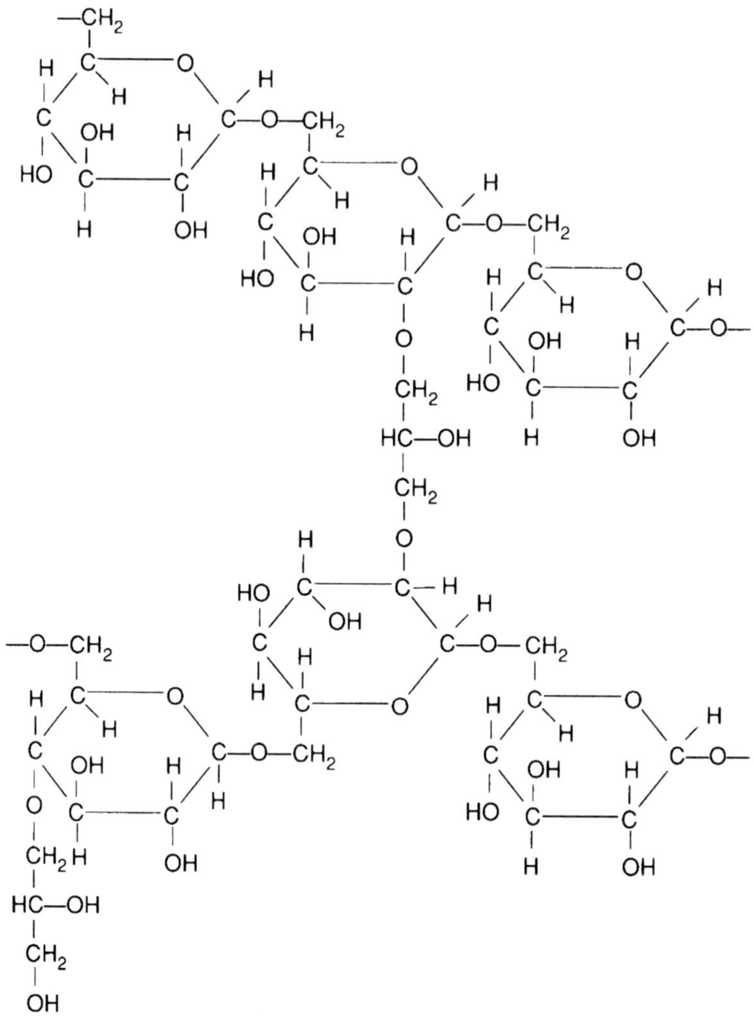

Figure 16 Structure of Sephadex

Practical details

When setting up a liquid chromatography column it is vital that the stationary phase is saturated with solvent, because any air present will interrupt the smooth flow and will result in inefficient or incomplete separation.

An inert material such as ceramic wool is normally inserted to stop any solid from clogging the tap at the bottom of the column. The stationary phase is then added. A common way of separating solutes is to allow the solvent and mixture to descend the column under gravity. For example, alumina can be mixed with a portion of solvent to form a slurry. This is then poured into the column, and any excess solvent is run out of the bottom.

If the column is not saturated with solvent at all times then air bubbles in the column can cause non-uniform separation. The sample is added to the column as a concentrated solution and is absorbed onto the top of the column. The column is then topped up with solvent and the solvent is allowed to flow through.

Unilever

THE ROYAL SOCIETY OF CHEMISTRY

An alternative technique is 'flash' chromatography. The conditions are similar to those used in liquid chromatography but the solvent is forced through the column at a faster rate by applying pressure from an inert gas such as nitrogen. The alumina or silica can be saturated by pouring solvent over the dry powder in the column, then applying pressure from a nitrogen cylinder.

Once the column is set up, the mixture to be separated is carefully added to the top of the stationary phase and some solvent poured in. A valve is attached to the top of the column and nitrogen is introduced at a pressure of up to 2×10^5 Nm^{-2} (2 atmospheres). The tap at the bottom of the column is opened and the solvent allowed to pass through. This is collected in tubes and the fractions identified. As the amount of solvent above the stationary phase decreases it can be replaced by closing the tap at the bottom of the column, and releasing the pressure in the column before topping up the head of solvent. The valve is then replaced and the gas supply turned on again before the tap at the bottom of the column is opened.

If the components of the mixture are coloured collecting them is easy. If they are colourless small volumes of the liquid leaving the column (the eluate) can be collected in tubes, and then the solutes identified by another method – *eg* fluorescence under ultraviolet light, or by TLC.

In both liquid chromatography and flash chromatography the solutes can be extracted from the liquid collected (the eluate) by evaporating off the solvent and if necessary they can then be identified by running a simple TLC experiment.

A more elaborate variation on liquid chromatography is high performance liquid chromatography (HPLC).

High performance liquid chromatography (HPLC)

The efficiency of a separation increases if the particles in the stationary phase are made smaller. This is because the solute can equilibrate more rapidly between the two phases. However, if the particles are made smaller, capillary action increases and it becomes more difficult to drain the column under gravity. Consequently, a high pressure has to be applied to the solvent to force it through the column. A schematic representation of the process is shown in *Fig. 17*.

The stationary phase normally consists of uniform porous silica particles of diameter 10^{-6} m, the surface pores having a diameter of 10^{-8}–10^{-9} m. (This gives the solid a very high surface area.) The particles can be bonded with a non-volatile liquid that allows interactions of solutes with different polarities. These liquids are held on the silica particles by covalent bonds – *eg* the surface of one polar resin has the structure

$$O - CH_2 - CH_3$$
$$/$$
$$\text{Particle} - Si - O - Si - CH_2 - CH_2 - CH_2 - CH_2 - NH_2$$
$$\backslash$$
$$O - CH_2 - CH_3$$

Interaction is then possible between the lone pair of electrons on the nitrogen atom and the solute molecule.

The stationary phase particles are packed into the HPLC column and are held in place by glass fibres coated with inert alkyl silane molecules. The separation in HPLC is normally so efficient that a long column is not necessary. (If the column was too

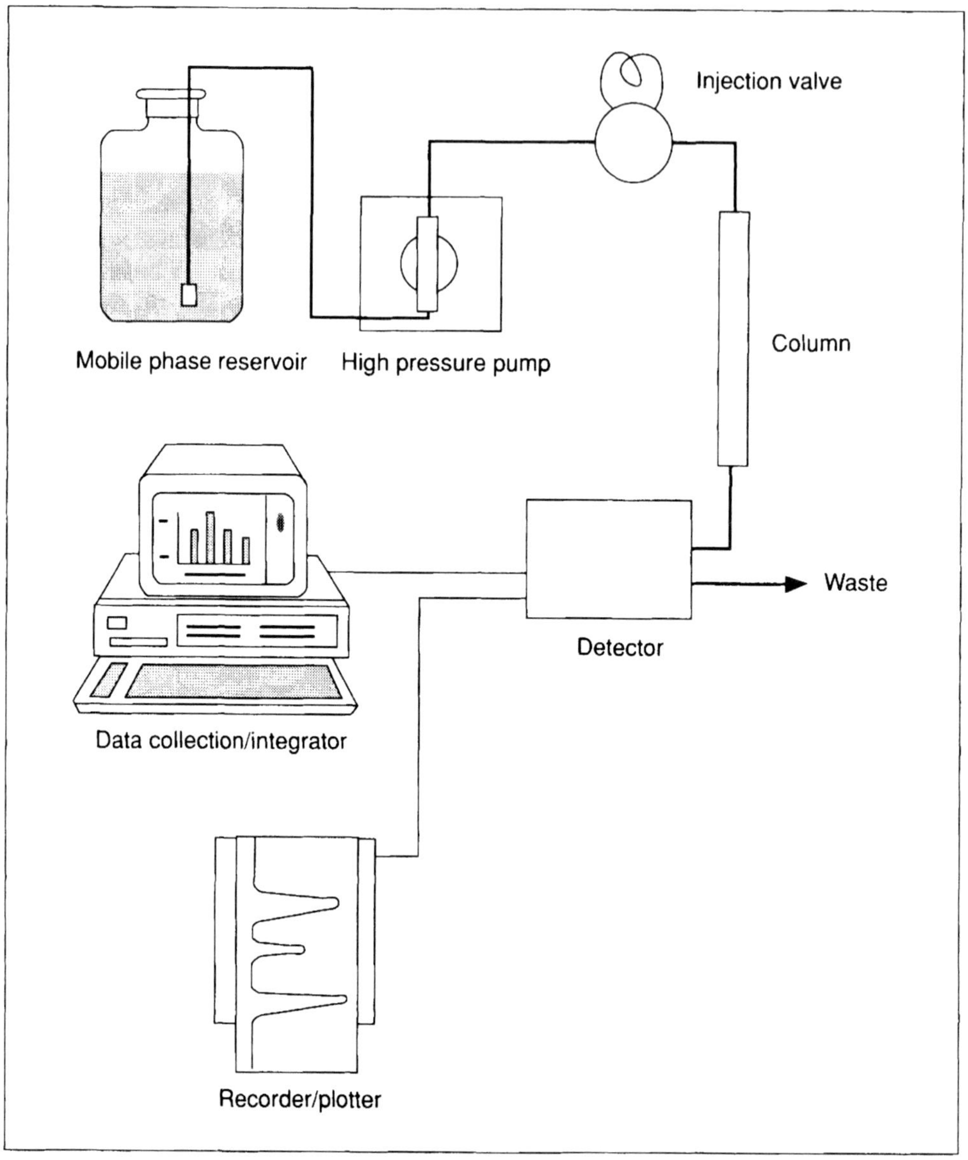

Figure 17 Schematic representation of an HPLC system

long the pressure needed would be excessive.) Columns are typically 10–30 cm long, with an internal diameter of 4 mm.

Reproducibility is essential, and this is only possible if a constant flow rate is maintained. This means that the pump used must be capable of generating a uniform pressure; twin cylinder reciprocating pumps are typical. This type of pump has two chambers with pistons 180° out of phase, and can generate pressures up to 10 MNm^{-2} (10 MPa/100 atmospheres). The high pressures involved mean that the instrumentation has to be very strong, and the 'plumbing' is usually constructed from stainless steel. The pump and the piping must be inert to the solvent and solutes being passed through them.

The flow rates of HPLC columns are slow – often in the range 0.5–5 $cm^3\ min^{-1}$. The volumes of the columns are very small, and this means that the injection of the sample must be very precise and it must be quick without disturbing the solvent flow. Sample volumes are small – 5–20 mm^3 is usually sufficient.

The passage of solutes through a GC can be speeded up by increasing the

Unilever

THE ROYAL SOCIETY OF CHEMISTRY

temperature of the column. The same effect in HPLC is achieved by changing the composition of the mobile phase – *ie* there is a concentration gradient in which the proportion of methanol, say, in a methanol/water system is increased linearly from 10 per cent methanol to 60 per cent methanol, during the separation.

The amounts passing through the column are usually too small to extract from the solvent before identification, so the solutes in solution are analysed as they leave the column. Most compounds separated by HPLC absorb ultraviolet light. The eluate is passed along a small cell so that ultraviolet radiation can be passed through the liquid (*Fig. 18*).

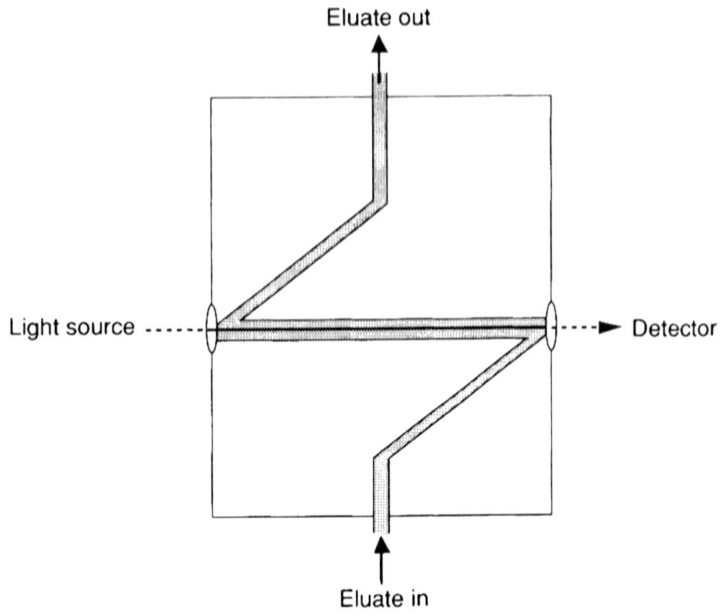

Figure 18 A micro cell for identifying solutes in the liquid leaving an HPLC column

The relative use of HPLC and GC varies from industry to industry and very much depends on the compounds to be separated. Many compounds decompose at the temperatures required for efficient GC separation while HPLC separation can be achieved readily. However, GC is particularly useful in detecting residual solvents in formulations and is also invaluable in looking for degradation products. Amines and acids are not separated well by GC because they tend to be too polar.

Other methods used in conjunction with HPLC for determining the presence of solutes are based on:

1. mass spectrometry;
2. infrared spectroscopy;
3. visible spectroscopy;
4. ultraviolet spectroscopy;
5. fluorescence spectroscopy;
6. conductivity measurement; and
7. refractive index measurement.

Whichever method is used it is vital that the volume of liquid used is very small, otherwise the sharpness of the separation peaks will disappear and the resolution of the final chromatogram will be lost.

Once the retention time of a solute has been established for a column using a set of operating conditions, that solute can be identified in a mixture from its retention time (assuming that another component with the same retention time is not also present).

Ion exchange chromatography

Ion exchange chromatography is used to remove ions of one type from a mixture and replace them by ions of another type.

The column is packed with porous beads of a resin that will exchange either cations or anions. There is one type of ion on the surface of the resin and these are released when other ions are bound in their place – *eg* a basic anion exchange resin might remove nitrate(V) ions (NO_3^-) from a solution and replace them with hydroxide ions (OH^-).

Many of the resins used are based on phenylethene (styrene) polymers with cross-linking via 1,4-bis-ethenylbenzene (divinylbenzene, *Fig. 19*).

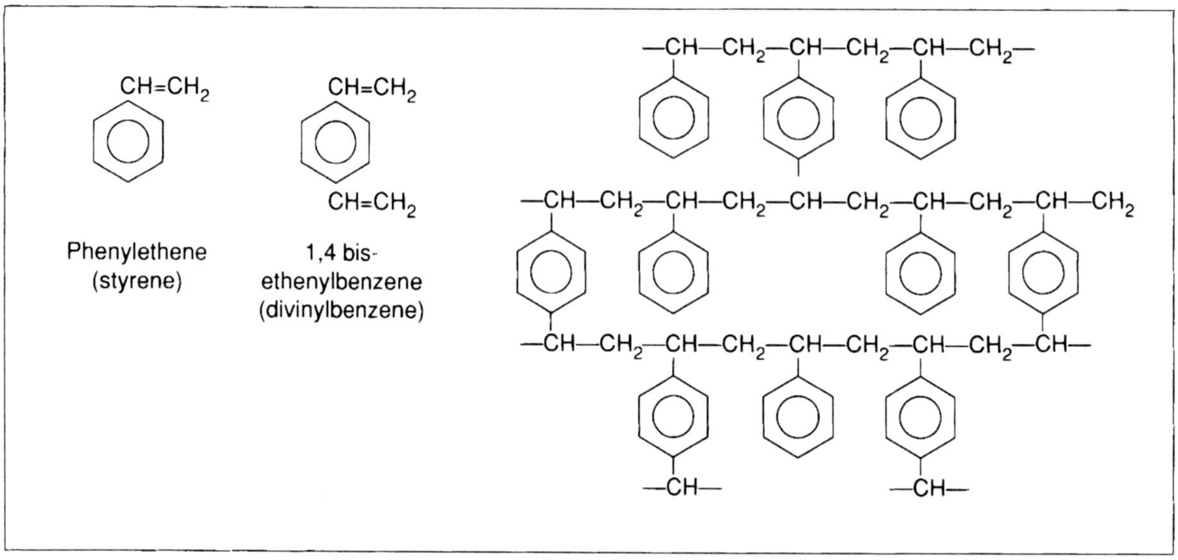

Figure 19 Copolymer of cross-linked styrene–divinylbenzene

If the ion is a quaternary ammonium group the resin is strongly basic (*eg* $-CH_2N(CH_3)_3^+\ OH^-$) then the resin will selectively remove the ions: $I^- > NO_3^- > Br^- > NO_2^- > Cl^- > OH^-$ ($> F^-$), thus liberating hydroxide ions while nitrate(V) ions, for example, are removed. The exchange site can be strongly or weakly acidic or basic depending on the group present. Examples of such groups are shown in Table 2.

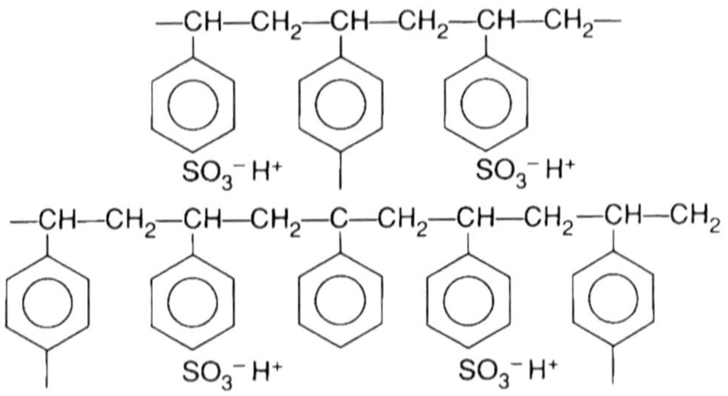

The presence of ionic groups bonded to the polymer of phenylethene and 4-ethenylphenylethene provides the ability to exchange ions. For example, the inclusion of $-SO_3^-\ H^+$ groups in the 4–position of phenylethene gives a strongly acidic resin which will selectively remove the ions $Ag^+ > Rb^+ > Cs^+ > K^+ > NH_4^+ > Na^+ > H^+$. This means that both potassium and sodium ions will be removed from a solution containing both, until all the H^+ ions have been discharged. Then potassium ions from the solution will replace the sodium ions on the column.

Figure 20 A strongly acidic cationic exchange resin

Table 2 Acid/base character of exchange site groups

Character	
Strongly acidic	$-SO_3^-\ H^+$
Weakly acidic	$-R-COO^-\ H^+$
Weakly basic	$-CH_2NH(CH_3)_2^+Cl^-$
Strongly basic	$-CH_2N(CH_3)_3^+\ OH^-$

The degree of cross-linking of the copolymer can have a significant effect on the behaviour of the resin, and can change selectivity. The range of cross-linking in commercial resins is approximately 4–12 per cent of the monomer units and the characteristics of resins having low and high degrees of cross-linking can be compared (Table 3).

Table 3 Characteristics of resins having low and high degrees of cross-linking

Low degree of cross-linking	**High degree of cross-linking**
Less rigid	More rigid
More porous	Less porous
Rapid exchange of ions	Slower exchange of ions
Greater swelling in water	Little swelling in water
Lower exchange capacity	Higher exchange capacity
Lower selectivity	Greater selectivity

The selectivity for certain ions can change with the degree of cross-linking because a highly cross-linked resin has smaller pores than one with little cross-linking, and bulky ions cannot get into the smaller pores, so they are not exchanged.

Ion exchange equilibrium

If two ions are competing for the binding sites of a resin an equilibrium will be established. For example, a cation–exchange resin used to remove ammonium ions and replace them with sodium ions would have the equilibrium:

$$Resin^{-}Na^{+} + NH_4^{+} \rightleftharpoons Resin^{-}NH_4^{+} + Na^{+}$$

This equilibrium can be described by an equilibrium constant (*K*) which is called the selectivity coefficient:

$$\text{Selectivity coefficient} = \frac{[Resin^{-}NH_4^{+}][Na^{+}]}{[Resin^{-}Na^{+}][NH_4^{+}]}$$

The equilibrium position can be changed in either direction by changing the concentration of ions in the aqueous phase. A column might remove relatively small amounts of ammonium ions almost quantitatively from a solution, but they can be liberated from the column equally efficiently by passing an excess of a concentrated solution of sodium ions through the column. The removal occurs despite the fact that ammonium ions are preferentially bound to the resin, and this principle is useful in regenerating the column. Once a column's exchange capacity is approached, it is regenerated simply by passing a concentrated solution of sodium ions through it. This shifts the equilibrium back to the left hand side.

Under most circumstances absorption and release of ions is effectively quantitative, so it is possible to remove selected ions and determine their concentrations by titration – *eg* if calcium ions are removed from a sample of hard water and are replaced by hydrogen ions from the acidic form of a resin, then the concentration of hydrogen ions can be determined by titrating the eluate with a standard alkaline solution.

Macromolecules can also be exchanged by using resins with larger pore sizes. The resins are generally derived from proteins or polysaccharides such as cellulose, and the exchange sites are bound to the OH groups of the saccharide units. The sites can be either weakly basic or weakly acidic, with polar covalent groups (*eg* amines); or they can be strongly acidic or strongly basic, with ionic groups (*eg* carboxylic or sulphonic acids).

Ion exchange applications

The solubility product of some partially soluble substances can be determined by using exchange methods. For example, a saturated calcium sulphate solution can be passed down a cation exchange column that replaces the calcium ions with hydrogen ions:

$$Ca^{2+}_{(aq)} + 2(Resin^{-}H^{+}) \rightarrow Ca^{2+}(Resin^{-})_2 + 2H^{+}_{(aq)}$$

The hydrogen ions liberated can then be determined by titration using a standard sodium hydroxide solution (phenolphthalein as indicator). Once $[Ca^{2+}]$ has been calculated from $[H^{+}]$, the solubility product of the salt can be calculated.

By passing a water sample through two columns, one a cationic exchanger and the other an anionic exchanger, it is possible to remove any ionic impurities. Tap water, for example, can be passed through one column where the metal ions present, such as Mg^{2+} and Ca^{2+}, are replaced by H^{+}; then the anions present, such as F^{-}, Cl^{-} and NO_3^{-}, are replaced by OH^{-} in the second column. The product is deionised water.

The dishwasher

A more sophisticated application of ion exchange resins is the water softener installed in many modern automatic dishwashers. The cleaning efficiency of dishwashers using detergents increases if the water is heated. This causes problems if the water supply has temporary hardness because when it is heated the hydrogencarbonate ions decompose to form carbonate ions and these are deposited on the heating element, reducing its efficiency. Further problems arise if the ions causing hardness are not removed, because inside the dishwasher they can deposit a film on the items that are meant to be cleaned. To overcome these problems calcium and magnesium ions are removed from the water entering the dishwasher by a cationic exchange resin. They are replaced by sodium ions – *eg*:

$$(Na^+)_2\,Resin^{2-} + Ca^{2+}{}_{(aq)} \rightleftharpoons Ca^{2+}\,Resin^{2-} + 2Na^+{}_{(aq)}$$

The exchange capacity of the resin is regenerated at the start of the next washing cycle by passing saturated sodium chloride (salt) solution through the resin. The high concentration of sodium ions favours the back reaction in the above equilibrium. The unwanted calcium and magnesium ions and the excess sodium chloride are pumped out of the dishwasher during the next drain cycle. It is important that no sodium chloride remains because it is corrosive.

The sodium chloride solution is formed by pumping fresh water through the salt chamber that the user periodically refills. The type of salt put into the dishwasher is important – table salt contains magnesium chloride to keep it free-flowing and the magnesium ions can both cause hardness and interfere with resin regeneration. The small crystal size of table salt also means that it can easily be washed into the resin chamber and cause a blockage. Granular or dendritic salt crystals are recommended to eliminate this risk (minute amounts of potassium hexacyanoferrate(II), $K_4Fe(CN)_6$), are added to salt solution to encourage the growth of dendritic crystals).

To minimise unnecessary consumption of salt some dishwashers can be set to take account of the relative hardness of the supply in the area where the dishwasher is used. Hotpoint dishwashers, for example, pump different volumes of salt solution through the resin according to the hardness of the water.

Other uses

Ion exchange resins can be used in HPLC columns, and a variety of such resins exist. It is possible to monitor the solutes leaving the HPLC column by measuring the conductivity of the solution, and extremely small ion concentrations can be determined by using this method. This is routinely done to monitor water purity in environmental analysis. An anionic exchange resin is used to separate ions. The sample to be analysed is added to an aqueous mobile phase containing sodium carbonate and sodium hydrogencarbonate, which is passed through an anionic exchange resin in an HPLC column. The ions to be detected exchange with carbonate ions on the resin surface. The carbonate ion concentration in the mobile phase is high enough to elute them off again, but at different rates. The anions in the sample are recognised by their retention times on the column, and their concentrations are determined from the conductivity of the eluate. Anions typically elute from the column in the order F^-, Cl^-, NO_2^-, Br^-, NO_3^-, PO_4^{3-}, SO_4^{2-}. Concentrations in the order of 10^{-3} g dm^{-3} (ppm/μg cm^{-3}) are routinely measured. Measuring concentrations down to 10^{-6} g dm^{-3} is required in the nuclear power industry, where the leak of radioactive contaminants must be detected (and stopped!) immediately.

Many EDTA (ethylenediaminetetraacetate) complexes absorb ultraviolet light, so transition metal concentrations can be determined by making complexes with EDTA. Alternatively, EDTA can be estimated by adding transition metal ions to its solution. This is done in the food industry, where canned shellfish have added EDTA. Crustacean blood is based on a Cu-haem complex (and not an Fe-haem complex as in humans). On cooking a blue/black colouring appears where the blood was and this colouring spreads into the flesh making it look unattractive. If EDTA is added it competes for the copper ions and complexes them so that discolouration does not occur. To conform to legislation the total EDTA concentration must be within certain limits. Excess copper ions are added to ensure that all the EDTA has complexed as $(Cu(EDTA))^{2-}$, and an ion exchange HPLC column is used to separate all the anions present according to their charge. The eluate containing all the EDTA (identified by its retention time on the column) is then passed through a cell where its ultraviolet absorption at 280 nm is measured; and from this measurement the EDTA concentration can be calculated.

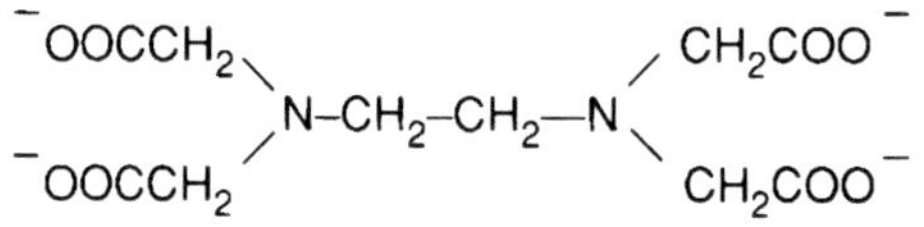

EDTA

Gel filtration or gel permeation chromatography

The separation of large molecules, often in biochemical situations, can be achieved in a column which works on the basis of molecular exclusion. The mixture of solutes is carried through the column by a solvent. The stationary phase (the gel) typically consists of particles of a cross-linked polyamide which contains pores. Separation occurs according to molecular size – the larger molecules passing through the column fastest (*Fig. 4*).

Different gels are available that allow the separation of proteins with relative masses ranging from a few hundred to in excess of 10^8. The greatest resolution is achieved by using very small gel particles, but the flow rate through the column then becomes much slower.

Applications of chromatography

Chromatography is used to separate compounds in reaction mixtures both in the laboratory and on the industrial scale. However, technology has now advanced sufficiently to allow chromatographic techniques to be interfaced directly to other analytical methods. For example, gas chromatographs are routinely linked to mass spectrometers, and HPLC columns are linked to ultraviolet/visible spectrometers.

Determination of caffeine and theobromine

Caffeine is found in both tea (leaf buds and young leaves of *Camellia sinensis*) and coffee (roasted seeds of *Coffea arabica* and *Coffea robusta*). In tea, caffeine constitutes 2–3.5 per cent on a dry leaf basis, though blended teas typically contain 3 per cent caffeine. Roasted coffee beans generally contain 1.1–1.8 per cent caffeine, while instant coffee powders average 3.0 per cent.

Caffeine is widely accepted as being a stimulant, although other physiological effects (such as its diuretic effect) are also observed. Some studies have suggested that there might be a link between caffeine and certain disorders – *eg* heart disease and kidney malfunction. Some common concerns that people have about caffeine are its stimulant effect and its addictiveness, and this has led to an increased market demand for decaffeinated products. Chlorinated solvents such as trichloroethene

Unilever

THE ROYAL SOCIETY OF CHEMISTRY

(trichloroethylene) and trichloroethane were often used to extract caffeine but the solvent residues may possibly be more harmful than the caffeine itself. Modern techniques for removing caffeine include using supercritical liquid carbon dioxide as a solvent and reverse osmosis. However, the methods for caffeine removal need careful control to avoid excessive loss of desirable flavour constituents.

Theobromine is a metabolite of caffeine – *ie* it is produced by the biological processes of the body. It also occurs naturally in the cacao beans (*Theobroma cacao*) that are used to make cocoa and chocolate. Approximately 1 per cent of the mass of the cocoa bean is theobromine, a proportion that rises to 2.5 per cent in de-fatted cocoa powder. As this amount is relatively constant it is possible to calculate the cocoa solids content of chocolate products such as confectionery, chocolate coatings, drinking chocolate and ice creams, from their theobromine content.

The chemical structures of caffeine, theobromine (which does not contain bromine) and other related xanthines are shown in *Fig. 22*.

To satisfy legal and labelling requirements it is essential that the amount of caffeine present, both initially and after processing, can be determined accurately. The method described here has the advantage that it can also be used to determine theobromine levels simultaneously.

In the first stage, a known mass of the coffee, tea, or chocolate product is gently boiled with deionised water. The boiling water extraction is essential to ensure complete dissolution of the xanthines, because they are not all readily soluble. The cooled solution is then clarified if necessary, with the aid of Carrez solutions (see Box). These solutions precipitate proteinaceous materials along with any soluble starch components.

> There are two Carrez solutions
>
> Carrez Solution 1:21.9 g of zinc ethanoate dihydrate dissolved in deionised water containing 3.0 g of ethanoic acid, made up to 100 cm^3 with deionised water.
>
> Carrez Solution 2:10.6 g of potassium hexacyanoferrate(II) (ferrocyanide) trihydrate dissolved in deionised water and made up to 100 cm^3.
>
> These two solutions are added to the extract in equal proportions to clarify the solution.

The resulting solution is made up to a known volume, mixed thoroughly, and filtered. The clear filtrate remaining is diluted if necessary to obtain xanthine concentrations within a standard calibration range. A schematic plan of the sample preparation is shown in *Fig. 21*.

An aliquot of the sample extract (typically 0.020 cm^3/20 μl) is then analysed. It is necessary to determine whether caffeine, theobromine, or both are present, and in what concentrations. The caffeine and theobromine are separated on a 15 cm HPLC column (see Box).

THE ROYAL SOCIETY OF CHEMISTRY

Unilever

The HPLC column contains a 5 x 10^{-6} m (particle size) silica based support with its surface chemically modified with bonded C_{18} groups. The solvent system (mobile phase) contains 12 per cent ethanonitrile (acetonitrile) in a 0.5 per cent aqueous solution of ammonium nitrate(V).

Systems such as this one, in which the mobile phase is the more polar, are known as reversed phase chromatography. (In industry reversed phase chromatography is more commonly used than normal phase chromatography.)

In normal phase chromatography the stationary phase is more polar – *eg* an unmodified silica stationary phase might be used – the surface hydroxy groups giving it its polar nature. Mobile phases used in such separations would tend to be less polar. Some solvents in order of increasing polarity are:

1:1 v/v Ethoxyethane:petroleum ether
3:1 v/v Ethyl ethanoate:petroleum ether
9:1 v/v Dichloromethane:methanol

In normal phase chromatography the least polar solute comes off the column first – in reverse phase chromatography the most polar solute comes off first. In each case the eluting order depends on the strength of the interaction between the solutes and the two phases.

(1) Weigh 0.5–1.0 g of finely ground sample into a 250 cm^3 flask

↓

(2) Add 100 cm^3 of deionised water and a few anti-bumping granules

↓

(3) Place on a hotplate and boil gently for 15 minutes

↓

(4) Allow to cool

↓

(5) Tea

(5) Chocolate and coffee samples → To precipitate, suspended proteins and starch, add small portions of each Carrez solution, mixing between additions

↓

(6) Transfer to a volumetric flask and dilute to 200 cm^3 mark, mixing thoroughly

↓

(7) Filter the solution

↓

(8) Inject 0.020 cm^3 onto a HPLC column to separate

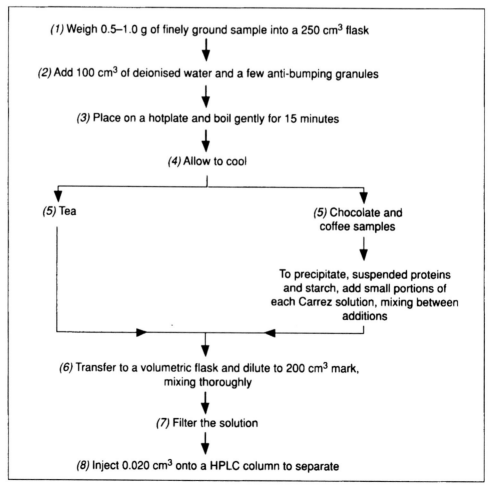

Figure 21 Schematic plan of sample preparation

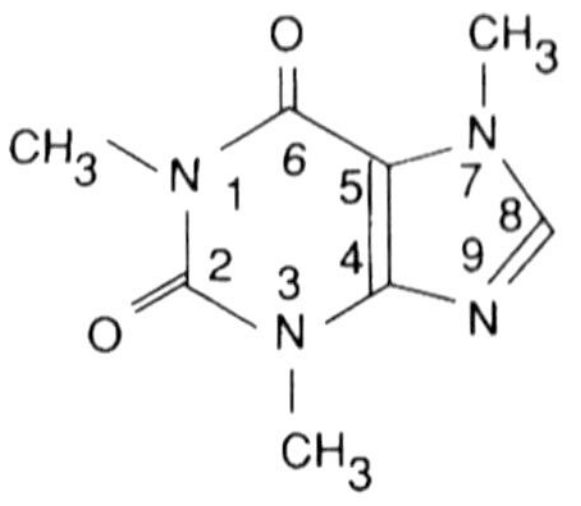

1,3,7-Trimethylxanthine
Caffeine
(a stimulant of the CNS; respiratory stimulant; diuretic)

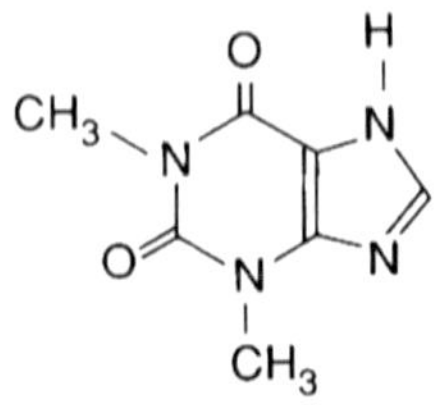

1,3-Dimethylxanthine
Theophylline
(a stimulant bronchodilator)

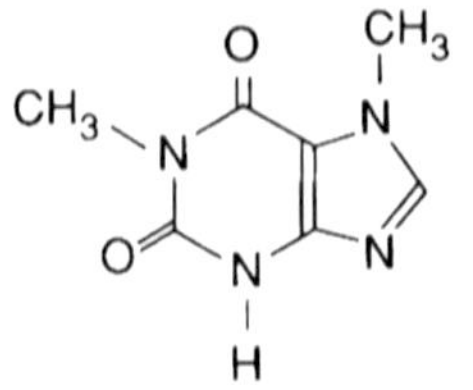

1,7-Dimethylxanthine
Paraxanthine
(physiological properties not yet studied)

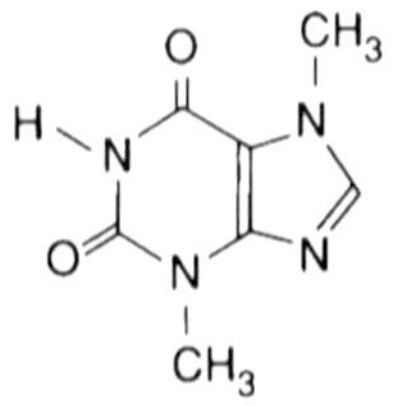

3,7-Dimethylxanthine
Theobromine
(a mild stimulant; diuretic; bronchodilator)

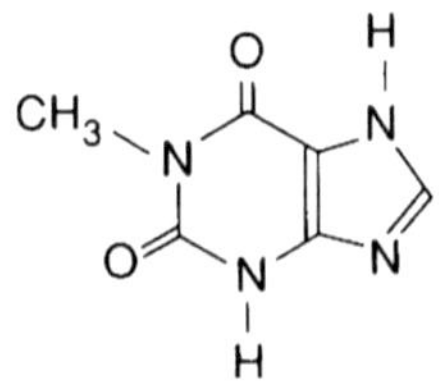

1-Methylxanthine

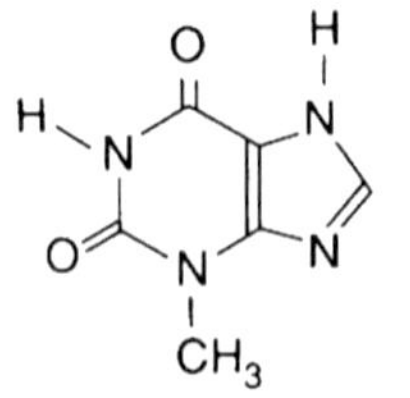

3-Methylxanthine

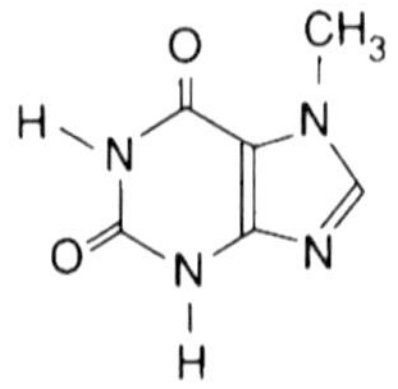

7-Methylxanthine

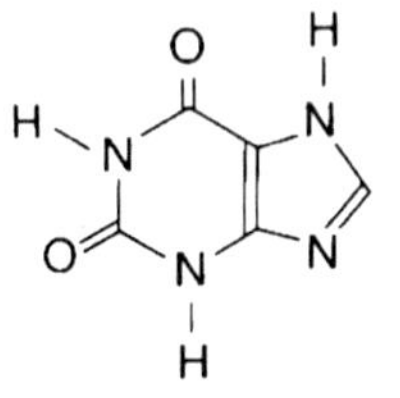

Xanthine

All the dimethyl and methyl xanthines appear as metabolites of caffeine in human urine

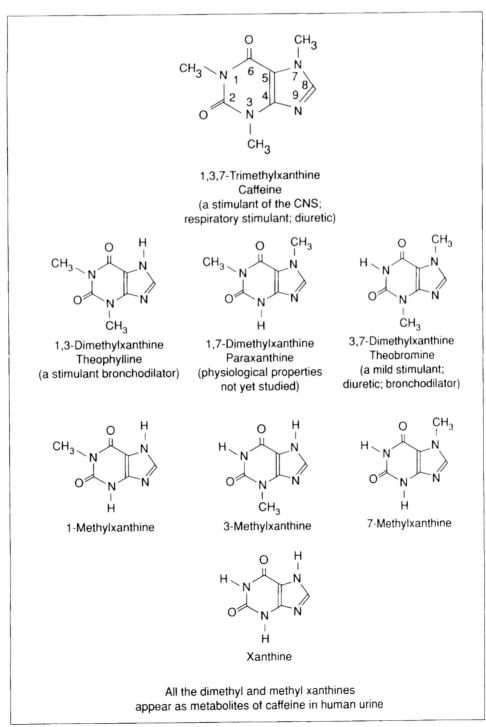

Figure 22 Chemical structures of xanthine and related substances

The flow rate is typical of HPLC, 1 cm^3 per min, and this lends itself nicely to interfacing with ultraviolet absorption detection. The eluate from the HPLC column passes through a 0.008 cm^3 (8 μl) flow cell made with ultraviolet transparent quartz windows. A variable wavelength spectrometer is set at 274 nm because both caffeine and theobromine absorb at this wavelength (*Fig. 23*).

The HPLC/UV system is calibrated so that the data emerging from it can be interpreted. Standard solutions of caffeine and theobromine of different concentrations pass through the column and ultraviolet detector so that the retention time of the solutes can be determined as well as the areas of the peaks corresponding to known concentrations (*Fig. 24*).

It is normal to calibrate the instrument by using two standards, and to ensure that the concentration of the solution analysed falls within the range provided by the standards. The standards used are:

	Theobromine μg per cm^3	**Caffeine μg per cm^3**
Standard 1	10	20
Standard 2	20	40

If the solution being analysed is too concentrated it is simply diluted by an appropriate factor to bring it into the required range. Once the concentrations of theobromine and caffeine have been determined their masses in the boiled solution can be calculated, and hence their percentages by mass.

This method is essentially straightforward and allows the food industry to maintain quality control of its products, as well as researching into the potential of new production methods.

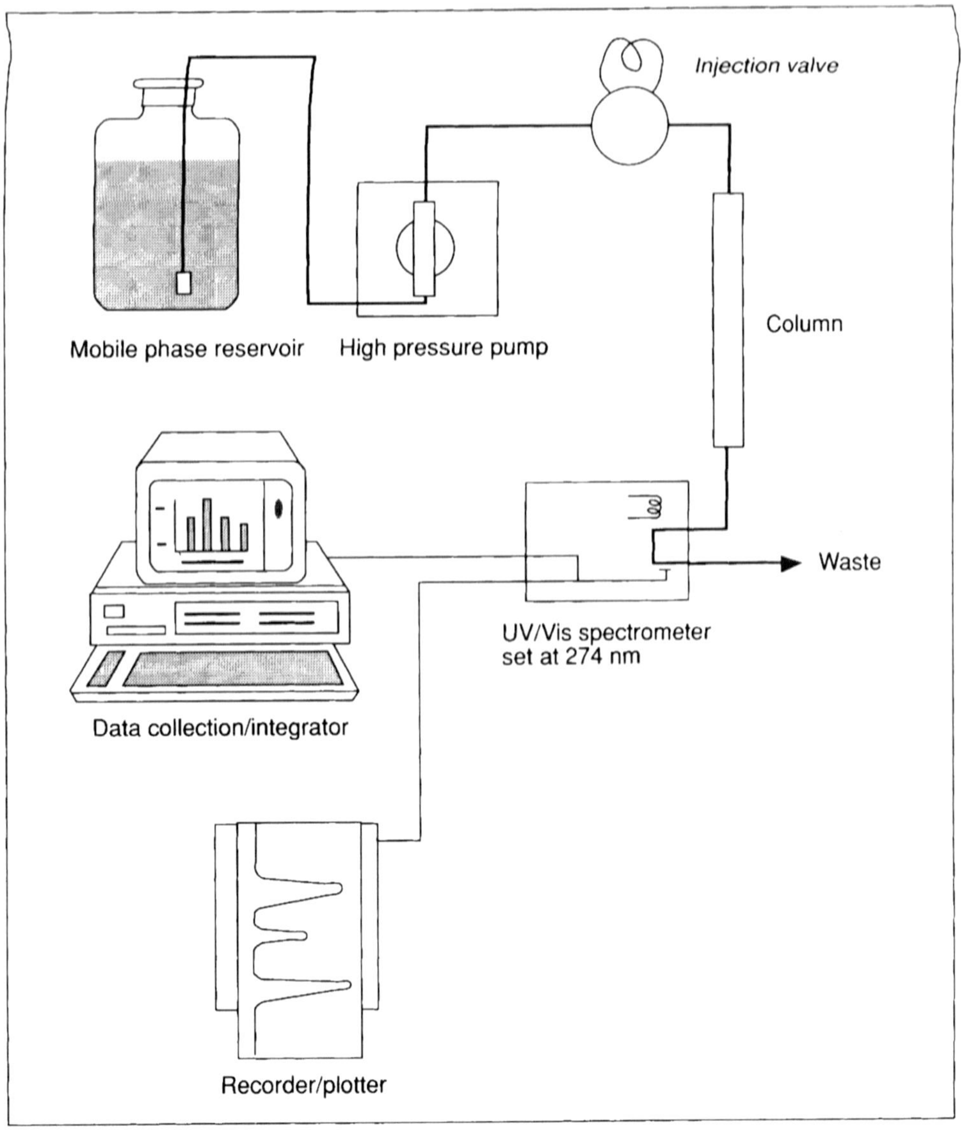

Figure 23 Separation using HPLC and detection using ultraviolet light at 274 nm

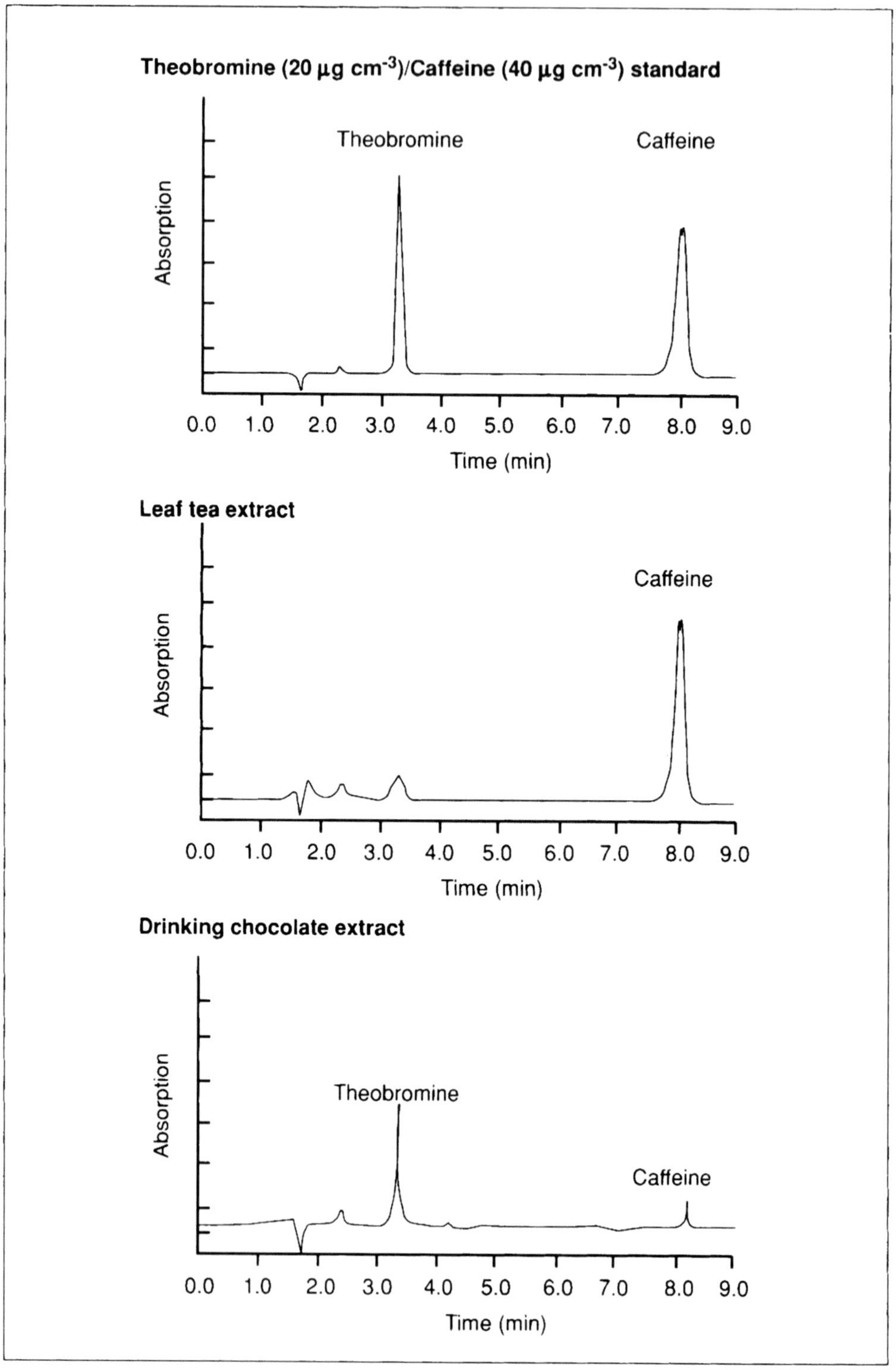

Figure 24 Chromatograms from the analysis of tea and drinking chocolate

Unilever

THE ROYAL SOCIETY OF CHEMISTRY

Screening for photoallergens

Irritant or allergen?

Some compounds – *eg* sodium hydroxide – cause irritation if applied to the skin. Each time the irritant is applied a reaction occurs. This is because the chemical affects the skin tissue. Photoirritants also exist, such as 8-methoxypsoralen (the active component of 'giant hogweed'), which cause irritation through a photochemical reaction.

Symptoms of irritancy might be erythema (reddening of the skin), itching or cracking of the skin. The same response is seen if an allergen (a substance causing an allergy) is applied to the skin. However, no response is observed on first exposure – indeed it can take several exposures of low level contact before an allergic reaction is seen. The allergic reaction is a learned response and is not observed in all subjects, whereas irritants cause a reaction in all cases – *eg* low levels of phenol cause skin irritation in all people, but not everybody shows an allergic reaction to cats.

In some cases allergic reactions occur because a photochemical reaction has taken place – this is known as photoallergy and is not observed on first exposure to the allergen.

To be a photoallergen a chemical must have two properties:

1. it must absorb wavelengths of light present in sunlight (if photoallergy is seen these are usually in the ultraviolet region); and
2. after absorbing light the chemical must produce a reactive species capable of binding covalently (*ie* irreversibly) to protein molecules in the skin.

Most of the chemicals identified so far as photoallergens in man are germicides and fungicides which were inadvertently used in soaps or medical preparations. These days such chemicals do not pass through screening, so are not marketed.

Photoallergens have a range of structures. A number of them contain halogenated aromatic rings. When ultraviolet light is absorbed halogen radicals (such as Cl• or Br •) are lost. It is the organic radical that remains which binds to skin proteins. Once binding has taken place a conjugate known as a 'complete antigen' is formed.

> The complete antigen is recognised as 'foreign' by the Langerhans cells, which leads to the formation of memory cells in the draining lymph nodes. When challenged, the conjugate is recognised by the memory cells which undergo transformation and proliferation. This leads to the release of soluble factors which produce responses such as itching of the skin and is known as a 'learned response'.

Screening

To avoid the possibility of exposing the public to photoallergens it is necessary to test new chemicals before they reach the market place. Testing on humans is regarded as unethical, and traditionally testing has been done on experimental animals. The test material is applied to the backs of the animals. The animals are exposed to simulated sunlight, and the test material is re-applied several weeks later, with further irradiation. If some of the animals show an allergic response the chemical is classed as a photoallergen and it is taken that similar results will be seen on a human population. By using appropriate controls it is possible to tell whether a reaction is due to irritation to the chemical or due to photoallergy.

A new way of testing is to use an *in vitro* method. A model protein, human serum albumin (HSA) extracted from blood, is used to test whether binding of the chemical occurs. HSA is the most abundant protein in human serum, and is also present in the skin. Binding of the chemical is detected by monitoring changes in the ultraviolet spectrum of the protein.

The technique depends upon two factors:

1 separating the protein (together with any substances bound to it) from the remainder of the test chemical; and

2 that the ultraviolet absorption spectrum of the conjugate formed between the protein and the test chemical is different from the protein alone.

The ultraviolet spectra of a solution of the protein and the test chemical are taken separately for reference, and then the two components are mixed. The ultraviolet spectrum of the mixture is taken and will usually be the sum of the two individual spectra unless a reaction has taken place. The mixture is then irradiated with an ultraviolet wavelength that is known to be absorbed by the test substance (for a photochemical reaction to occur light must be absorbed, *Fig. 25*).

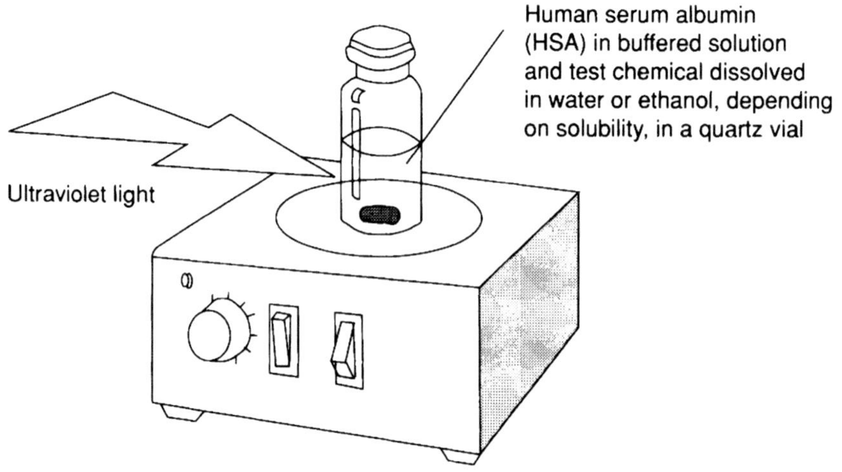

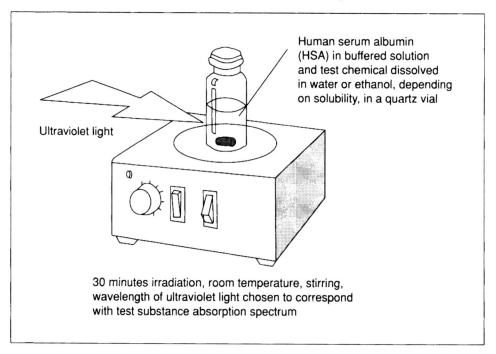

Human serum albumin (HSA) in buffered solution and test chemical dissolved in water or ethanol, depending on solubility, in a quartz vial

Figure 25 Irradiation of the HSA/test chemical mixture

If the test substance reacts with the protein, the ultraviolet spectrum of the sample will be different after irradiation. If it is the same after irradiation it is likely that binding has not taken place. However, the spectrum could also be different after irradiation because the test chemical has decomposed or because a decomposition product has reacted with the protein. To determine whether or not binding has occurred the protein fraction is separated from the other components of the mixture.

Separation is achieved by using a gel permeation column packed with Sephadex beads. (The principle of gel permeation is given on page 118 and the structure of Sephadex on page 130.) Human serum albumin absorbs radiation at 280 nm, so the eluate is passed through an ultraviolet beam set at this wavelength *(Fig. 26)*. The test material can also be detected if it absorbs radiation at this wavelength (*Fig. 27*). The different fractions are identified and collected for spectroscopic analysis.

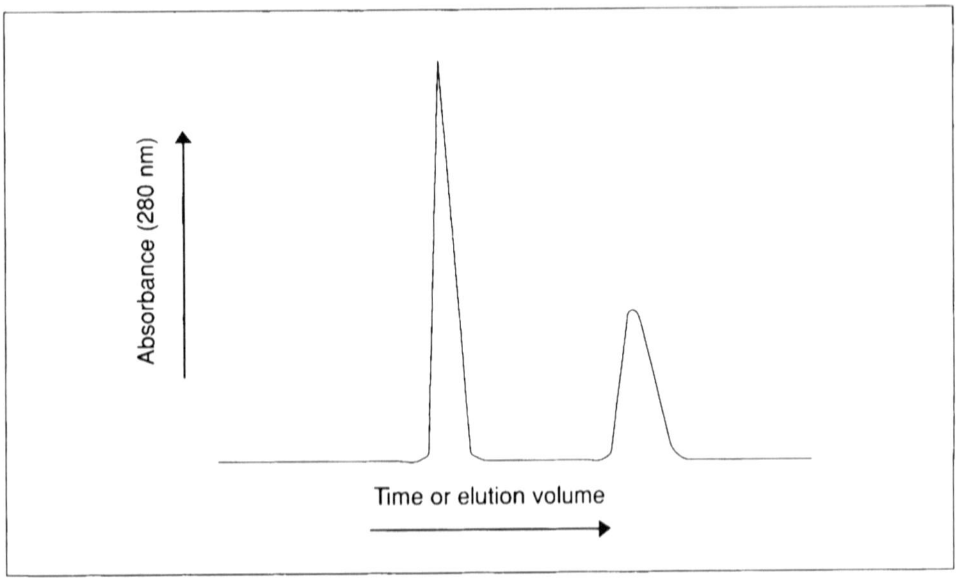

Figure 26 Detection of different fractions using ultraviolet light of wavelength 280 nm

The full spectrum of the protein fraction is then run on a variable wavelength ultraviolet spectrometer (*Fig. 27)* with a diode array detector (*Fig. 32*). If the spectrum is identical to that of HSA alone, it is probable that no binding has occurred. If the spectrum is different it is possible that there has been some binding, though this does not always indicate that the test chemical is a photoallergen. A non-covalent interaction, such as that between hydrophobic groups on the two substances might have occurred or alternatively, the test material might have decomposed when irradiated with ultraviolet light and one of the products may be bound non-covalently to the protein.

To eliminate a false-positive result arising from a non-polar interaction, a less polar solvent can be used to try to separate the molecules. If complete separation results in an ultraviolet spectrum of HSA alone, photochemical binding has not occurred.

To decide whether a photodecomposition product is giving a false-positive result, the test compound alone is irradiated with ultraviolet light before being mixed with HSA. The solution is then separated as before. By mixing the compound with HSA and passing it through the gel permeation column without first irradiating the solution, it is also possible to find out whether binding is taking place without absorbing ultraviolet light. Only if the last two tests give the ultraviolet spectrum of unbound HSA alone is it then possible to conclude that a photochemical reaction has taken place. An overview of the method is given in (*Fig. 29*).

A compound that causes photoallergy is bithionol (bis-(2-hydroxy-3,5-dichlorophenyl) sulphide). This acts as both a fungicide and a germicide. The change observed in the ultraviolet spectrum of HSA when bound to this chemical is shown in *Fig 27*.

Another example of a photoallergen is T_4CS (3,3',4',5-tetrachlorosalicylanilide *Fig. 28*), a germicide, added to toilet soaps in the 1950s and 1960s. A number of people suffered photodermatitis following repeated exposure to soap containing T_4CS, together with exposure to sunlight.

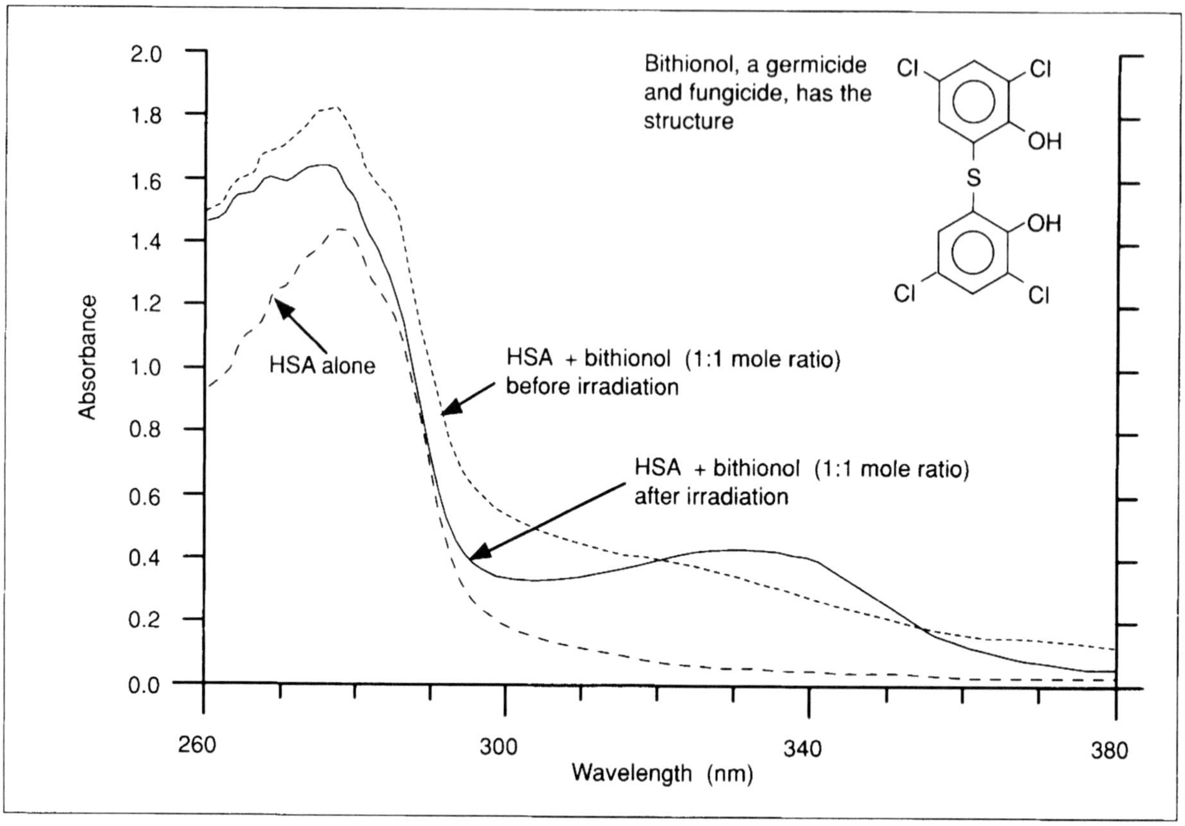

Figure 27 The ultraviolet spectra of HSA and bithionol, a known photoallergen, before and after irradiation with ultraviolet light. (Spectra reproduced with permission R U Pendlington and M D Barratt, Int. J. Cosmetic Science, 1990, 12, 91-103, Chapman & Hall)

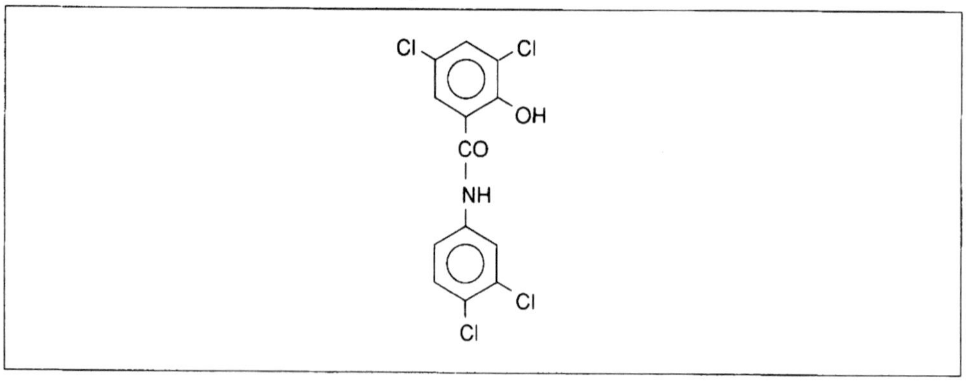

Figure 28 Structure of T_4CS; a germicide

Although testing with HSA is not the same as testing on humans directly, it does have some very significant advantages over tests using guinea pigs:

1 it does not involve the use of any animals;

2 it is much faster, the whole test routine takes a few days whereas animal tests can take weeks; and

3 it is much cheaper.

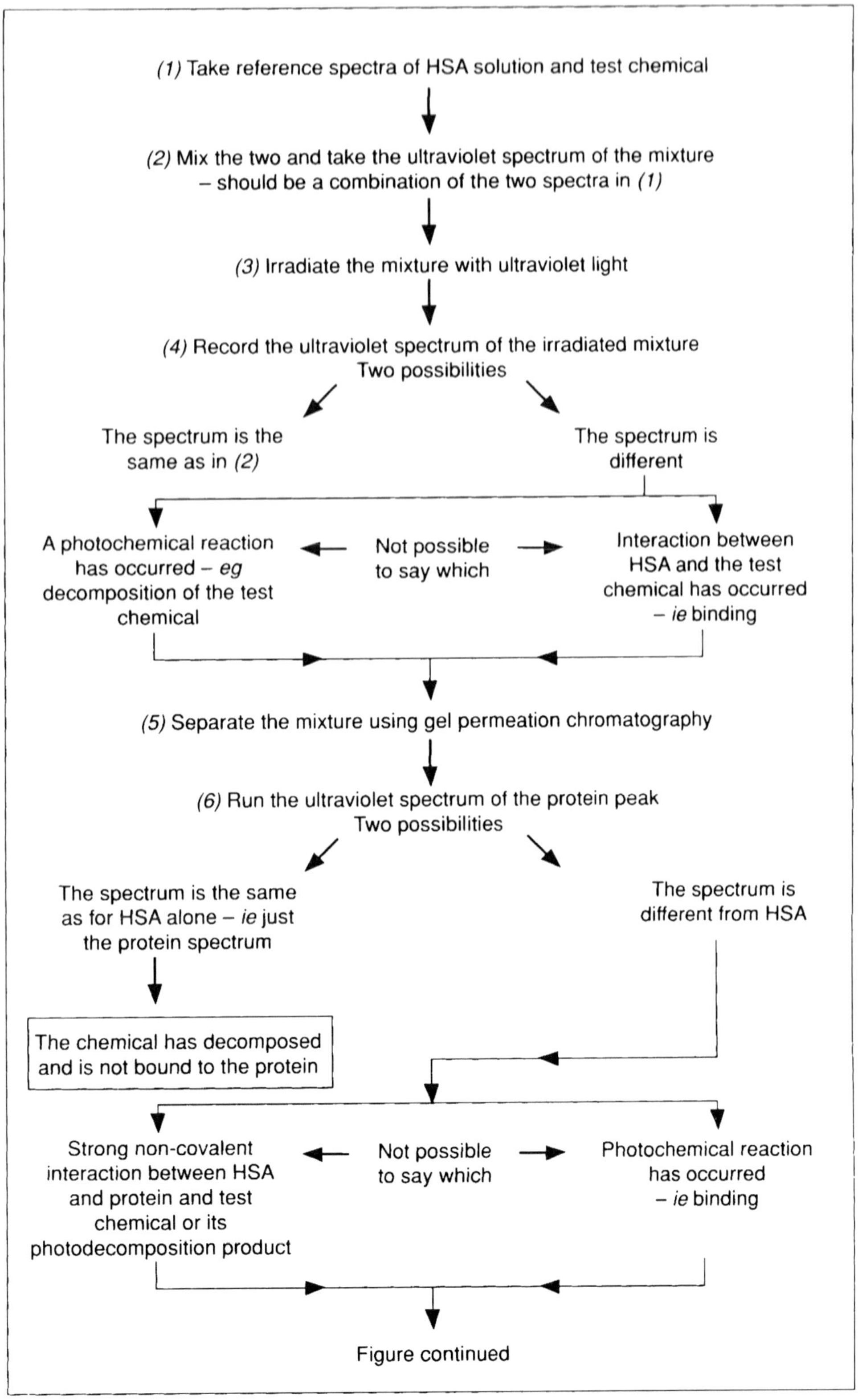

Figure 29 In vitro screening techniques (continued overleaf)

THE ROYAL SOCIETY OF CHEMISTRY

Unilever

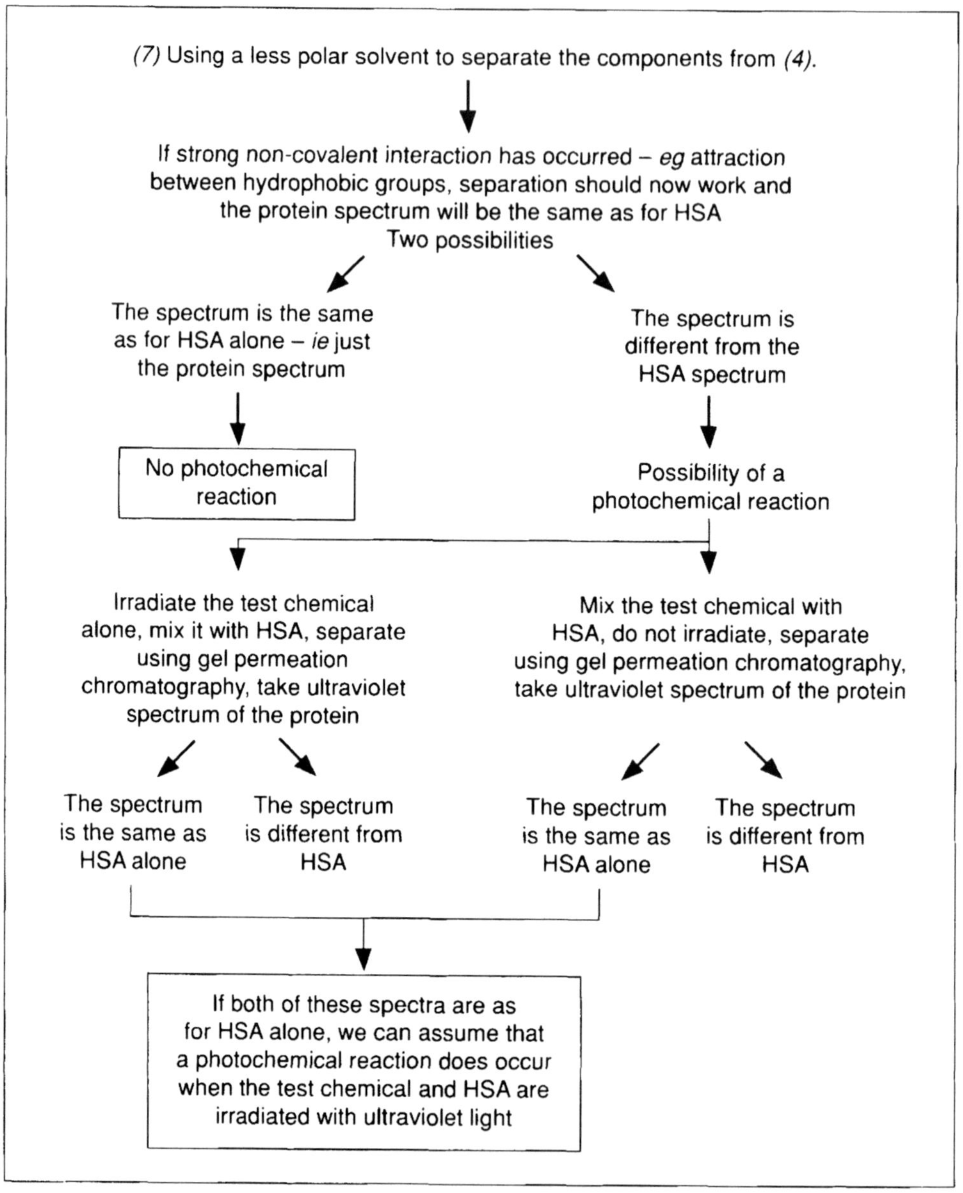

Figure 29 continued In vitro screening technique

Dope testing of horses

The Jockey Club of Great Britain adheres to a number of internationally agreed rules on the use of drugs in horseracing. This ensures that when horses run, they win or lose on their own merits and not because of illegal substances. Approximately 10 per cent of all runners are routinely tested for drugs, 60–70 per cent of which will be race winners. It is the local stewards who decide exactly which horses should be tested, usually after taking advice from a Jockey Club veterinary surgeon. Any horse that performs particularly well, or badly – such as a favourite that is left behind – will usually be tested.

The aim is to have at least one horse from each race tested, though in the case of classic races several runners will be tested. All the UK testing, and some overseas testing is done at the Horseracing Forensic Laboratory in Newmarket. Approximately 0.3 per cent of all domestic tests are positive, but this figure is small compared with the 2–3 per cent in some countries (although in exceptional cases up to 10 per cent have failed).

If a horse fails a drug test, it is the trainer who is held responsible, and if the horse has been placed it automatically loses the prize money. There are several ways in which a horse can have illicit substances in its body – *eg*

1 through medication. Even if given by a vet to treat a condition, if the substance is not permitted the horse loses the race, the trainer is responsible;

2 through negligence. A drug taken orally might be given to the wrong horse; again the trainer is responsible because his stables have not been run efficiently;

3 by deliberate doping. This is the most serious scenario, and can be done either to enhance a horse's performance or to slow it down *ie* if it is clear that only two horses stand a realistic chance of winning a race, the favourite might be drugged so that the second favourite can be backed heavily and win; and

4 through contaminated feedstuff. Although this might be outside the control of the trainer a horse might lose a race because his feed contained theobromine (which is not allowed) from cocoa wastes.

If a horse fails a drug test the Jockey Club will hold an enquiry. Depending on the circumstances, the trainer might be fined (fines of several thousand pounds have been recorded) or lose his licence if the doping was deliberate. The rules to which the Jockey Club work are those of the sporting body and not criminal law, unless 'nobbling' is involved. In these cases a criminal offence has been committed (fraud) and the police become involved.

In routine testing the Horseracing Forensic Laboratory will look for any substance that might affect a horse's system – to improve or impair performance, or to conceal an injury or illness. These might include diuretics; drugs to fight infection; anabolic steroids that build up muscle fibre and increase the aggression of the horse; sedatives, which can be used to impair a horse's performance and reduce its chances of winning; painkillers and local anaesthetics that allow a horse to run while not really in a fit state to do so; and stimulants that increase a horse's chances of winning.

The presence of these drugs or their metabolites is determined by testing the horse's urine. After a race a veterinary officer's assistant has to wait with the horse until it passes water. The sample collected is split into two portions of approximately 250 cm^3. They are kept in two identical sealed bottles which are bar coded so that the testing is anonymous and the analysts cannot be bribed to falsify the results. One portion is tested the day it is received, and the other is frozen and kept. If the tests are negative the second sample is destroyed.

If the tests are positive the trainer and/or owner can elect to have the second sample analysed, so long as they do so within 21 days of being notified that the horse has failed the drug test. The second sample has to be analysed by one of five laboratories – the UK laboratory at Newmarket, or at laboratories in Ireland, France, Germany or Spain. (This is different from human blood alcohol testing, where car drivers are given a sample of their blood and can take it to any certified analyst for independent analysis). To allay any doubts about collusion or falsification of results the owner/trainer is permitted to have a witnessing analyst present when the second sample is tested.

An enormous range of compounds is present in equine urine, and it is first necessary to separate the mixture according to acid/base character (Table 4).

Table 4 Acid/base character and examples of drugs found in horse urine

Character	Examples of types of drug detected in this fraction
Strong acid	Diuretics, anti-inflammatory drugs
Weak acid	Painkillers, anti-inflammatory drugs
Neutral	Diuretics, xanthines (*eg* caffeine, theobromine)
Basic	Local anaesthetics

Four internal standards are added to a 20 cm^3 aliquot of the sample (one for each fraction) so that any foreign substances can be measured against them if necessary. These standards are compounds with structures similar to those of the drugs being tested for, so they will behave in a similar way in the test. The solutes are then totally absorbed on a diatomaceous earth support in a small chromatography column, and a mixture of organic solvents is run through the column (a liquid–liquid extraction). This takes three of the fractions through, and leaves one (strong acids) behind (*Fig. 30*). The acids are eluted off the column by acidifying the organic solvent to suppress the ionisation of the acids and make them soluble.

The solution containing the three remaining fractions is concentrated and separated using solid–liquid chromatography. A buffer/methanol solvent system is used to take two of the fractions through a bonded silica chromatographic cartridge (a small disposable column), and leave one behind (this can be eluted by changing the buffer in the solvent). This is repeated on a different column to separate the two remaining fractions (*Fig. 30*). The four fractions are then treated separately. An overview is given in *Fig. 31*.

The strongly acidic fraction is spotted onto an ultraviolet sensitive TLC plate consisting of silica particles bound to an aluminium support. A dichloroethane/ethanoic acid solvent system is used to separate the components, which show up as bright spots under ultraviolet light. Any spots other than the internal standard are investigated further because they suggest that illegal substances have been administered.

Unilever

THE ROYAL SOCIETY OF CHEMISTRY

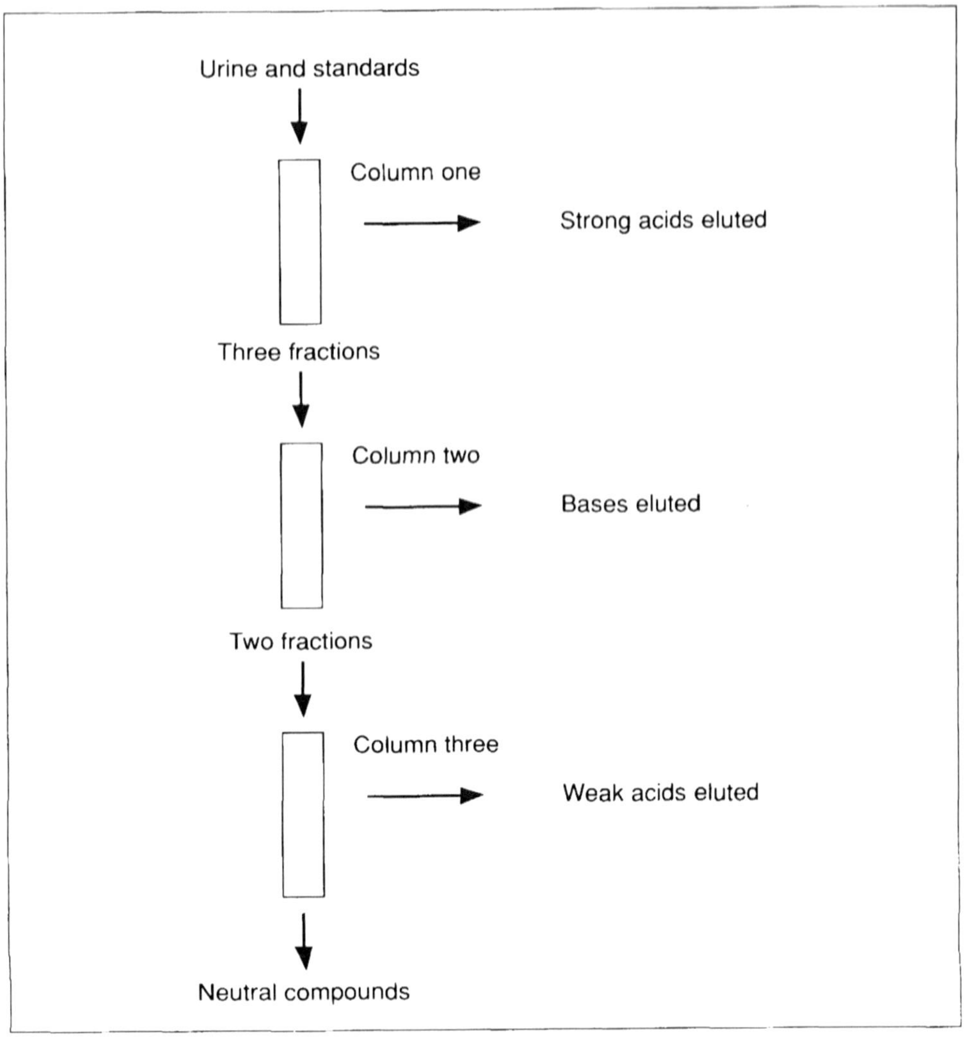

Figure 30 Separation of a sample according to acidity

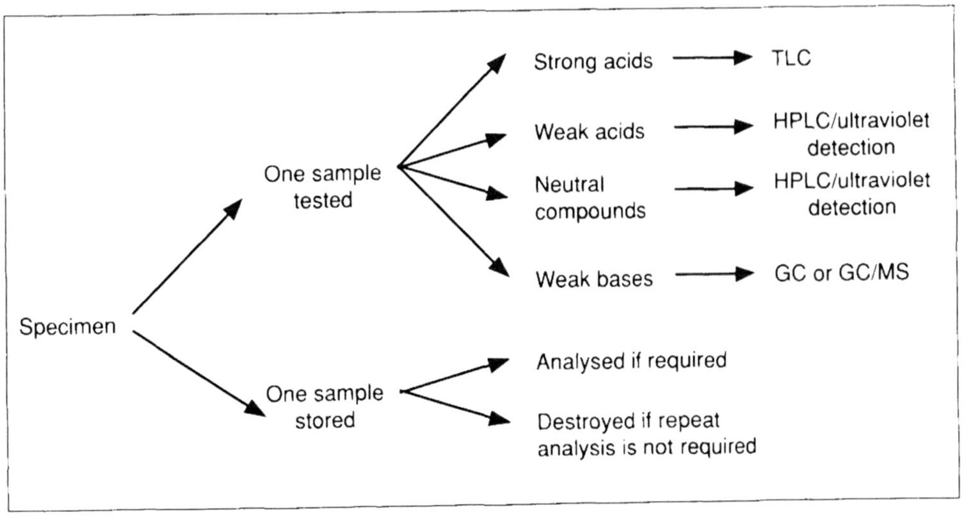

Figure 31 Overview of the analytical techniques used for the dope testing of horses

The weakly acidic and neutral fractions are treated similarly. The 0.02 cm^3 (20 μl) portions are injected onto standard octadecylsilane (C_{18}) HPLC columns, and are eluted with a methanol/ethanoic acid solvent. The relative proportions of the two liquids in the solvent are changed during the separation to achieve maximum separation – *ie* there is a concentration gradient. The mixture normally takes 15 min to separate, using a flow rate of 1.5 cm^3 min^{-1}. The eluate from the HPLC column passes through a 0.008 cm^3 (8 μl) cell and the components are detected from their ultraviolet spectra using a diode array detector (*Fig. 32*). All the data generated are stored on a computer. Once the run is complete each peak is compared with a library of roughly 40 drugs, according to the retention time and ultraviolet spectrum of each peak. Stimulants such as caffeine are frequently detected by using this method. Examples showing HPLC chromatograms of mixtures of some drugs are shown in *Figs. 33* and *34*.

Anything unusual is investigated further. The mass spectrum of a sample of the unknown compound might be obtained first. If any test is positive the whole procedure is repeated by a single person who also runs a blank sample through the system to ensure that there is no doubt about the accuracy of the instrumentation.

The basic compounds are separated by GC. The basic fraction is injected onto a neutral bonded silica GC column by using helium as the carrier gas. (Some compounds are not very volatile, so the column is heated from 70 °C to 290 °C during the separation.) The presence of any basic solutes is monitored by a nitrogen specific detector (see page 127). The compounds are identified from their retention time on the column and by their mass spectra.

Horses' urine can contain complex water soluble conjugated metabolites, and these conjugates have to be broken down before the drugs, or their metabolites, can be extracted. The conjugated metabolites are hydrolysed by using enzymes before being separated using a solid–liquid chromatography column with a basic organic solvent. Chemically similar compounds are collected, and these are further separated by using GC before their mass spectra are obtained. However, the compounds tend to be too polar to separate efficiently (*ie* – having too many OH groups) so the polarity of the compounds is reduced by making their trimethylsilyl derivatives. The derivatives are then separated by using GC and their mass spectra are recorded.

The mass spectrum of each peak from the GC is stored on computer, however, it is not necessary to interpret all of the peaks of each spectrum. This is because different drugs of a given type have at least one fragment in common. By looking for such peaks it is possible to establish that a tranquilliser or a β-blocker, for example, has been administered. Consequently, five peaks might cover the possible presence of a hundred drugs.

Unilever

THE ROYAL SOCIETY OF CHEMISTRY

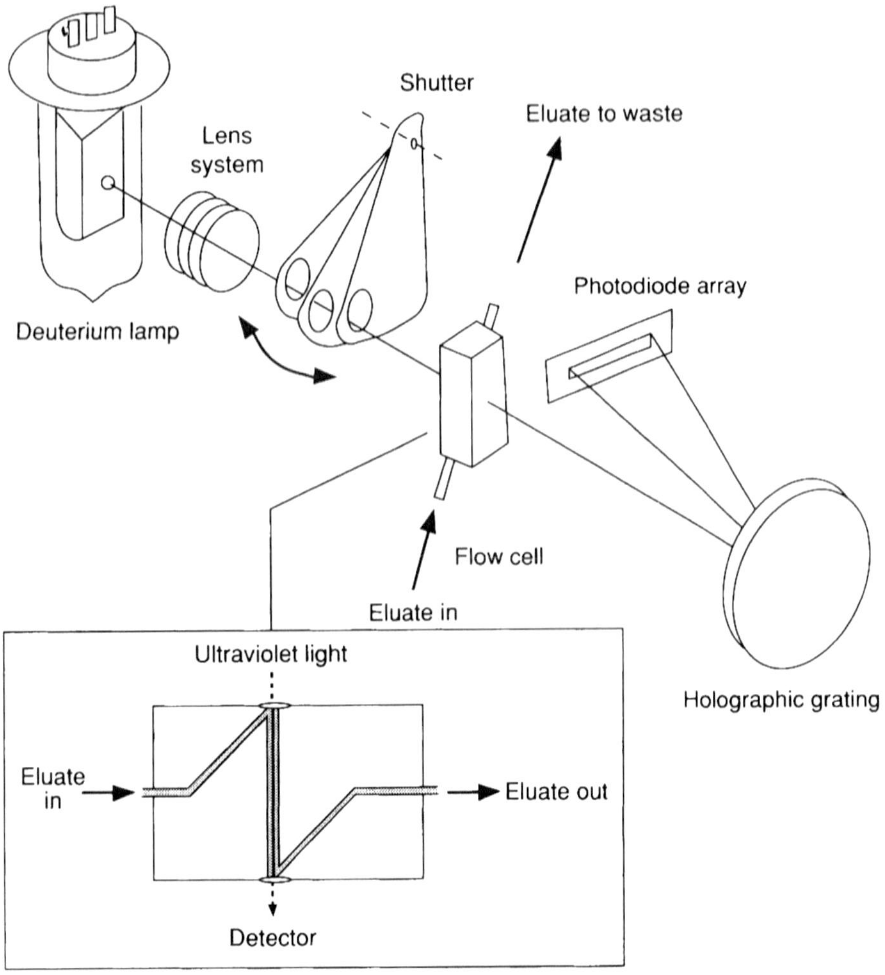

In the diode array detector light from the deuterium lamp is focused onto the flow cell. Having passed through the cell the light is dispersed into single wavelengths and these are focused onto a row of photosensitive diodes etched onto the surface of a silicon chip. This allows all wavelengths across the desired range to be monitored simultaneously. Such an arrangement does not have the same sensitivity as more conventional (variable wavelength) ultraviolet spectrometers. This is because the diode array detector does not measure single wavelengths at a time. A reference measurement is taken at the beginning of the run, and is stored by the computer.

Figure 32 The diode array detector

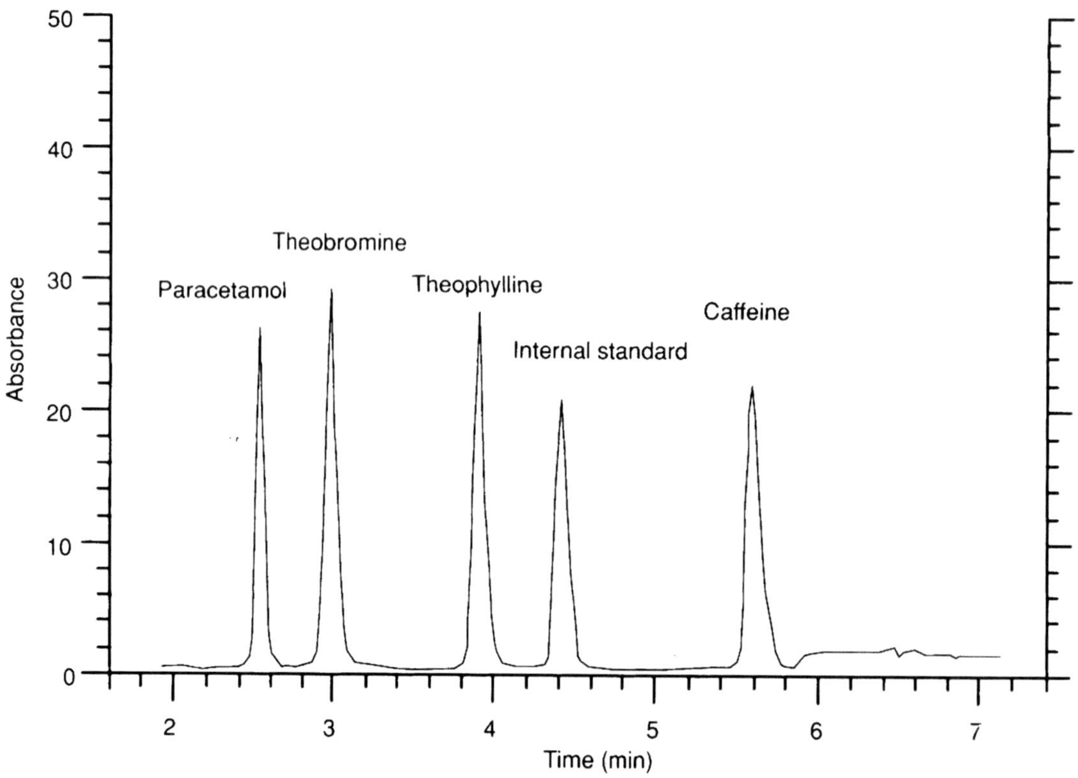

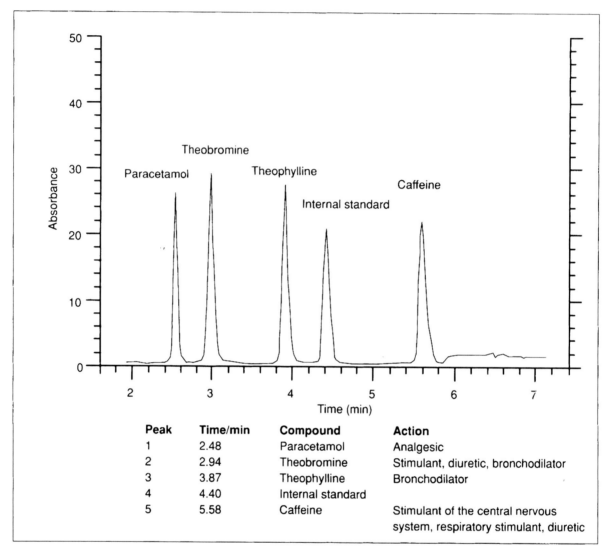

Peak	Time/min	Compound	Action
1	2.48	Paracetamol	Analgesic
2	2.94	Theobromine	Stimulant, diuretic, bronchodilator
3	3.87	Theophylline	Bronchodilator
4	4.40	Internal standard	
5	5.58	Caffeine	Stimulant of the central nervous system, respiratory stimulant, diuretic

Figure 33 HPLC chromatogram of a mixture of neutral compounds

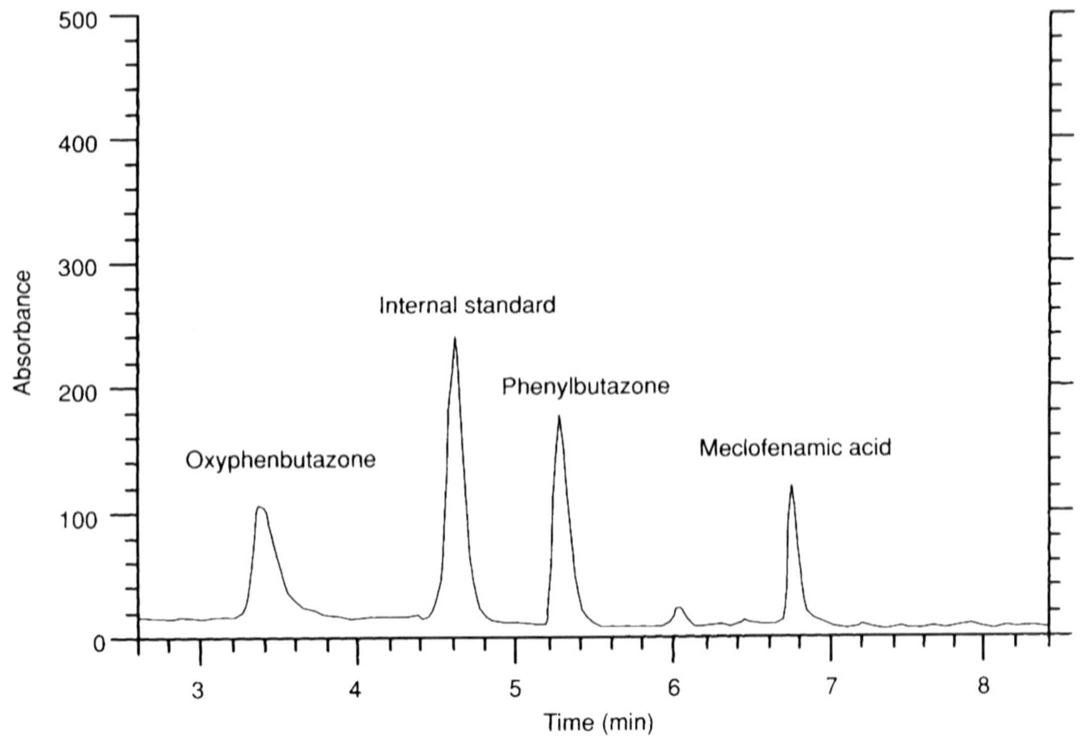

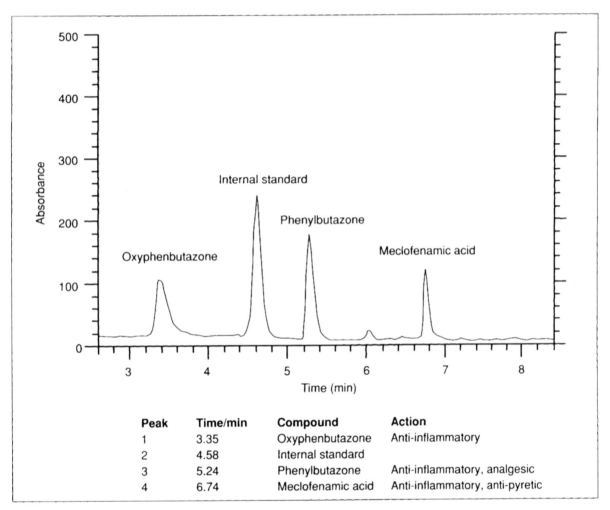

Peak	Time/min	Compound	Action
1	3.35	Oxyphenbutazone	Anti-inflammatory
2	4.58	Internal standard	
3	5.24	Phenylbutazone	Anti-inflammatory, analgesic
4	6.74	Meclofenamic acid	Anti-inflammatory, anti-pyretic

Figure 34 HPLC chromatogram of a mixture of weakly acidic compounds

If the output from the GC is less than 1 cm^3 min^{-1} it can be passed through a heated block before being transferred directly to the mass spectrometer. The heated block ensures that all components in the eluate from the GC are gaseous. The pumping system of the mass spectrometer will then remove enough of the gas to maintain the operating pressure of the spectrometer.

Anabolic steroids are tested for in a rather different way, by using a technique not covered in detail in this book – immunoassay. A sheep or a rabbit is harmlessly and painlessly injected with a form of the drug that is to be tested for. The animal will develop antibodies to the drug, which will appear in its blood shortly afterwards. The antibodies are separated from the blood, and are used as the reagent to react with any of the drug in the horse urine sample. An enzyme- or radio-labelled reaction is usually chosen.

THE ROYAL SOCIETY OF CHEMISTRY

Unilever

Once in the urine the antibodies will bind to the drug, and the bound complex can be separated. The actual technique used is a little more elaborate. A small labelled sample of the drug is first introduced into the urine, then, when an equal amount of the antibodies is added they will bind exclusively to the labelled drug and all the radioactivity detected will be in the separated complex. However, if a small amount of the drug is already present in the urine some will bind to the antibodies. There will not be sufficient antibodies for all the labelled drug to bind, and once the complex has been separated there will be a residual radioactivity in the urine sample. A strongly positive test will leave a large concentration of the labelled drug in solution, and hence a higher radioactive count rate (*Fig. 35*).

The presence in horse urine of substances not permitted by Jockey Club rules does not automatically mean that an offence has been committed. For example, theobromine and theophylline are metabolites of caffeine. Caffeine is a stimulant and is completely banned. Theobromine is present in chocolate and cocoa residues, so a urine sample will reveal if the horse has been given food containing cocoa residues. As a result, small amounts are permitted. On the other hand, if theophylline is detected there is more of a problem because it is not only a metabolite of caffeine, but is also used in the treatment of asthma. In both cases the trainer can expect a visit from a Jockey Club representative!

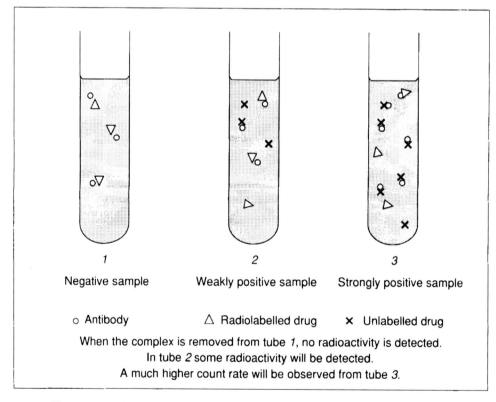

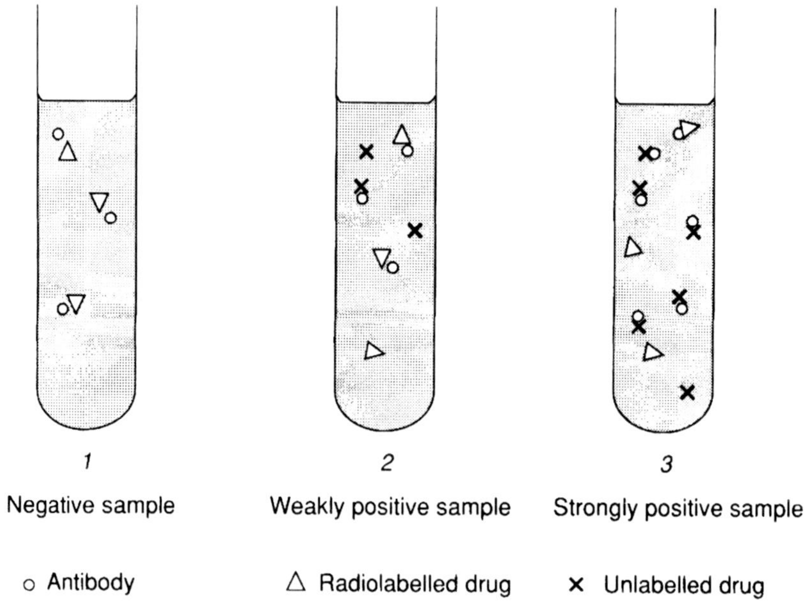

When the complex is removed from tube *1*, no radioactivity is detected.
In tube *2* some radioactivity will be detected.
A much higher count rate will be observed from tube *3*.

Figure 35 Immunoassay technique

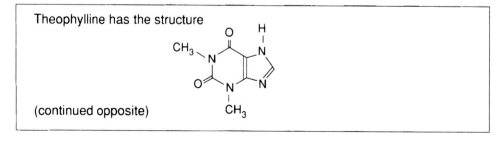

Theophylline has the structure

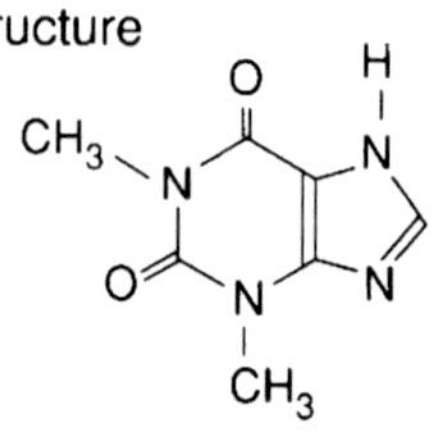

(continued opposite)

As such, it is similar to caffeine (see page 138, caffeine/theobromine analysis) but with one methyl group replaced by a hydrogen atom.

Theophylline and some of its derivatives are effective bronchodilators so they are frequently used in patients suffering from acute or chronic bronchoconstriction – *eg* asthmatics.

The compound responsible for the relaxation of smooth (involuntary) muscle in the respiratory system is destroyed by another substance in the body. Theophylline inhibits the action of the destroying substance, enabling the muscle to relax.

Recent trends

Chromatographic techniques have become very important in industry for the purification and separation of intermediates in multi-stage syntheses. (Such separations have to be done in batches rather than in continuous flow.)

In terms of scientific advances, one of the major innovations in the past five years has been the development of efficient columns capable of separating specific chiral compounds from a mixture. They work by the stereospecific adsorption of one enantiomer onto the surface of the stationary phase. The resin contains only one enantiomer, hence its stereoselectivity towards other chiral molecules. The cost of a column for chiral separations can be roughly three times (or even higher) the cost of a standard analytical column.

A similar type of chromatography, affinity chromatography, works on a similar principle – a substrate is covalently bound to a resin, and only enzymes with vacant sites in the correct orientation can interact with these sites.

Research has also shown that graphite-based HPLC packing materials can separate compounds with low or modest polarities, very efficiently. Studies of particle and pore sizes have shown that the surface area of graphite-based materials in HPLC columns can be as high as 1000 $m^2 g^{-1}$. Some classes of compound that have been separated by such columns include:

1. aromatic hydrocarbons;
2. alkylnaphthalenes;
3. methylphenols;
4. polychlorinated biphenyls;
5. steroids; and
6. some amino acids.

By using different solvent systems one area which has been found to have great potential is supercritical fluid chromatography (SFC). A supercritical fluid is one at a temperature and pressure above its critical point. Supercritical carbon dioxide (used to extract caffeine from coffee) in HPLC columns has the solvating properties of a liquid and the transport properties of a gas. Another advantage is that extraction of the solutes from the eluate is easy – the carbon dioxide is simply allowed to evaporate.

However, supercritical carbon dioxide would not be a good solvent for a chromatography column if the eluate were to be analysed by infrared spectroscopy, because of the strong spectral absorption of carbon dioxide which would obscure the absorptions of the eluted compounds. Supercritical xenon has been found to achieve good separation of compounds such as polyaromatic hydrocarbons and if the eluate flows through a microcell no absorption due to xenon is observed. This is a particular advantage if SFC is to be linked with infrared spectroscopy.

THE ROYAL SOCIETY OF CHEMISTRY

Unilever

6. Electron microscopy

There is a limit below which it is not possible to resolve an image in light microscopy. However, structural information on a specimen can still be obtained, by using electrons instead of light. The principles involved are similar, although the operational practicalities are somewhat different. An electron microscope can be used to obtain magnification in the range $10–10^6$ x. Gas molecules scatter (diffract) electron beams, so the vast majority of studies involving electron microscopy have to be at very low pressures, typically $1.33 \times 10^{-3}–1.33 \times 10^{-5}$ Nm^{-2}. This limits the range of materials that can be studied using this technique to dry, solid specimens that are stable at these very low pressures.

The theory

The interaction of materials with electrons is shown in *Fig. 1*.

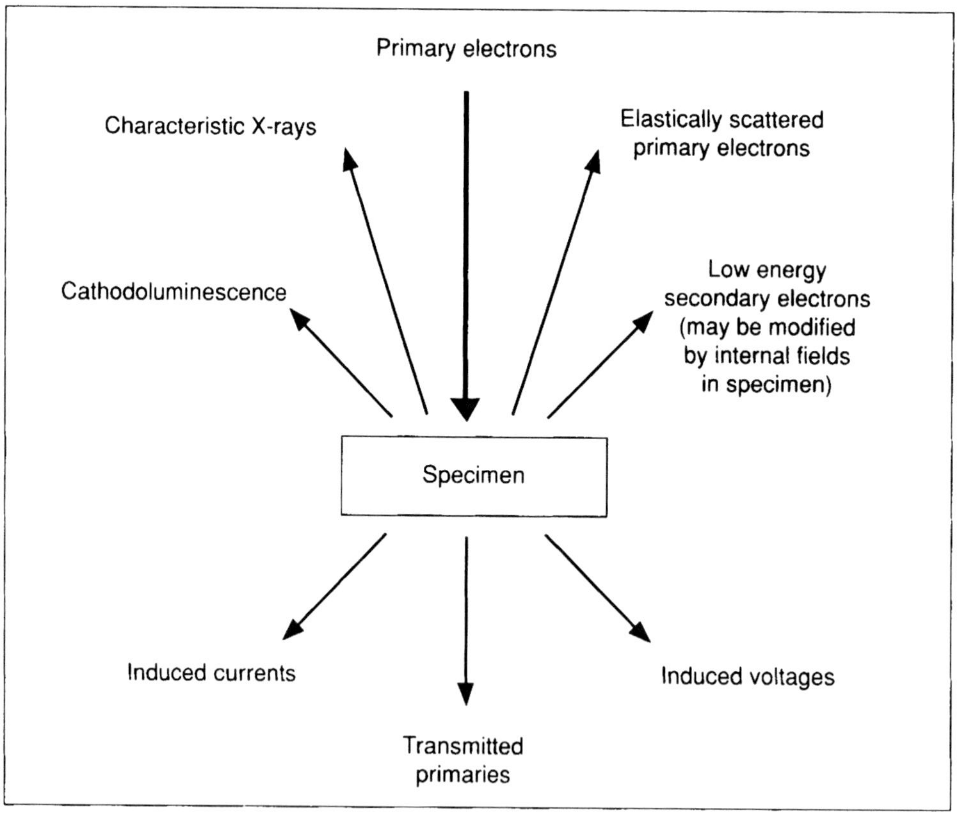

Figure 1 Interaction of electrons with materials

One major similarity between light and electron microscopy is that images can be formed from the radiation that is transmitted through the specimen or from radiation that comes back towards the radiation source, be it a lamp or an electron gun. In the case of electron microscopy, different conditions are necessary for generating and detecting the radiation. Scanning electron microscopes are useful for displaying images of surface structures, which are generated by secondary electrons. The transmission electron microscope relies on the primary electrons passing through the specimen to give high resolution images of internal structures of samples (which must be less than 1×10^{-7} m/0.1 μm thick).

X-rays are formed when a primary electron strikes an inner shell electron of an atom in the specimen and gives it sufficient energy to ionise. Once the inner shell electron has been removed an electron from a higher energy orbital will drop down to the lower level and emit its excess energy as an X-ray photon. The energies of the photons produced are characteristic of the elements from which they have been formed. (The gap left at the higher level can then be filled by an electron from a level higher still, so that a range of characteristic X-ray energies is observed.)

The scanning electron microscope (SEM)

Electron beam formation and focusing

The most important signals to consider in the SEM are:

1 secondary electrons;

2 backscattered electrons; and

3 X-rays.

Secondary electrons are usually used to provide the image because the electron beam is not spread out and resolution is often very high, usually in the range 5×10^{-9}–2×10^{-8} m (5–20 nm). This type of electron is generated as a result of inelastic scattering of the incident electrons (*Fig. 2*). The secondary electrons have low energies, typically 3–8 x 10^{-19} J (2–5 eV) although they can be as high as 8×10^{-18} J (50 eV). The inelastically scattered incident electrons can continue and cause other events.

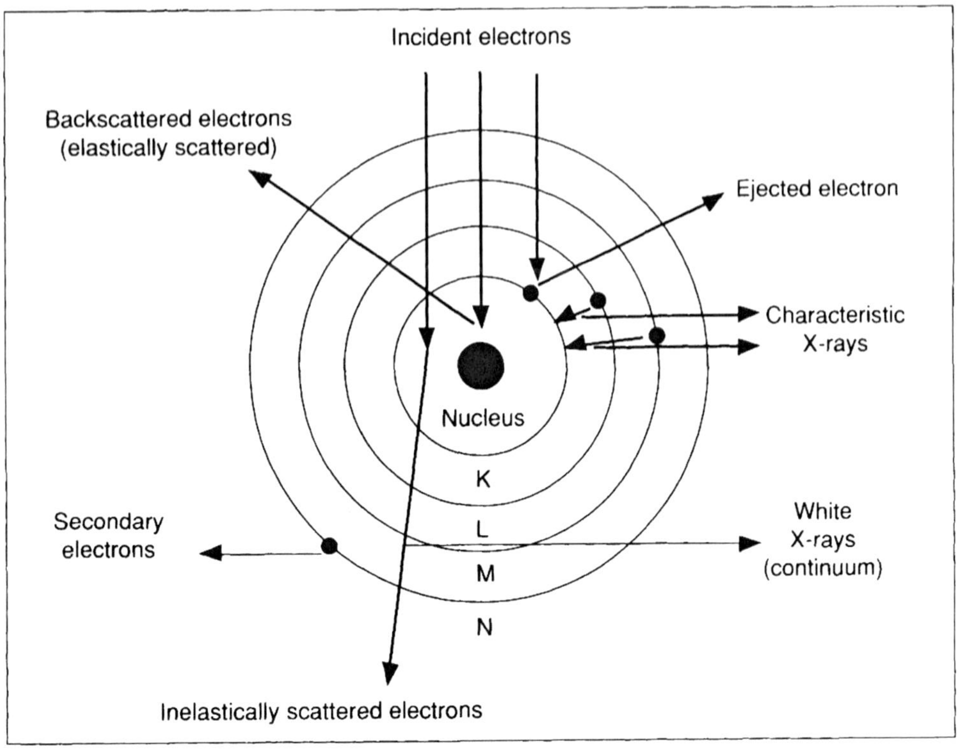

Figure 2 Scattering of electrons and X-ray formation

Backscattered electrons are the primary beam electrons that have been scattered elastically by the nuclei in the sample (*Fig. 2*). These electrons are useful for imaging the atoms in a specimen by atomic number contrast. This is because low atomic number samples give low emissions of backscattered electrons while high atomic number samples give high emissions of these electrons. The backscattered electrons have higher energies than secondary electrons – usually from approximately 8×10^{-18} J (50 eV) up to the energy of the primary beam electrons.

The electrons can be scattered from relatively deep positions within the sample – typically up to 1×10^{-7} m (100 nm), but because the spread of electrons is relatively large, the resolution of any image from these electrons is low (perhaps 2×10^{-8} m) compared with secondary electron images.

The incident electrons are usually generated by passing an electric current through a tungsten filament at the top of a column (other methods exist such as applying a potential to a lanthanum hexaboride single crystal). A voltage, usually in the range 300 V to 40 kV, is applied between the electron source (the cathode) and the rest of the column (the anode). This voltage accelerates the electrons down the column, towards the specimen. Whereas light rays are focused in a light microscope by glass lenses, electrons are focused in an electron microscope by electromagnetic lenses (*Fig. 3*).

The condenser lens is used to collimate the electron beam which the objective lens focuses onto the specimen, producing a 'probe' of diameter *ca* 1×10^{-8} m (10 nm). Scanning coils are then used to direct the beam across the specimen in a series of parallel lines so that when the parallel scans are put together a two dimensional image is obtained (similar to a domestic television set). Scan rates can be as fast as 25 frames per second for immediate study, or as slow as several minutes per scan if more clearly defined images are required for a photographic record.

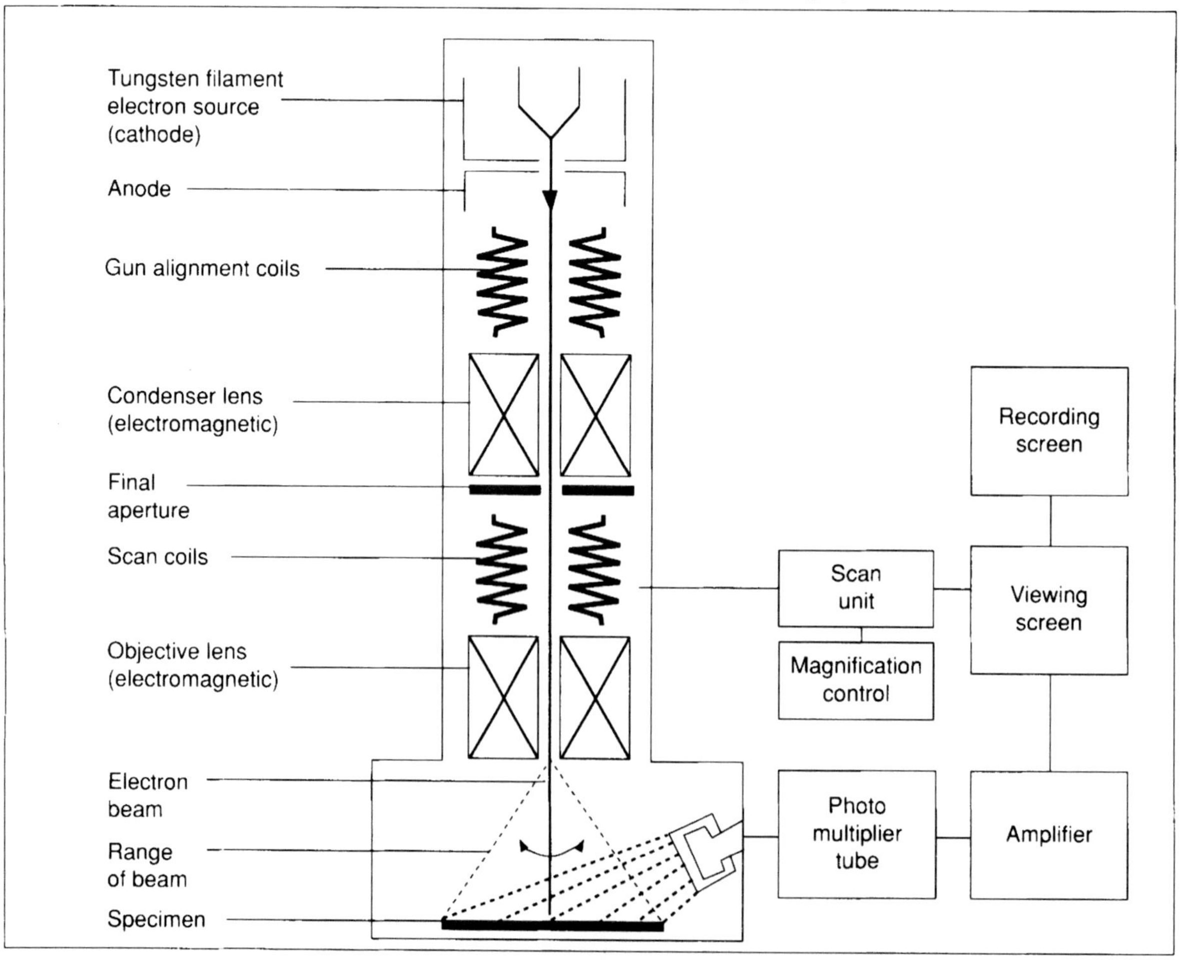

Figure 3 The scanning electron microscope (SEM)

The sample is held on a movable stage in a chamber at the base of the column. The stage enables the specimen to be moved in the *x*, *y* and *z* directions, and also allows for tilt and rotational adjustments to be made. (Some electron microscopes have air locks so that the sample can be changed while keeping the remainder of the column under vacuum.)

One limitation is that samples that are non-conductors of electricity have to be treated before they can be studied with the electron microscope. A plasma of gold ions is sputtered onto the sample at very low pressure and a thin film of gold forms on its surface. This coating inhibits image distortion by sample charging, and does not normally affect surface detail because the gold coating can only be detected at relatively high magnifications. Gold is often used because it is an excellent electrical conductor and being a heavy metal, has a high secondary and back scattering electron yield.

Image formation

Secondary electrons are useful for high resolution imaging. They are attracted by a grid, typically set at +200 to +600 V potential, in front of a scintillation detector. They are further accelerated by a potential of about 10 kV onto the scintillation detector surface, where their energy is converted to visible light. The light emitted passes down a perspex light guide to a photomultiplier tube where it is converted to an electrical current. This signal can be amplified to produce an image on a cathode ray tube (a television screen). A large number of secondary electrons results in a bright image on the screen.

Photographic images are produced by placing a camera in front of a suitable screen and moving images can also be recorded by using videotape. Images can be clarified by removing unwanted background 'noise' with the aid of a computer.

The magnification can be changed by changing the area of the sample scanned while keeping the screen size constant. A large magnification is achieved by scanning a very small area of the sample. The images obtained have an advantage over light microscopy images because they have a 'three dimensional' quality and have an appreciably greater depth of field ca 300 x better (see diagram of polymer bead, page 166).

Chemical analysis

The X-rays produced when primary electrons interact with the sample have energies characteristic of the elements contained in the specimen. A solid state detector can be used to measure the energy of the X-rays formed and, when used in conjunction with a computer, can be used to identify the atoms present. The systems are capable of identifying elements with atomic numbers 5–92 (boron to uranium) simultaneously (*Fig. 4*). The sample used was copper mounted on an aluminium base using a silver based adhesive.

Once the elements in a sample have been identified by the energies of the X-rays emitted from it, it is possible to programme a computer to display the location of the different elements in different colours on a screen. This can be done so that one element is shown per image, or many elements shown in the same image. Individual atoms cannot be 'seen', but their distribution in a sample can.

The transmission electron microscope (TEM)

Electron beam formation and focusing

Electron beam formation is similar to that in the SEM, but does not require the scanning coils (*Fig. 5*). The accelerating potential between the cathode (the electron source) and the anode (the rest of the column) tends to be higher than for the SEM –

THE ROYAL SOCIETY OF CHEMISTRY

Unilever

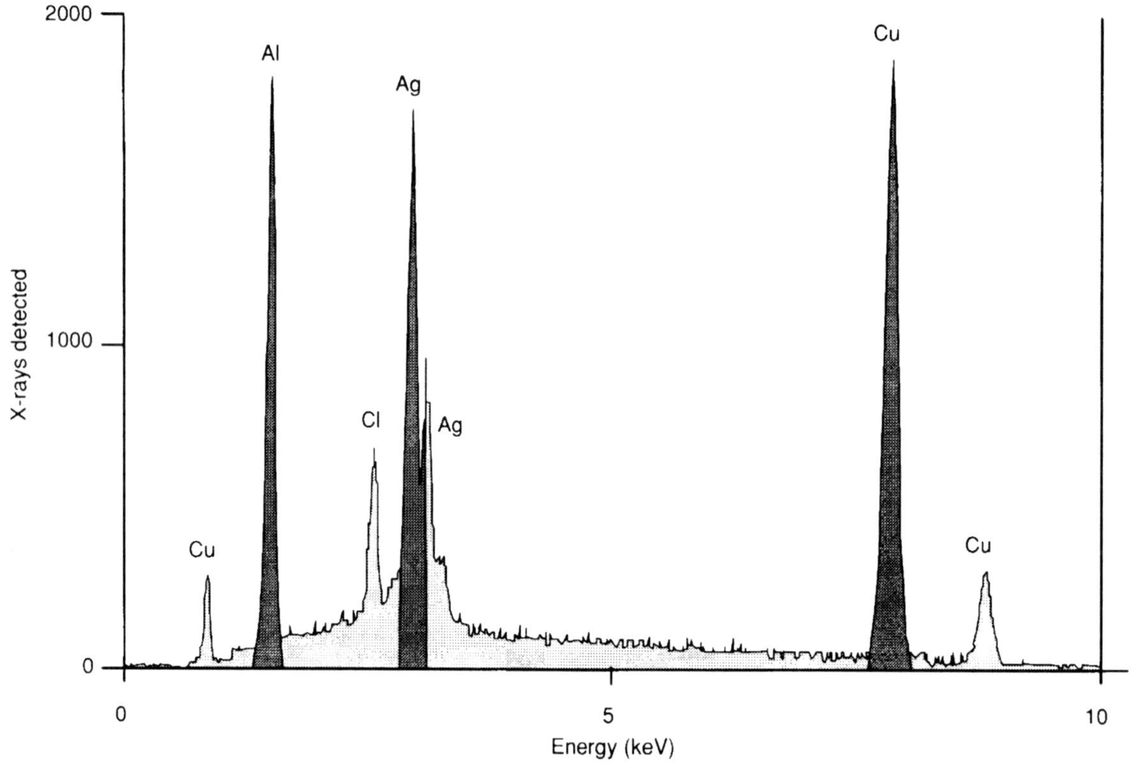

Figure 4 Identification of elements from their characteristic X-rays in a scanning electron microscope

usually 80–200 kV, although instruments with very high resolution might require a potential of 1 MV. Higher potentials are necessary to give the electrons sufficient energy to penetrate the specimen. The pressure inside the instrument also has to be lower to achieve high resolution images, typically 1.33×10^{-4}–1.33×10^{-5} Nm^{-2}.

Electromagnetic condenser coils collimate the beam, which then strikes the specimen. The specimen must be dry, solid, stable, and capable of withstanding the heating effect of the electron beam. It must also be extremely thin (of the order 1×10^{-7}m) and transparent or semi-transparent to the beam. Microtomed sections of samples can be used (a microtome is an instrument used for cutting thin slices from a sample), and techniques such as etching and chemical staining also can be used to visualise the detail in the final image.

Image formation

Once the electron beam has been passed through the specimen it is magnified and focused by an image forming electromagnetic lens (the 'objective'). It then strikes a fluorescent screen where the energy of the electrons is converted to visible light, forming an image. The image can be viewed through a lead glass window. Alternatively, the screen can be replaced by a camera so that a photographic image is recorded.

If scanning coils are used, high resolution images can be obtained from the secondary electrons detected. Alternatively, the X-rays emitted can be analysed. If the TEM is put into scanning mode (STEM) it is possible to deduce where in a bulk sample small amounts of material are. Under favourable conditions, as little as 10^{-16} g of a substance can be detected in STEM mode.

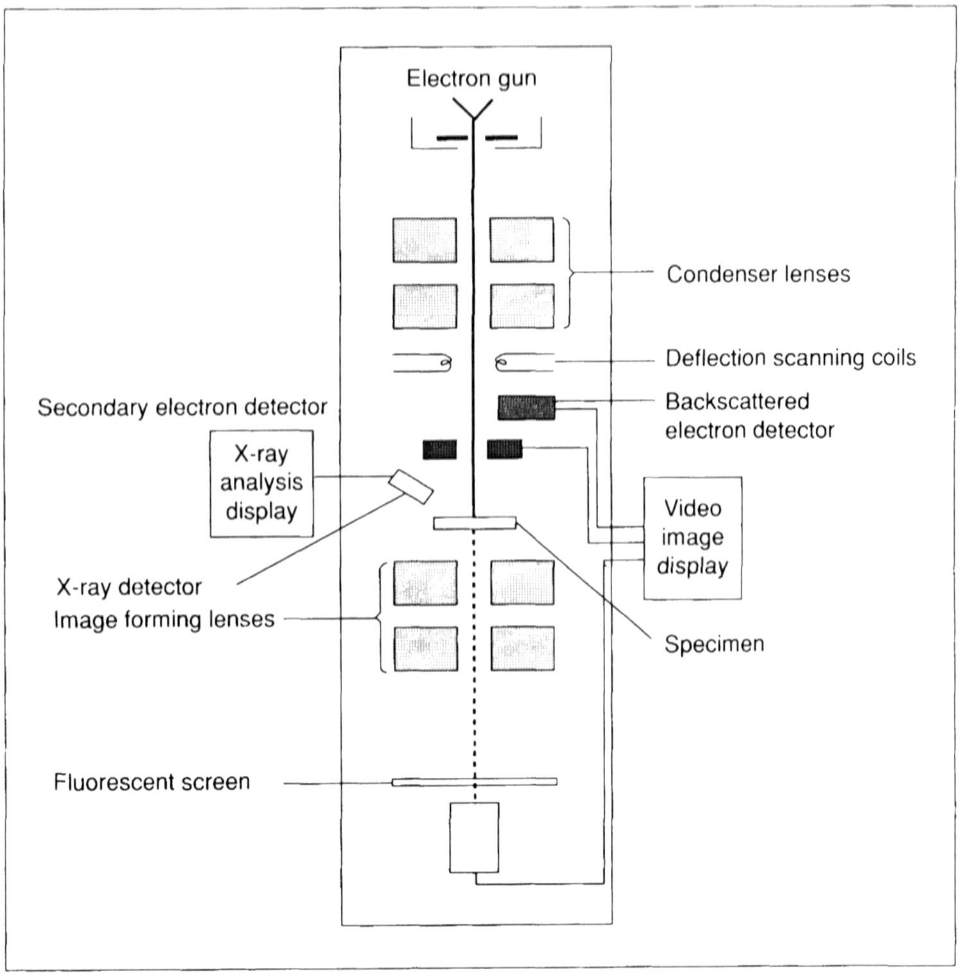

Figure 5 The transmission electron microscope (TEM)

Applications

Electron microscopy does have some limitations. Images of living systems cannot be obtained, and the preparation of a specimen can be complicated and expensive. Without programming a computer to assign different colours to different grey tones the images produced are always in monochrome (*ie* they are not in colour). However, a wide range of materials can be studied by electron microscopy. These include: powder particles; soaps; hair; teeth; bacteria; timber; plastics; metals and foils; plant and animal tissues; lotions, creams and emulsions (*eg* ice creams) [see Advanced Techniques, page 171]; foodstuffs and oils; and packaging.

THE ROYAL SOCIETY OF CHEMISTRY

Unilever

Imaging

Scanning electron microscope images

Electron microscopy has been used in industry to study the pore size in a Polyhipe (poly high internal phase emulsion). To the naked eye and under low magnification the polymer appears similar to expanded polyphenylethene (polystyrene). However, under the scanning electron microscope it is possible to see the fully interconnected open pore structure of the material – it is about 90 per cent air (*Fig. 6*). By measuring the pore sizes (about 5×10^{-6} m) and its structure, its properties can be determined.

Figure 6 Internal structure of 'polyhipe' ™

It is also possible to study the effect of shampoos on human hair. Under the electron microscope the structure of a hair can be seen clearly, and the changes in the structure of the hair before and after treatment with a shampoo can be followed. Other fibres, such as wool, dog hairs, nylon *etc*, can be identified from their structures and any scales (cuticle) covering them.

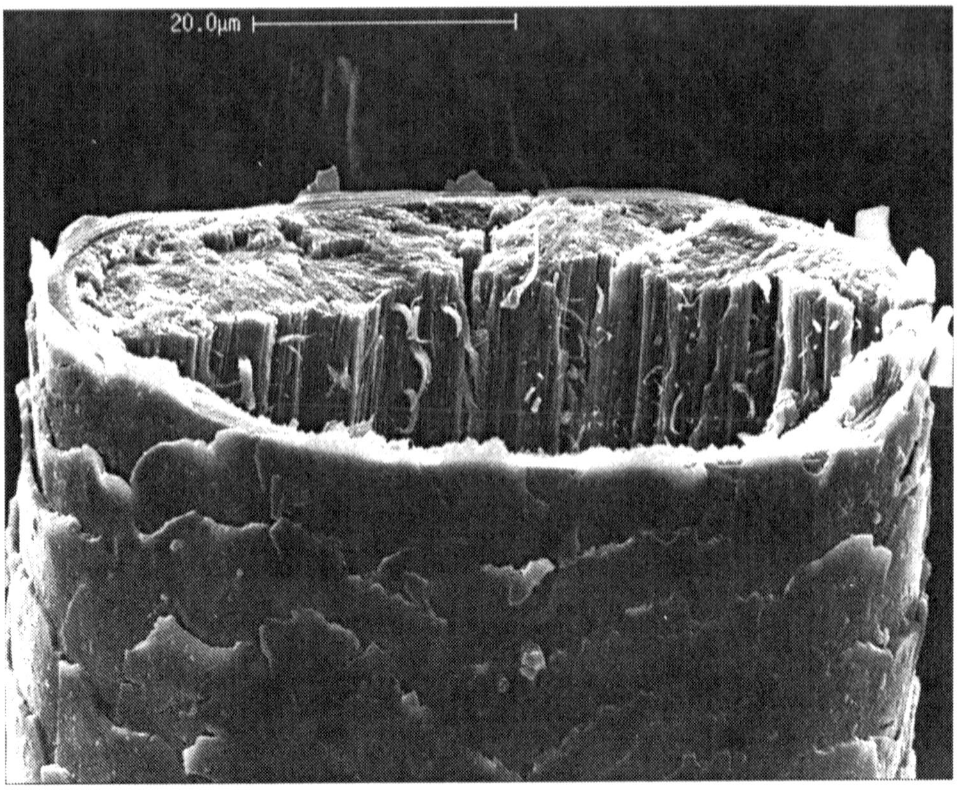

Figure 7 Human hair broken under tension and showing the outer cuticle and the inner cortex

Soiled fabrics can also be studied by using the electron microscope, because dirt particles which are bound to different fibres can be seen. The efficiency of different types of soap and detergent in removing the particles can then be determined.

THE ROYAL SOCIETY OF CHEMISTRY

Unilever

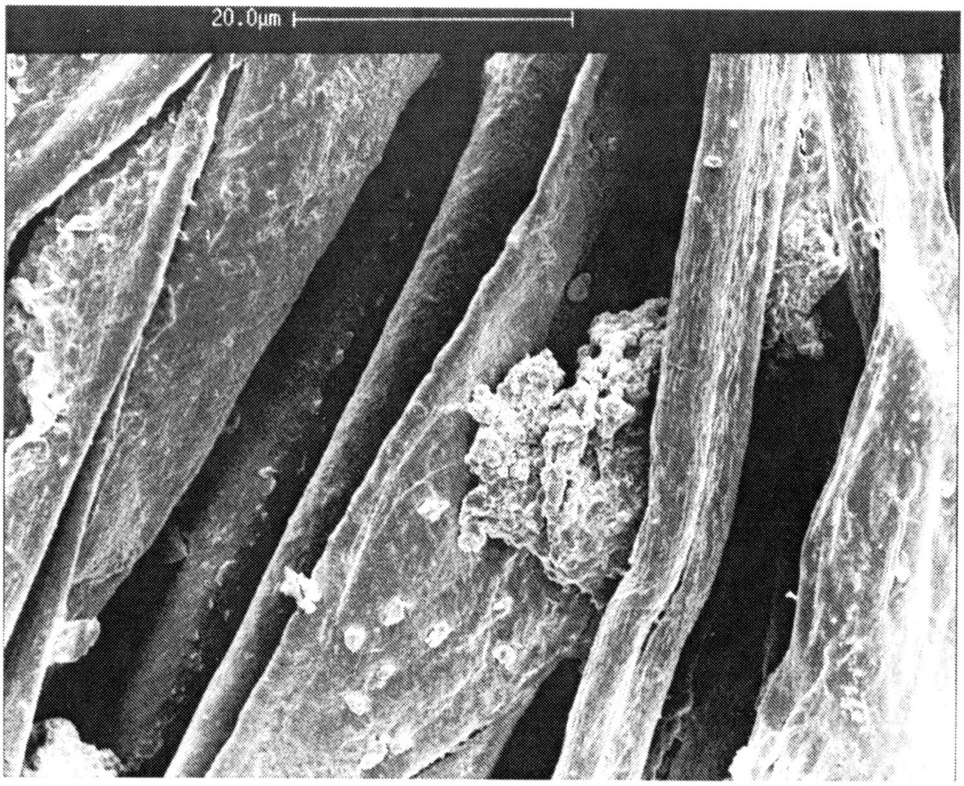

Figure 8 Cotton fibres with soil particle attached

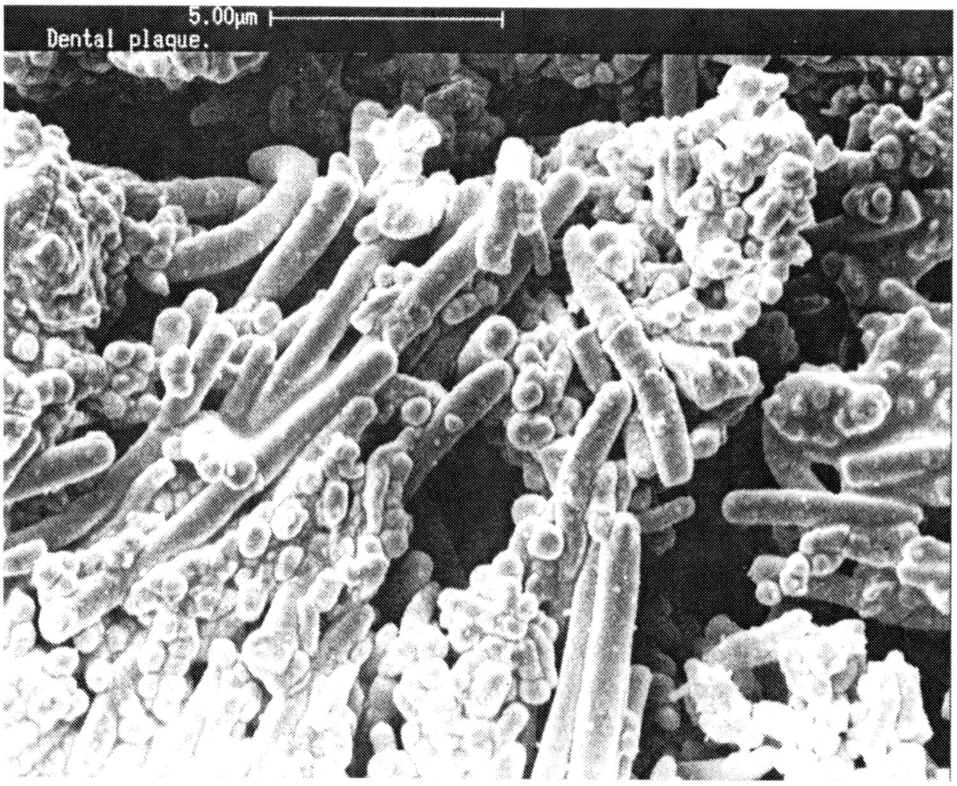

Figure 9 Dental plaque

Unilever

THE ROYAL SOCIETY OF CHEMISTRY

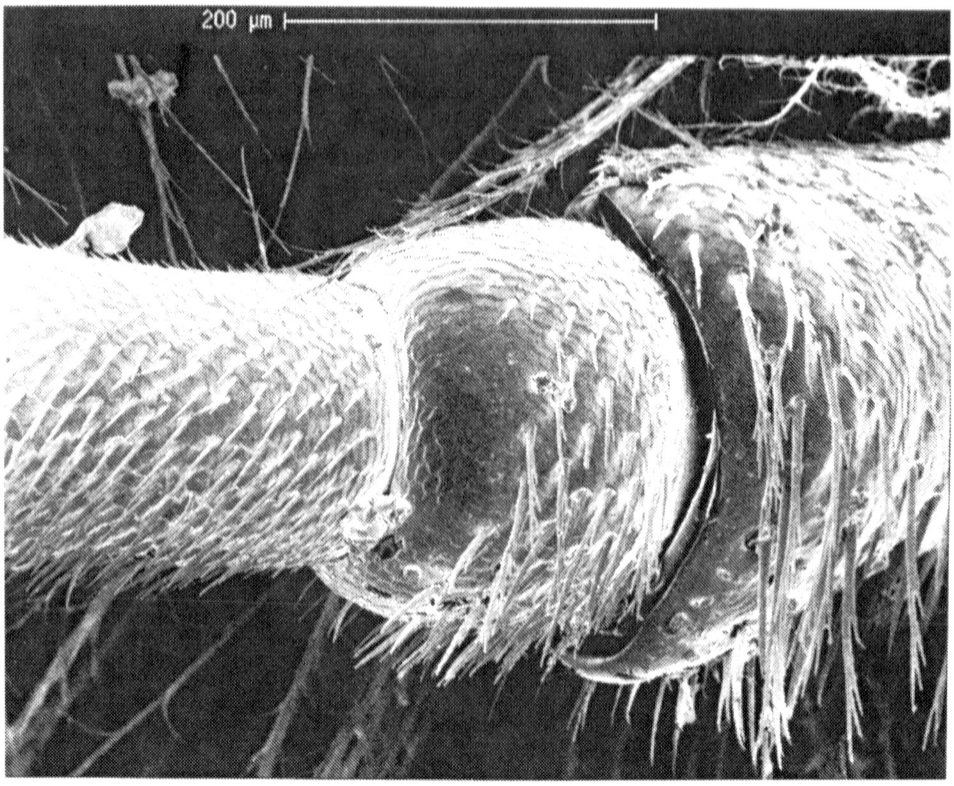

Figure 10 Knuckle joint on a bee's head where the antenna is attached

Transmission electron microscope images

Using the TEM images of carbon dioxide gas hydrate crystals can be obtained (*Fig. 11*). The image was obtained by forming a carbon replica of the surface and shadowing the replica with tungesten, and it is possible to see images of the crystals embedded with an ice (1h) matrix. (The magnification is x 10^4.)

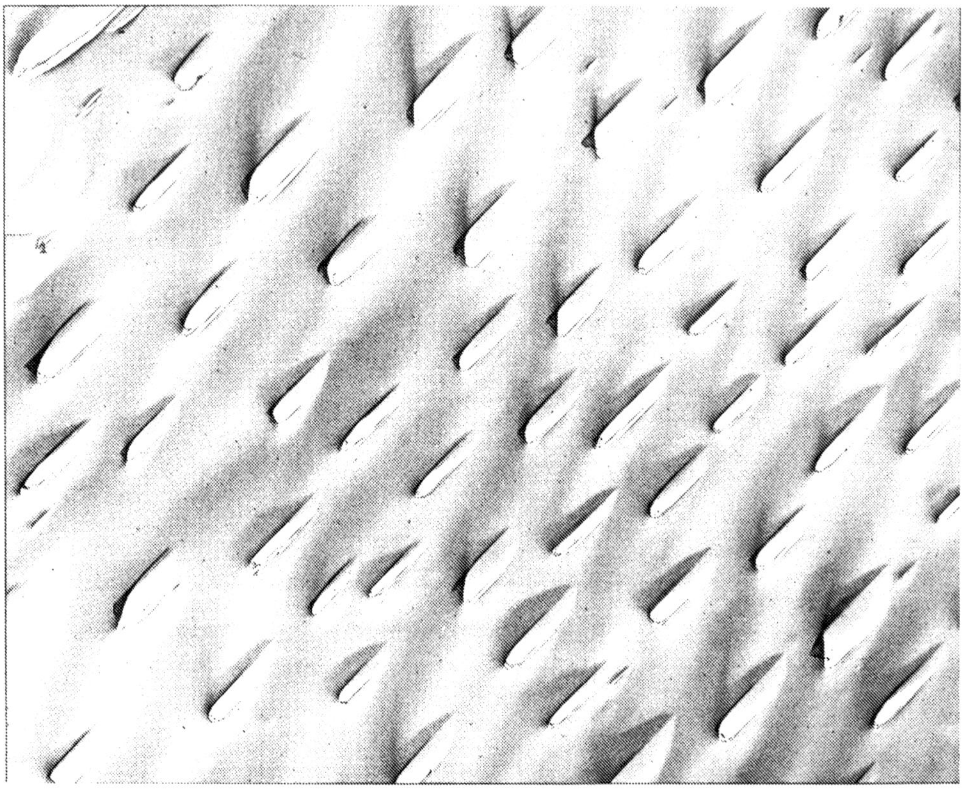

Figure 11 Carbon dioxide gas hydrate crystals

TEM can also be used to look at thin sections. Figure 12 is of Crambe abyssinica, a commercial oil seed. The thin section was obtained by fixing with potassium manganate (VII) and embedding in epoxy resin before sectioning with a microtome. Under magnification of 4 x 10^3 the dark areas mainly represent protein and cell walls, nuclei and starch bodies can be seen clearly.

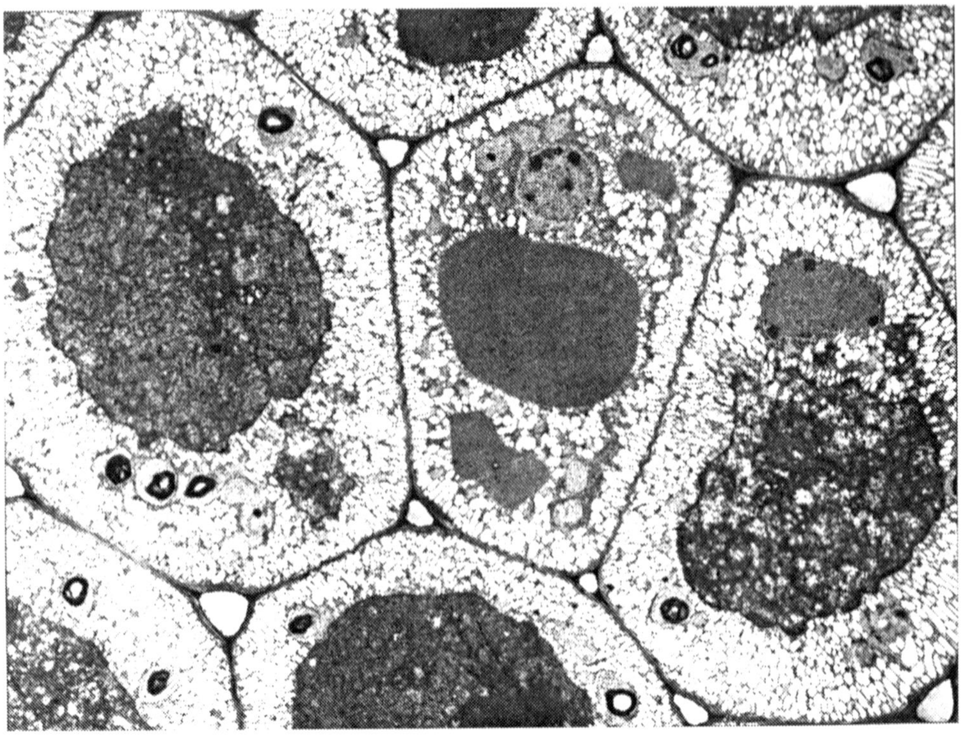

Figure 12 Crambe abyssinica thin section

Advanced techniques

Some materials do not lend themselves to electron microscopy because of their physical state or stability. However, useful images can still be formed. For example, replicas of liquids and soft solids are formed under high vacuum by evaporating platinum/carbon at an oblique angle onto a previously frozen sample. This produces 'shadows' which highlight changes in topography, and improve image contrast. Carbon is then evaporated from a gun mounted directly above the sample forming a continuous thin film layer. Later, at room temperature and pressure, the original sample material is dissolved away and the remaining carbon–platinum/carbon film (literally a 'carbon copy') is placed in the TEM for examination. Alternatively, it is sometimes possible to use the frozen sample as the specimen itself. In such cases they are typically frozen to –180 °C so that their vapour pressures are insignificant and the heating effect of the incident electrons is unlikely to cause them to melt.

It is sometimes inappropriate to put certain samples into an electron microscope – *eg* skin in the study of the effect of a moisturising cream. In this case a sample of skin would have to be removed from the subject before and after application of the cream. This is clearly unacceptable, so a negative of the skin is taken by painting a liquid polymer over the skin and peeling it off once it has cured to a rubber-like consistency. A positive is then made by pouring an epoxy resin (such as 'Araldite') into the polymer 'mould' and removing the polymer once the resin has set. The positive is then coated with gold. The result is a specimen that is easy to handle and is stable to the vacuum and the electron beam.

Scanning electron microscopes are available that are capable of extremely high resolution – down to approximately 8×10^{-10} m (0.8 nm). These require a very high vacuum – of the order of 1.33×10^{-8} Nm^{-2}. Such instruments are not cheap, however, and currently (1992) cost about £500 000. Transmission electron microscopes are capable of even greater resolution – to about 1.4×10^{-10} m (0.14 nm).

7. Following a synthetic route

The spectroscopic techniques discussed in this book can be used to assess the purity of compounds. It is also possible to monitor the progress of a reaction from the data obtained on the reagents and/or products of the reaction. Although this can sometimes be achieved by using one technique alone, more conclusive evidence can be gained from using several techniques.

The purpose of this chapter is to show how spectra change as a molecule is modified in a chemical synthesis, including the data from mass, infrared, and NMR spectral investigations. A book could be devoted to the interpretation of these spectra alone, so some of the important peak assignments have been given and a brief discussion has been included.

The target molecule is Ibuprofen (a drug invented and marketed by Boots Pharmaceuticals). Details of the synthesis are given in the Box.

All of the compounds are liquids at room temperature and pressure except Ibuprofen which is a solid. The mass spectra were all obtained by 70 eV (6750 kJ mol^{-1}) EI using a probe temperature of 150 °C; the infrared spectra were obtained from liquid films, except ibuprofen which was obtained from a KBr disc; and the NMR spectra were obtained from solution in deuterated trichloromethane, $CDCl_3$, using a 90 MHz instrument.

Isobutyl benzene

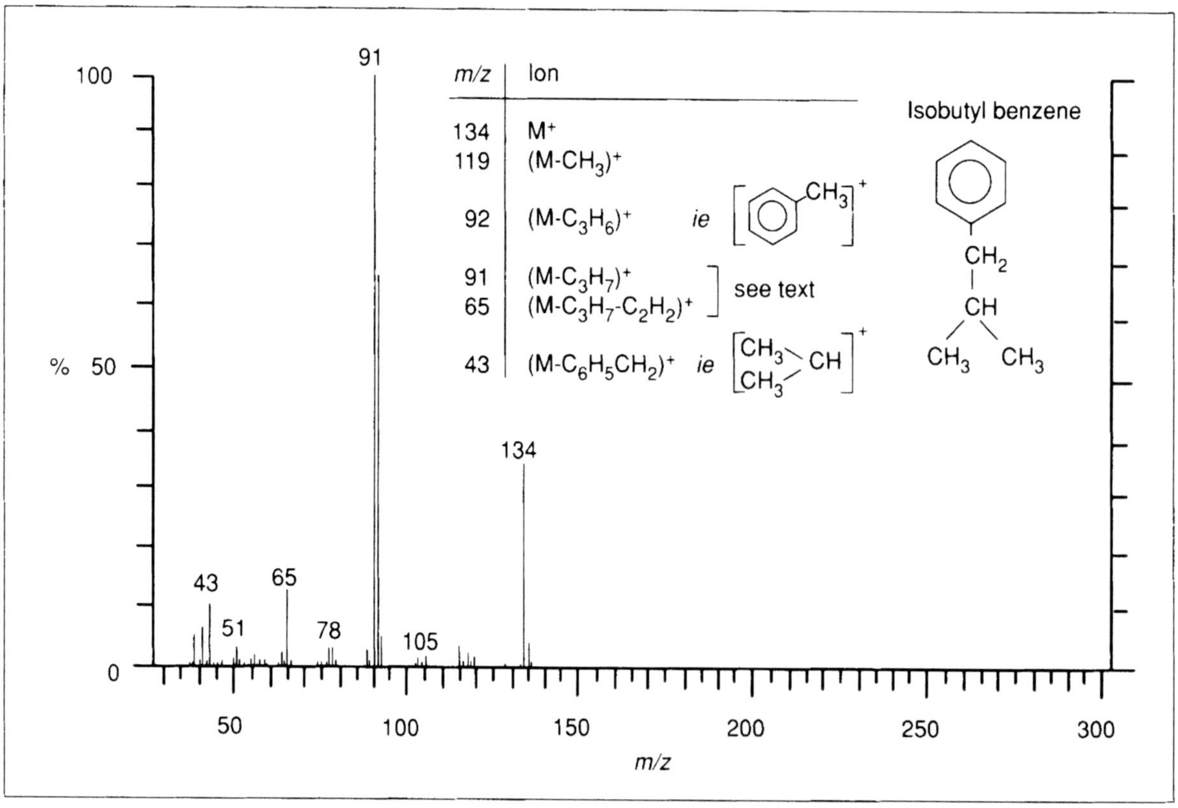

Figure 1 Mass spectrum of isobutyl benzene

Unilever

THE ROYAL SOCIETY OF CHEMISTRY

Methyl propyl benzene (isobutyl benzene) is used as the starting material.

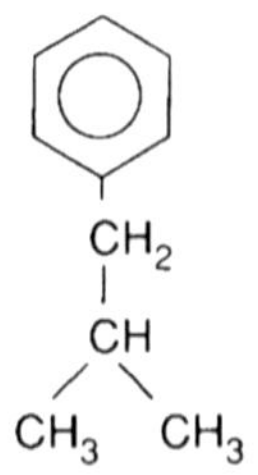

This undergoes a Friedel-Crafts acylation to give 4-isobutyl acetophenone.

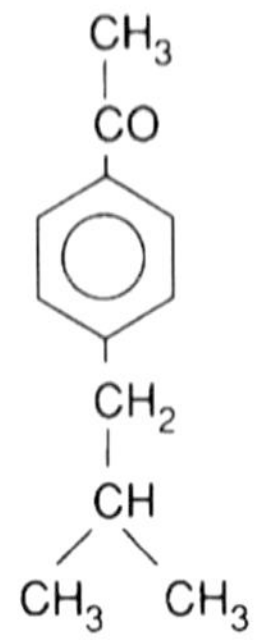

This is converted via a Darzen's condensation to 2,(4-isobutylphenyl) propionaldehyde.

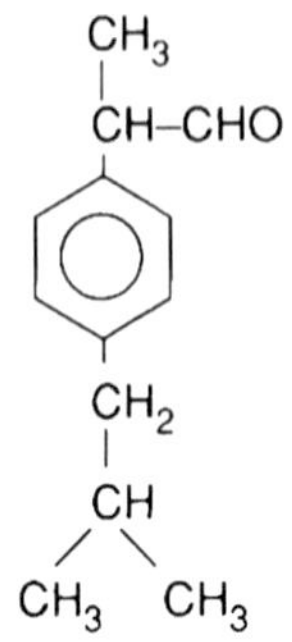

Which is oxidised to ibuprofen, 2,(4-isobutylphenyl) propanoic acid.

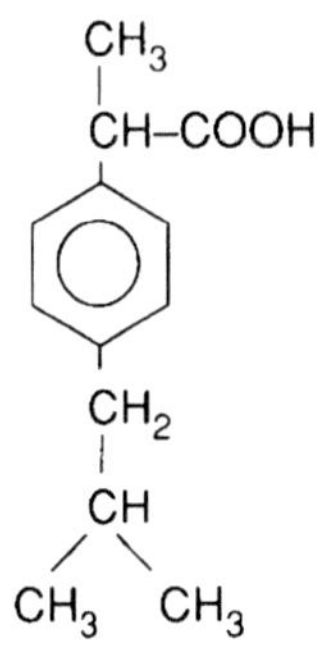

The major feature of this spectrum is the ease with which the link one bond away from the benzene ring is broken. Although methyl groups are fragmented, giving a peak at (M-15)=119, and there is loss of C_3H_6 (m/z = 92) – the most abundant peak is at m/z = 91. This corresponds to loss of the $(CH_3)_2CH$- group, to give $C_7H_7^+$ which rearranges as:

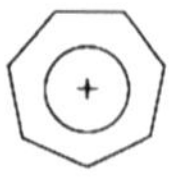

It is known as the tropyllium ion and loses C_2H_2 readily to give a relatively stable five membered ring with m/z = 65:

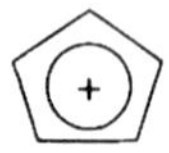

The ions of these seven- and five-membered rings are also detected in the mass spectra of ibuprofen and its intermediates.

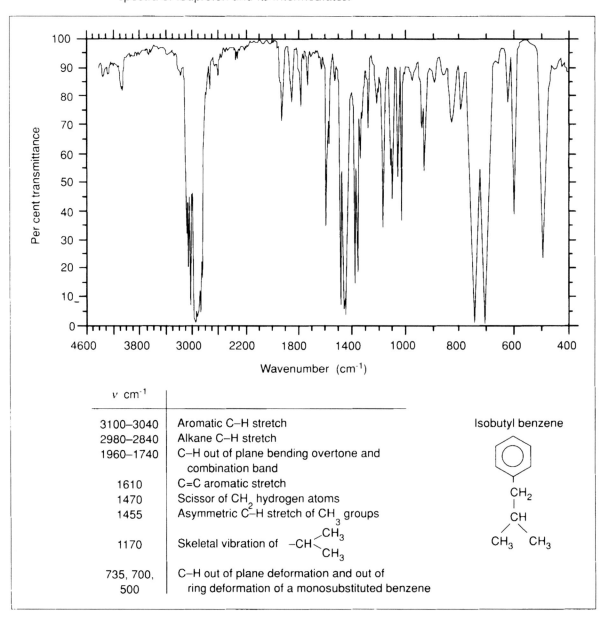

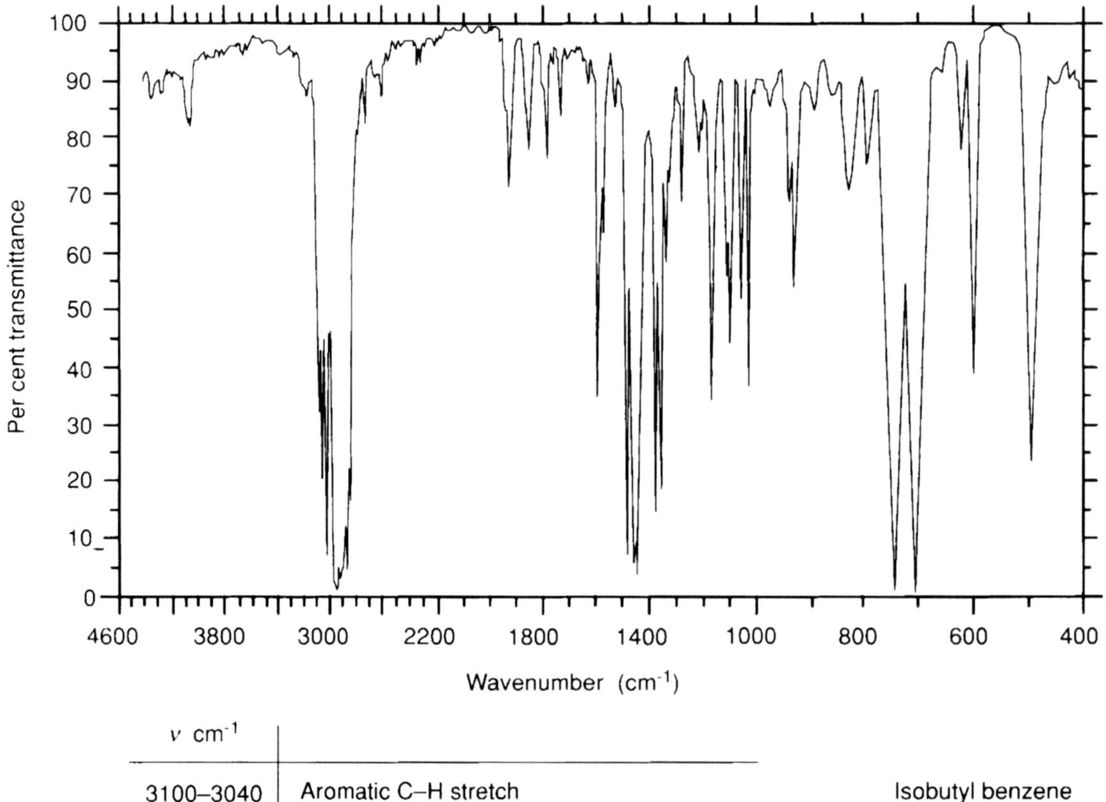

ν cm^{-1}	
3100–3040	Aromatic C–H stretch
2980–2840	Alkane C–H stretch
1960–1740	C–H out of plane bending overtone and combination band
1610	C=C aromatic stretch
1470	Scissor of CH_2 hydrogen atoms
1455	Asymmetric C–H stretch of CH_3 groups
1170	Skeletal vibration of $-CH(CH_3)_2$
735, 700, 500	C–H out of plane deformation and out of ring deformation of a monosubstituted benzene

Isobutyl benzene

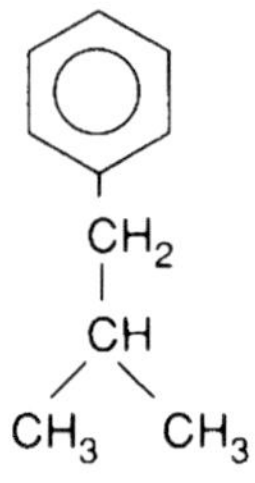

Figure 2 Infrared spectrum of isobutyl benzene

Unilever

THE ROYAL SOCIETY OF CHEMISTRY

Because this molecule is a hydrocarbon, it is not surprising that its infrared spectrum is dominated by the vibrational modes of the C–H, C=C and C–C bonds. Some of these are assigned on the spectrum.

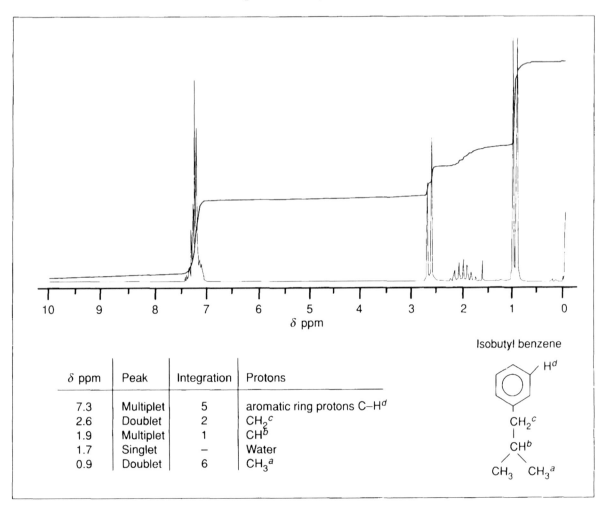

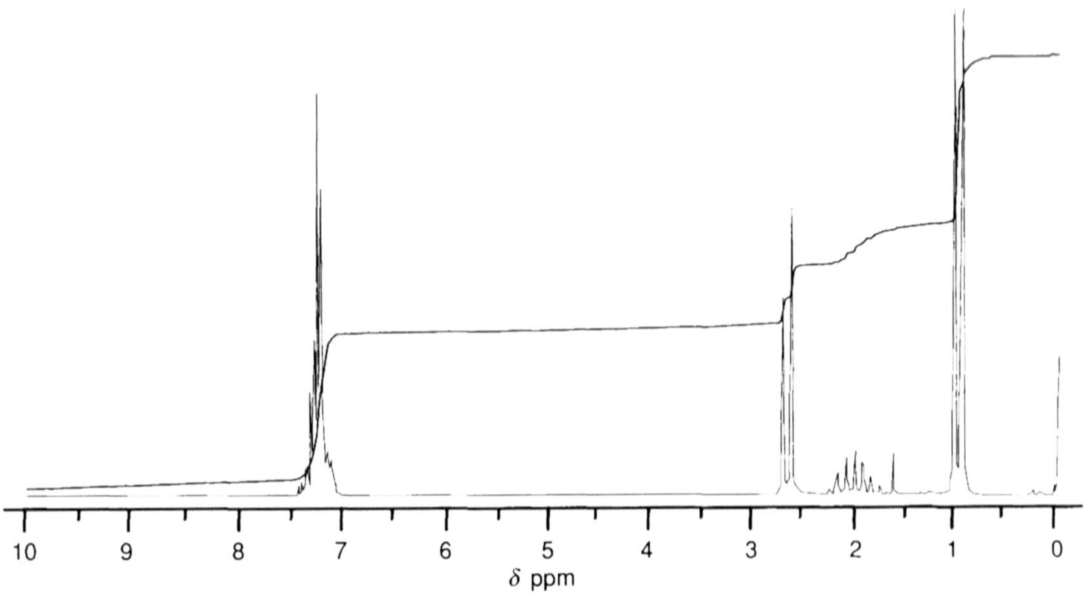

δ ppm	Peak	Integration	Protons
7.3	Multiplet	5	aromatic ring protons C–H[d]
2.6	Doublet	2	$CH_2{}^c$
1.9	Multiplet	1	CH^b
1.7	Singlet	–	Water
0.9	Doublet	6	$CH_3{}^a$

Isobutyl benzene

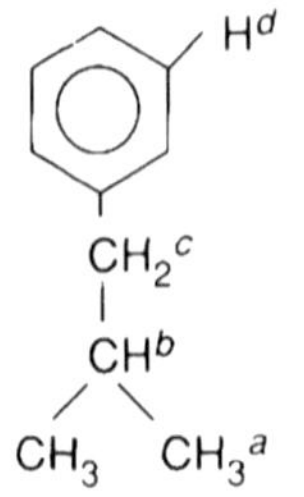

Figure 3 NMR spectrum of isobutyl benzene

The assignment of peaks is quite straightforward in this case. They are well separated and contain simple splitting patterns. The peak for the methine proton (*b*) is really a septet of triplets, because the proton couples with the two methylene protons (*c*) and the six (equivalent) methyl protons (*a*). The splitting is not clear owing to peak overlap. Because the alkyl side chain of the molecule is unchanged throughout the synthesis all of its protons should give the same peaks in all the NMR spectra. This is found to be the case. As expected, the resonance of the ring protons is seen at a chemical shift of *ca* $\delta = 7.3$.

An unexpected peak appears at δ = *ca* 1.7. This is due to water.

THE ROYAL SOCIETY OF CHEMISTRY

Unilever

4-isobutyl acetophenone

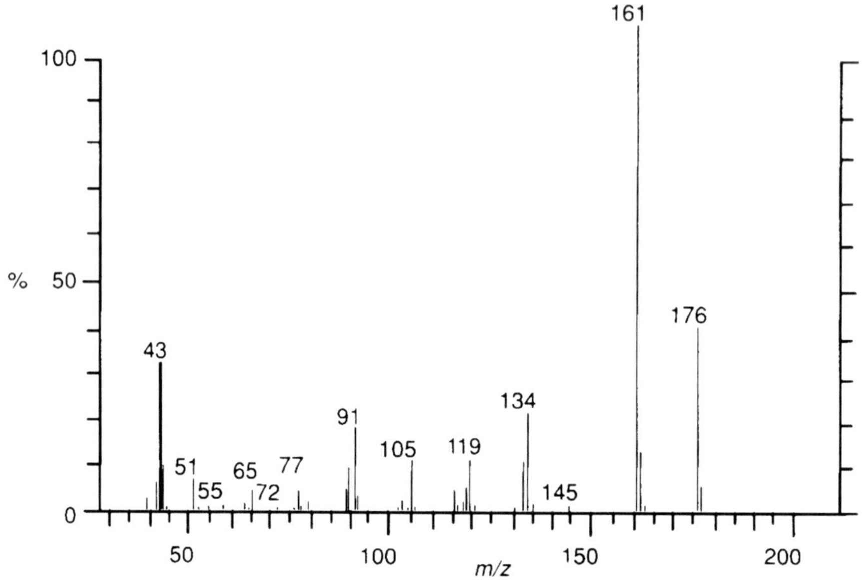

m/z	Ion
176	M^+
161	$(M\text{-}CH_3)^+$
134	$(M\text{-}C_3H_6)^+$ *ie* 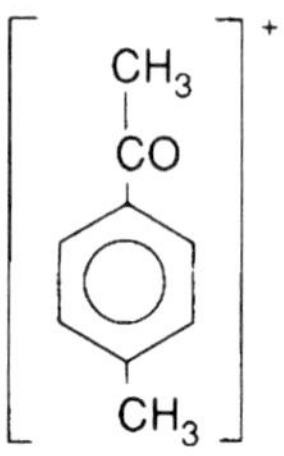
133	$(M\text{-}C_3H_7)^+$
119	$(M\text{-}CH_2CH(CH_3)_2)^+$
105	$(M\text{-}CH_2CH(CH_3)_2 - CH_3)^+$ *ie* 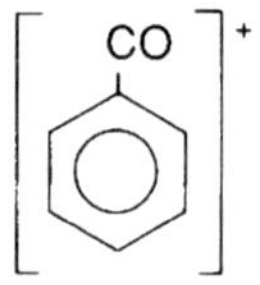
43 (2 peaks)	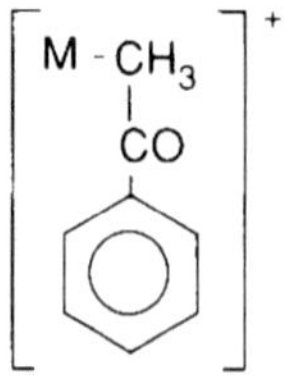 *ie* $(CH_3CO)^+$ (more abundant peak) 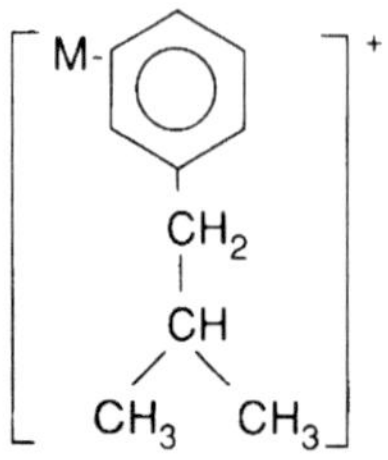 *ie* $(C_3H_7)^+$ (less abundant peak)

4-isobutyl acetophenone

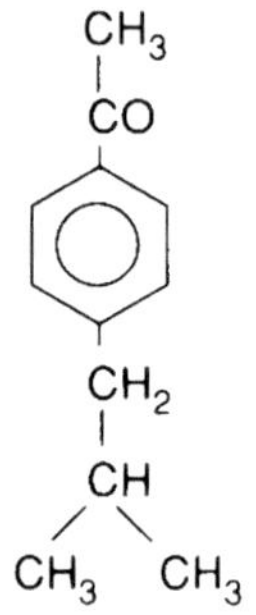

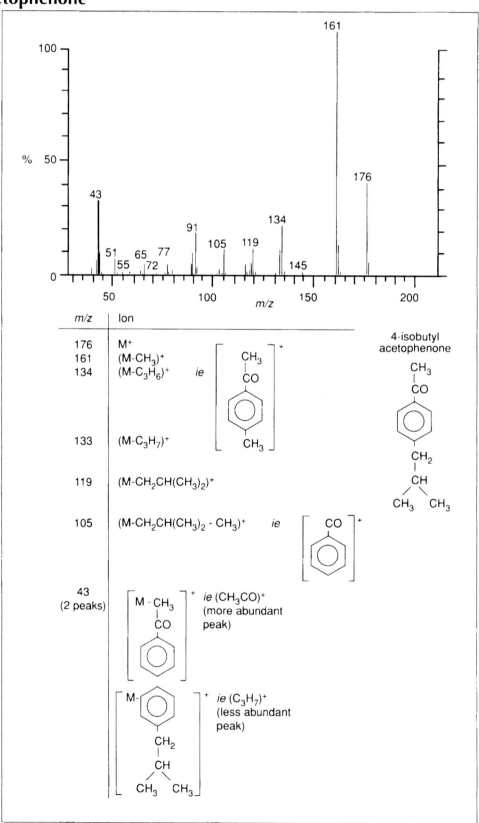

Figure 4 Mass spectrum of 4-isobutyl acetophenone

An ethanoyl group has been put in the ring position *para* to the alkyl side chain, and the mass spectrum now shows a particularly stable ion at $m/z = 161$ (M-15) if a methyl group is cleaved off.

The spectrum also shows that the bonds between the carbonyl group and the benzene ring; and the C–C bond between the CH_2 and CH groups are susceptible to cleavage. In both cases the charge can be on either fragment, so both are observed. The smaller fragment in each case has $m/z \approx 43$, and these are seen as separate peaks in the spectrum because of the slight differences in the accurate masses of the ions.

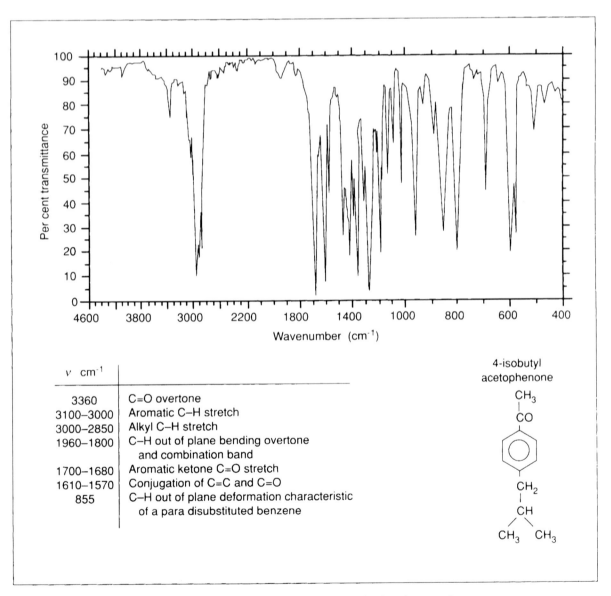

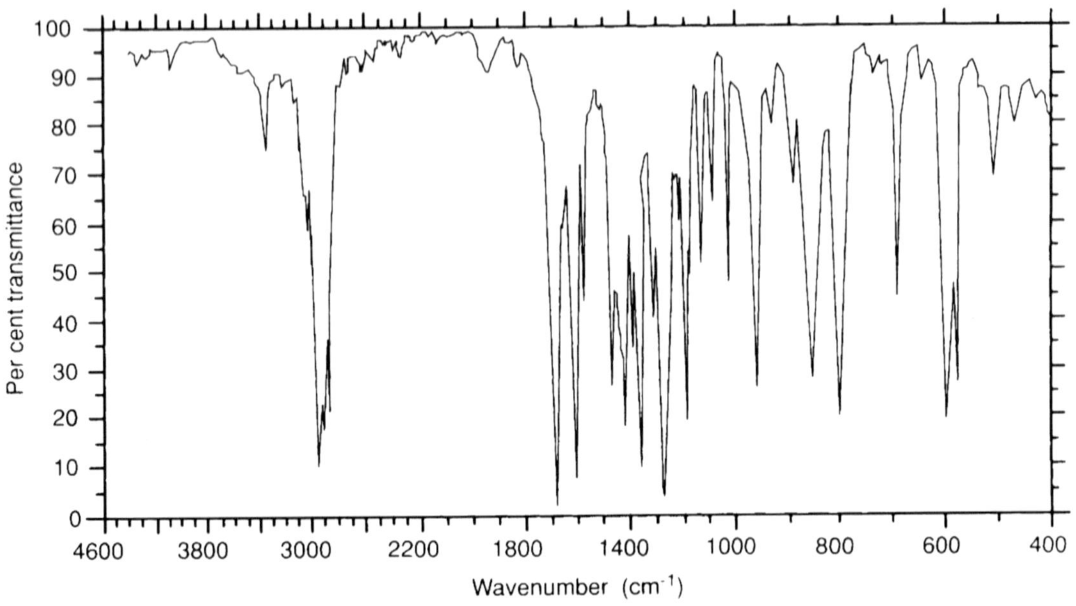

ν cm⁻¹	
3360	C=O overtone
3100–3000	Aromatic C–H stretch
3000–2850	Alkyl C–H stretch
1960–1800	C–H out of plane bending overtone and combination band
1700–1680	Aromatic ketone C=O stretch
1610–1570	Conjugation of C=C and C=O
855	C–H out of plane deformation characteristic of a para disubstituted benzene

4-isobutyl acetophenone

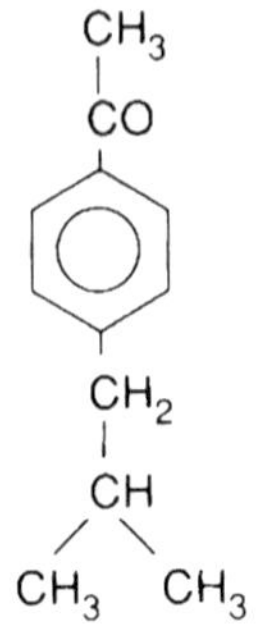

Figure 5 Infrared spectrum of 4-isobutyl acetophenone

The major difference between this spectrum and that of isobutyl benzene is the presence of the peak due to the aromatic ketone carbonyl at 1680 cm^{-1}. An overtone of this vibration is observed at twice this frequency, 3360 cm^{-1}. Two other peaks of interest are seen at 1570 and 1610 cm^{-1} – these are absorptions resulting from the conjugation of the carbonyl group with the unsaturated benzene ring.

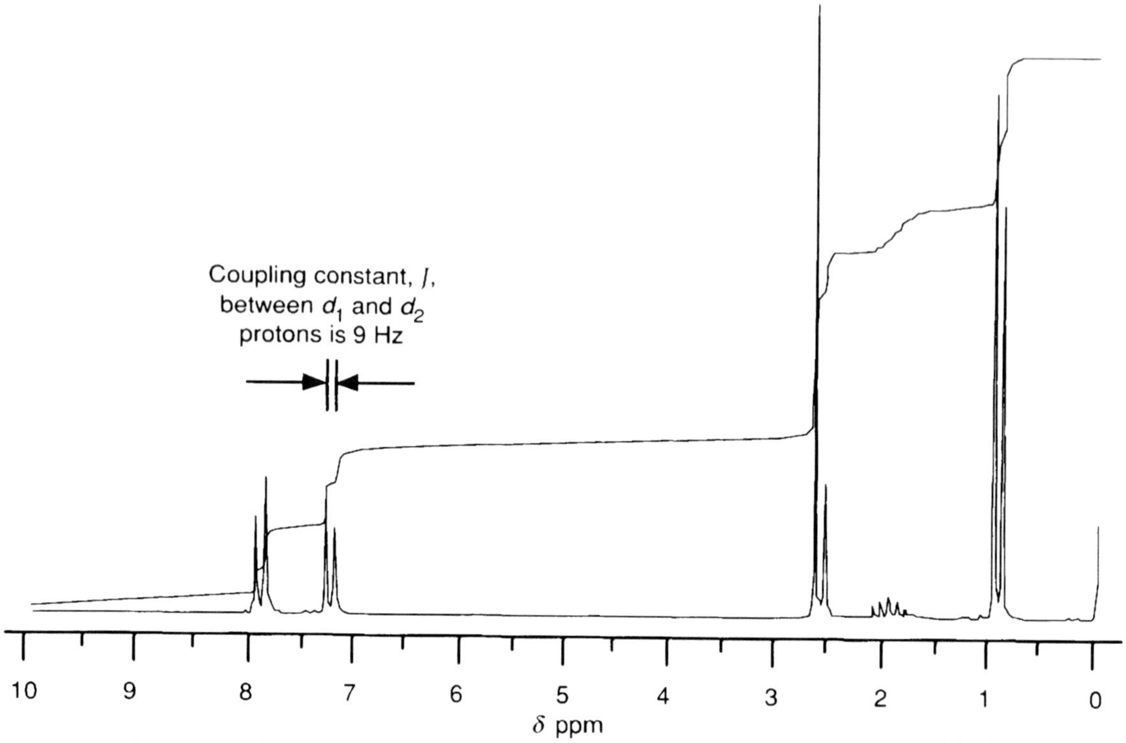

δ ppm	Peak	Integration	Protons
7.8	Doublet	2	Aromatic CH^{d2}
7.2	Doublet	2	Aromatic CH^{d1}
2.6	Singlet	3	CH_3^{e}
2.5	Doublet	2	CH_2^{c}
1.9	Multiplet	1	CH^{b}
0.9	Doublet	6	CH_3^{a}

4-isobutyl acetophenone

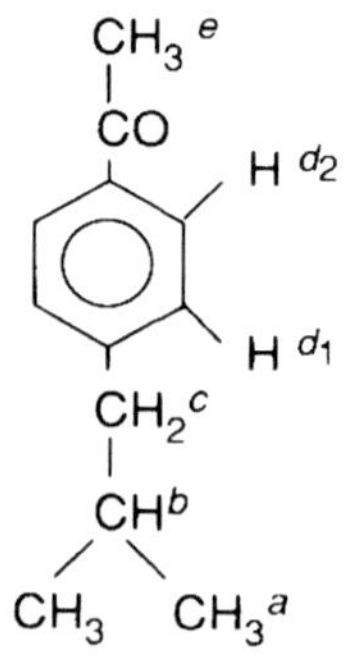

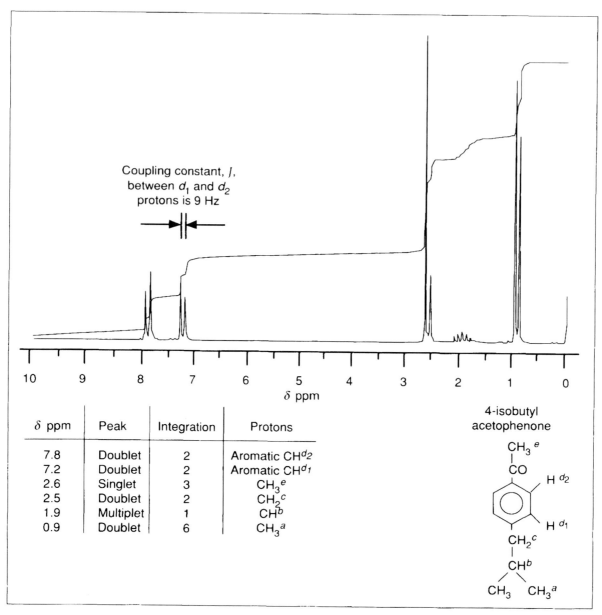

Figure 6 NMR spectrum of 4-isobutyl acetophenone

Adding the ethanoyl group to the molecule has two effects on the NMR spectrum:

1. a singlet appears at $\delta = 2.6$ – this is from the methyl group adjacent to the carbonyl and overlaps one of the lines of the methylene protons (*c*); and

2. the peak from the benzene ring protons is split into two doublets: an upfield doublet from the protons *ortho* to the alkyl side chain (d_1); and a downfield doublet at $\delta = 7.8$ from the protons ortho to the ethanoyl group (d_2) – these protons are more deshielded because of the presence of the electronegative oxygen. Measurement of the difference in frequency between the lines in each doublet gives the coupling constant between the d_1 and the d_2 protons. This pattern confirms that the ethanoyl group has been introduced in the para position.

Unilever

THE ROYAL SOCIETY OF CHEMISTRY

2,(4-isobutylphenyl) propionaldehyde

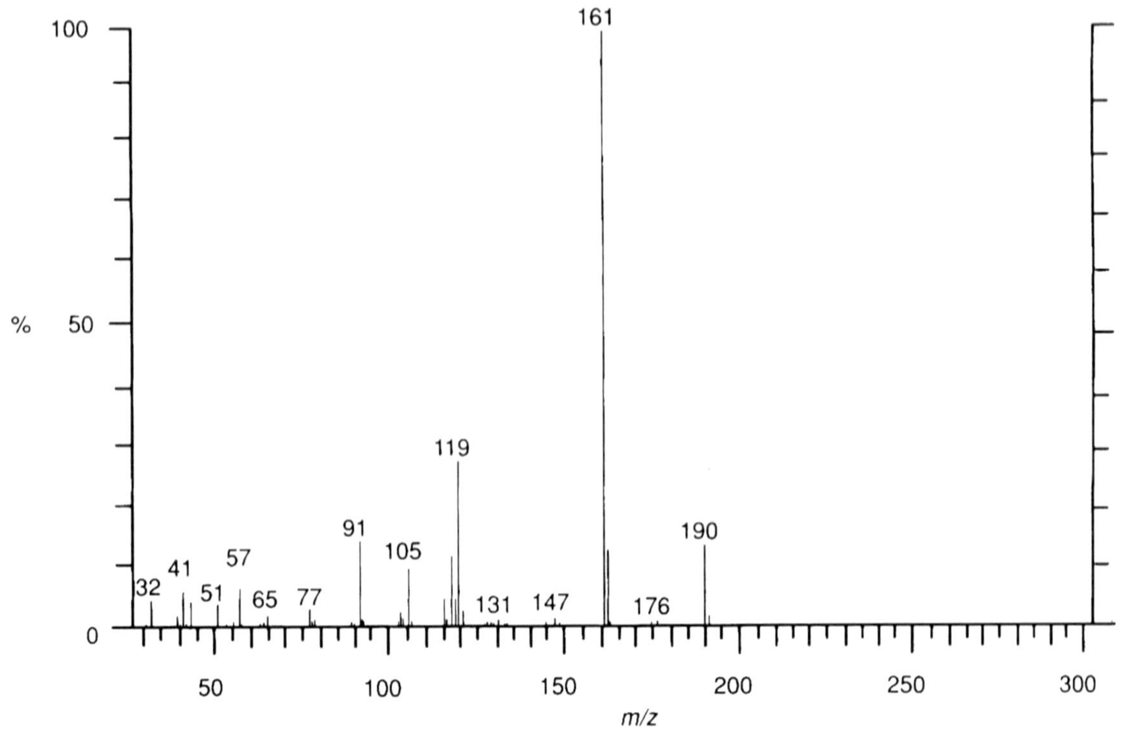

m/z	Ion
190	M^+
161	$(M\text{-}CHO)^+$
119	$(M\text{-}CHO\text{-}C_3H_6)^+$ *ie*
105	$(M\text{-}CHO\text{-}C_3H_6\text{-}CH_2)^+$ *ie*

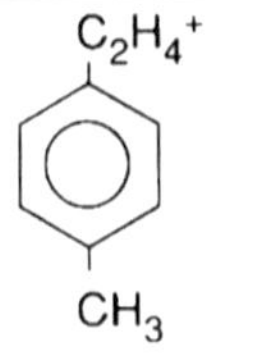

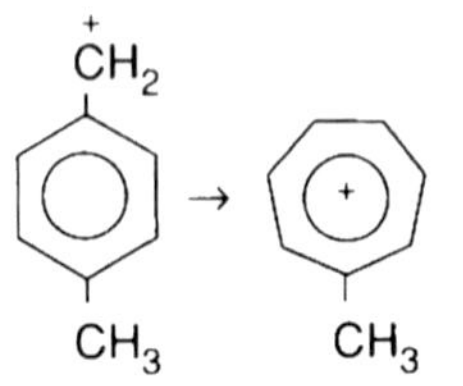

2, (4-isobutylphenyl) propionaldehyde

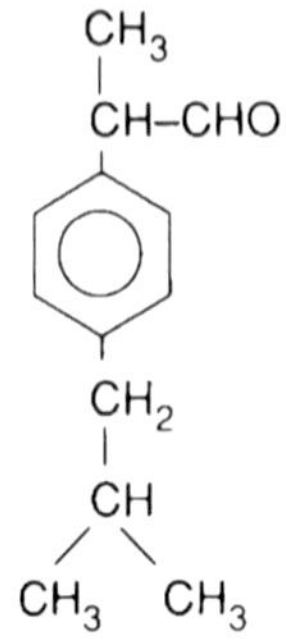

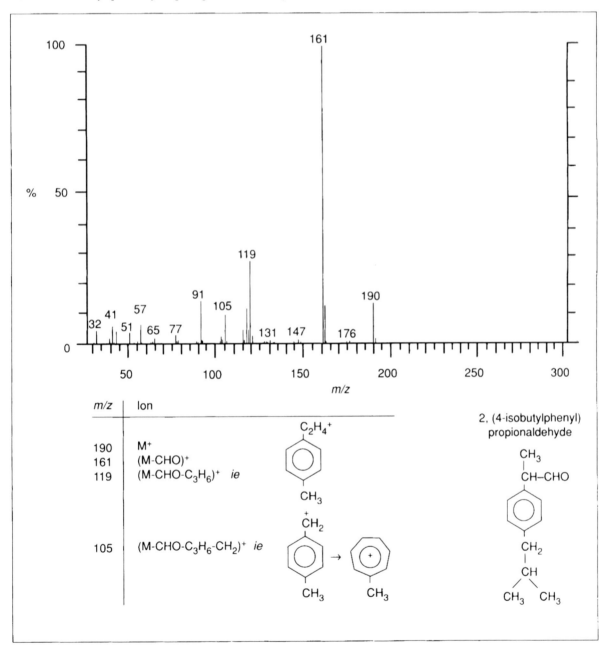

Figure 7 Mass spectrum of 2,(4-isobutylphenyl) propionaldehyde

After the Darzen's condensation the most significant peaks are due to the loss of the aldehydic group CHO at (M-29), and subsequent loss of a C_3H_6 group from the alkyl side chain at (M-71).

THE ROYAL SOCIETY OF CHEMISTRY

Unilever

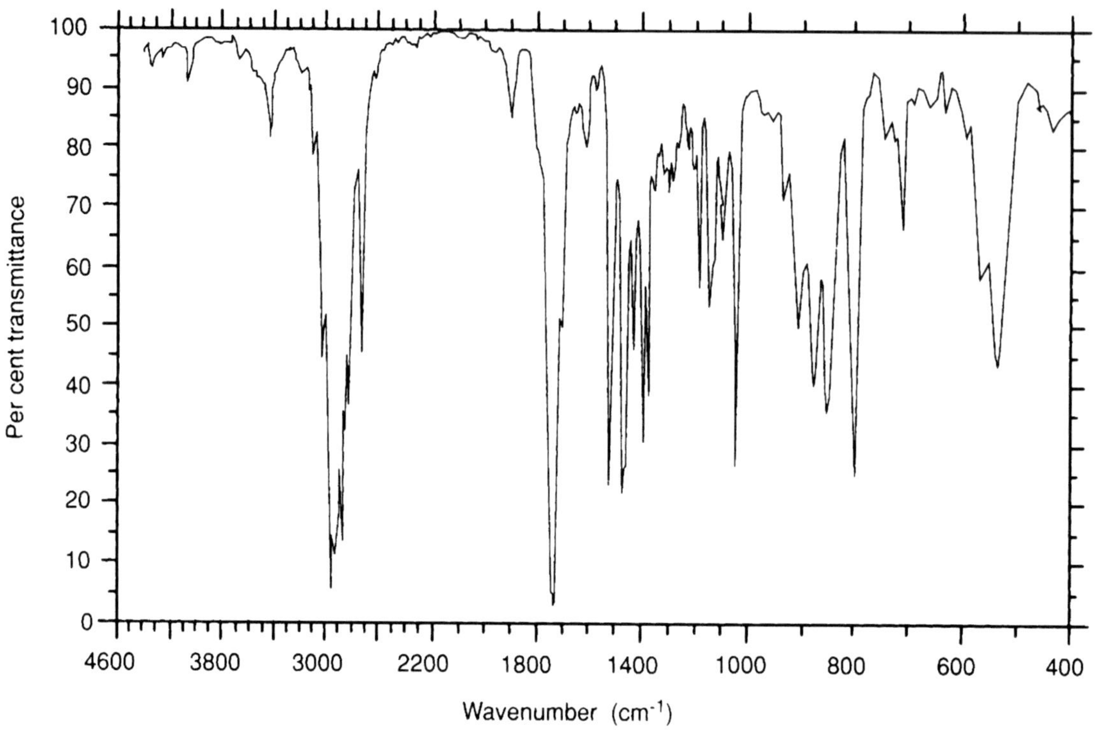

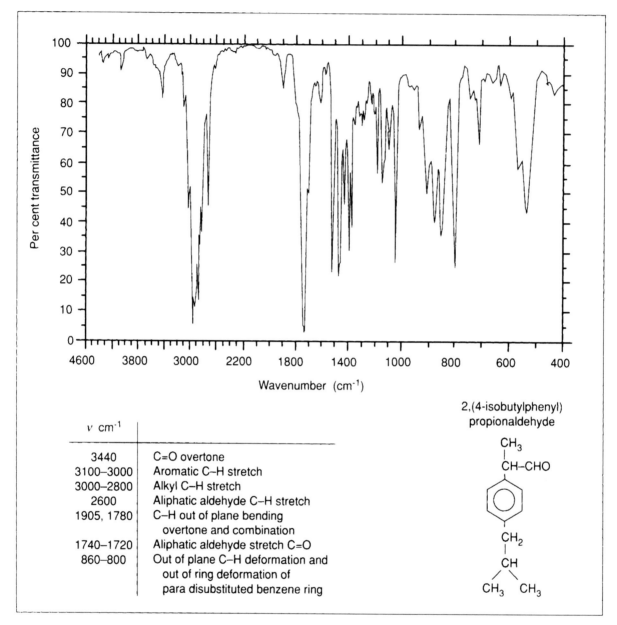

ν cm^{-1}	
3440	C=O overtone
3100–3000	Aromatic C–H stretch
3000–2800	Alkyl C–H stretch
2600	Aliphatic aldehyde C–H stretch
1905, 1780	C–H out of plane bending overtone and combination
1740–1720	Aliphatic aldehyde stretch C=O
860–800	Out of plane C–H deformation and out of ring deformation of para disubstituted benzene ring

2,(4-isobutylphenyl) propionaldehyde

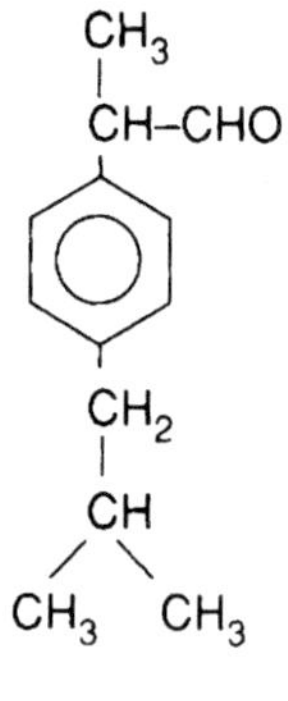

Figure 8 Infrared spectrum of 2,(4-isobutylphenyl) propionaldehyde

The absorptions due to the hydrocarbon vibrations are still present, but the carbonyl vibration has moved from 1680 cm^{-1} in the aldehyde to 1720 cm^{-1}. An overtone at twice this new carbonyl stretch frequency is again observed.

Unilever

THE ROYAL SOCIETY OF CHEMISTRY

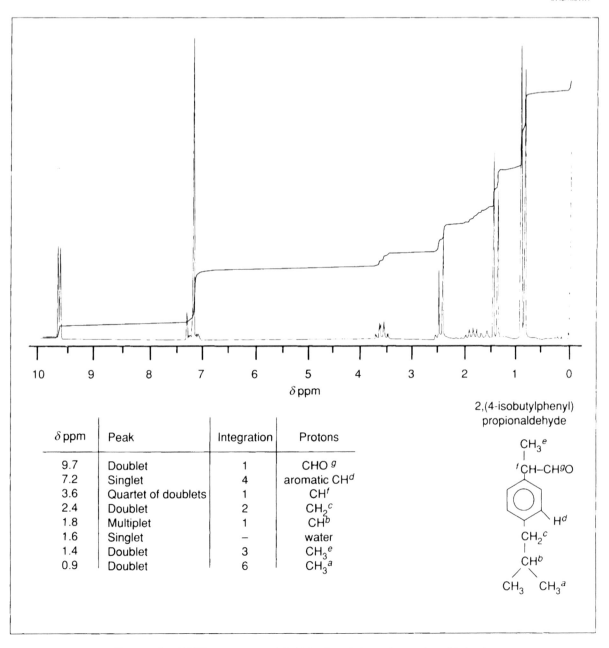

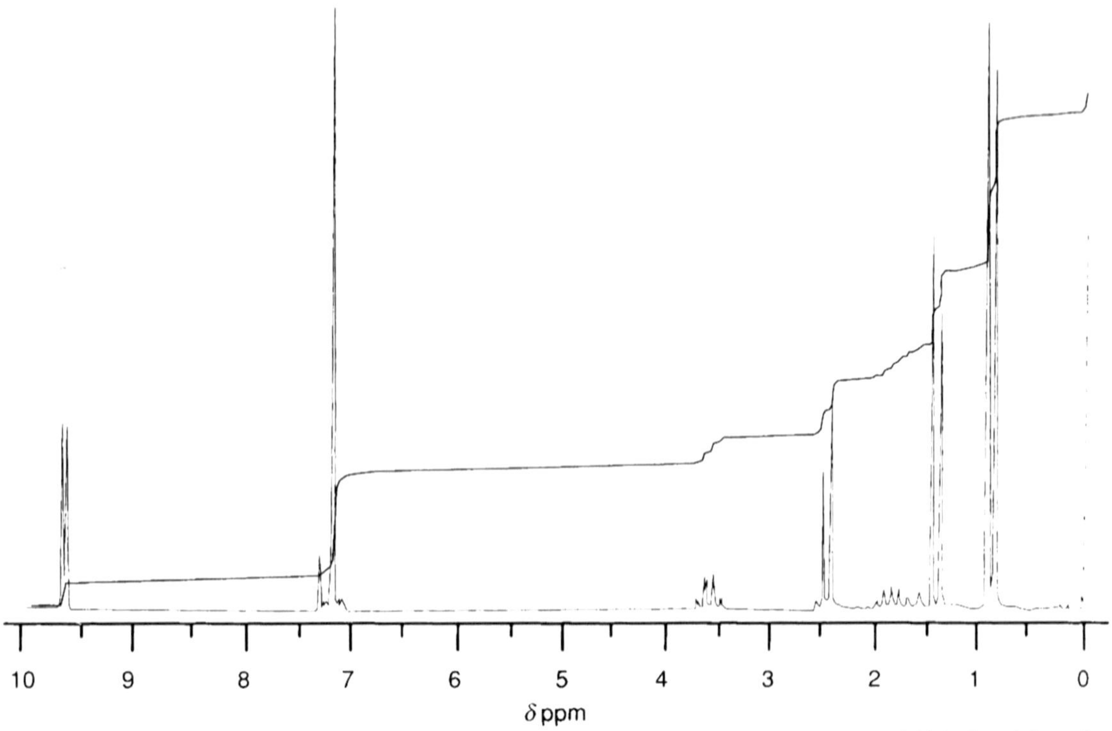

δ ppm	Peak	Integration	Protons
9.7	Doublet	1	CHO ^{g}
7.2	Singlet	4	aromatic CH^d
3.6	Quartet of doublets	1	CH^f
2.4	Doublet	2	$CH_2{}^c$
1.8	Multiplet	1	CH^b
1.6	Singlet	–	water
1.4	Doublet	3	$CH_3{}^e$
0.9	Doublet	6	$CH_3{}^a$

2,(4-isobutylphenyl) propionaldehyde

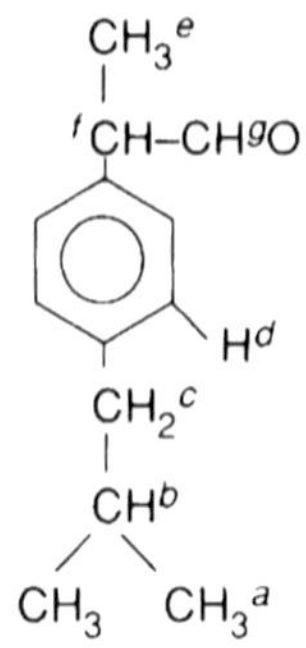

Figure 9 NMR spectrum of 2,(4-isobutylphenyl) propionaldehyde

The singlet from the methyl protons (*e*) has now expanded to a doublet and has moved upfield to $\delta = 1.4$ because of the proton (*f*) on the CH group now adjacent to it. The methine proton (*f*) gives a quartet of doublets as it couples with both the methyl protons (*e*) and the aldehydic proton (*g*). The aldehydic proton itself appears downfield as a doublet at $\delta = 9.7$ because it couples with the methine proton (*f*).

The ring protons do not give the expected 1,4- disubstituted splitting pattern at $\delta = 7.2$ because there is insufficient difference in the environments of the protons previously labelled (d_1) and (d_2).

A small peak due to water is seen at $\delta = 1.6$. Another small peak (not seen before) appears at $\delta = 7.3$. This is from small amounts of $CHCl_3$ in the $CDCl_3$ solvent, and a similar peak is observed in the NMR spectrum of ibuprofen. It was not detected in the previous spectra because the signal from the ring protons was on top of it.

THE ROYAL SOCIETY OF CHEMISTRY

Unilever

Ibuprofen

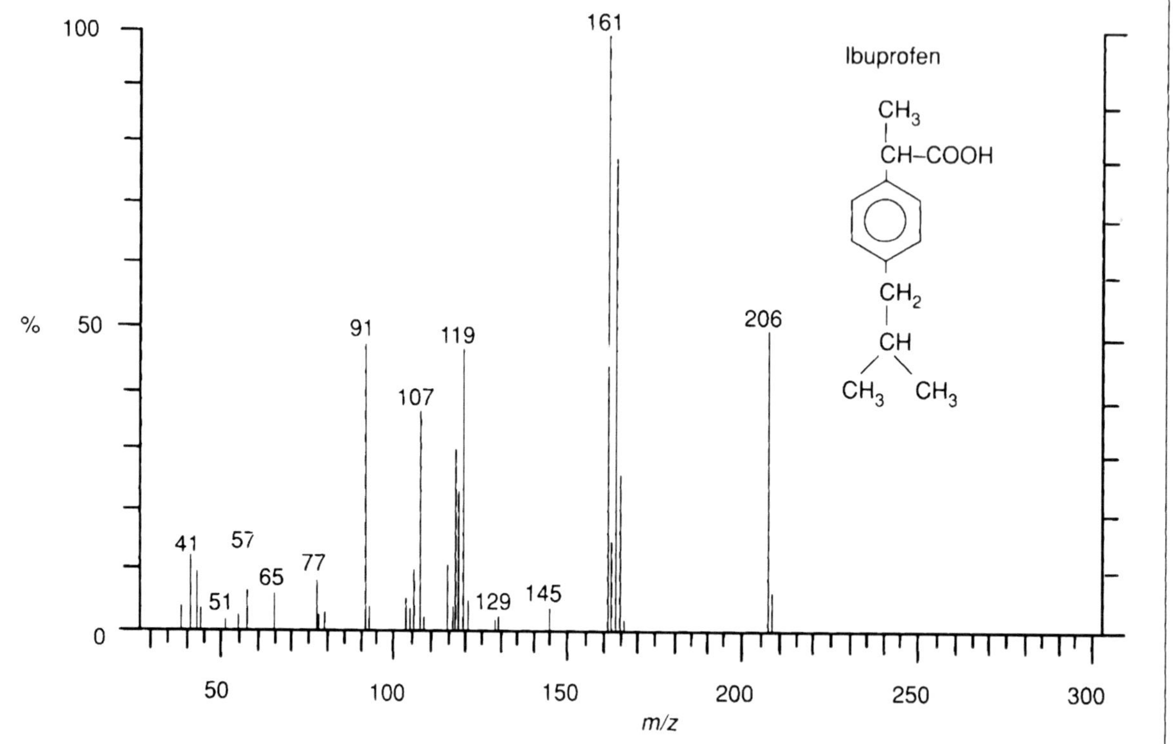

m/z	Ion	
206	M^+	
164	$(M\text{-}C_3H_6)^+$ *ie*	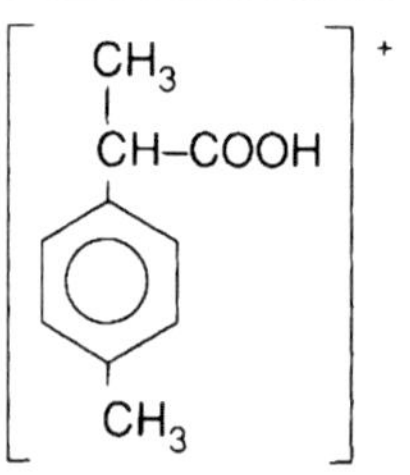
163	$(M\text{-}C_3H_7)^+$	
145	$(M\text{-}C_3H_7\text{-}H_2O)^+$ *ie*	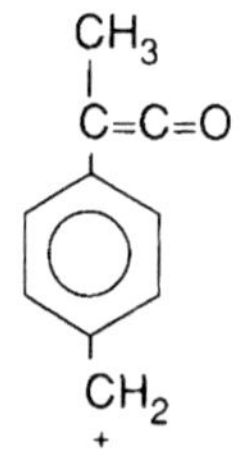
107	$C_7H_7O^+$	

m/z	Ion	
206	M^+	
161	$(M\text{-}COOH)^+$	
119	$(M\text{-}COOH\text{-}C_3H_6)^+$ *ie*	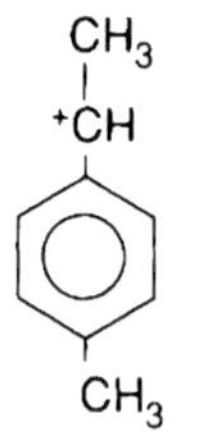
91	$(M\text{-}COOH\text{-}C_3H_6\text{-}C_2H_4)^+$ *ie*	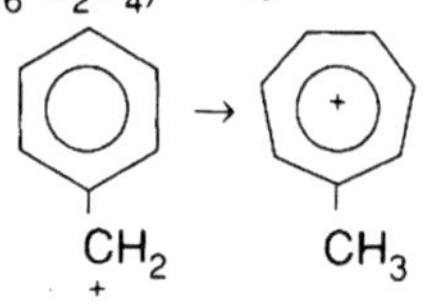
65	$(M\text{-}COOH\text{-}C_3H_6\text{-}C_2H_4\text{-}C_2H_2)^+$	

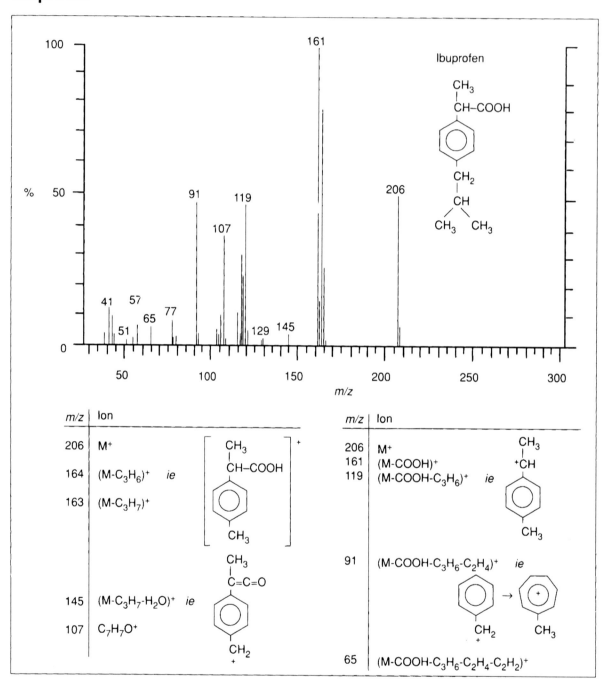

Figure 10 Mass spectrum of ibuprofen

Whereas the molecular ions of some carboxylic acids are not observed in their mass spectra because they decarboxylate when heated at low pressure (*ie* the condition of the source of the mass spectrometer) this is not the cause with Ibuprofen, as witnessed by its abundant ion at m/z = 206. However, it does lose water in one of its fragmentation routes. The two routes shown indicate that fragmentation occurs from both side chains, losing the carboxylic acid group is the most common first fragmentation. As with the other compounds formation of the tropyllium ion is observed.

Unilever

THE ROYAL SOCIETY OF CHEMISTRY

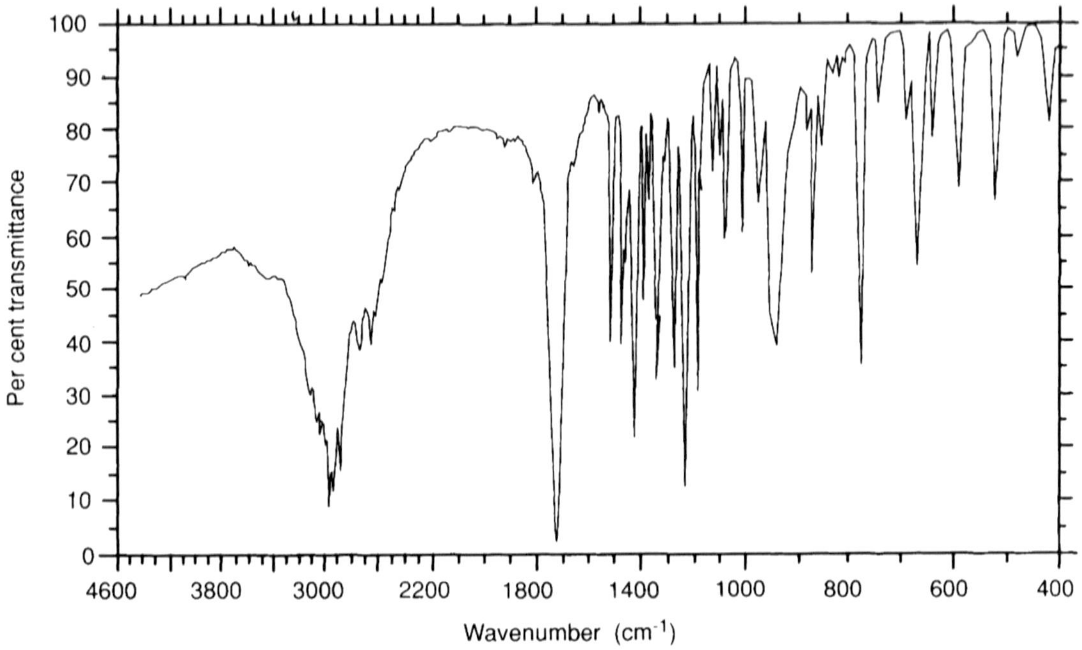

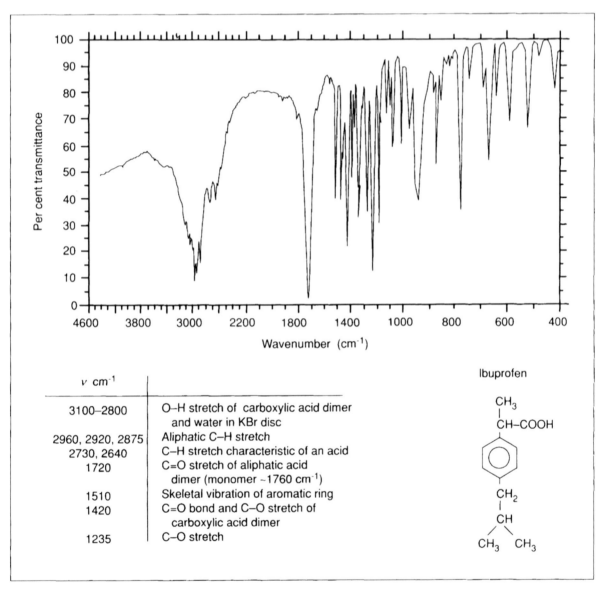

ν cm^{-1}	
3100–2800	O–H stretch of carboxylic acid dimer and water in KBr disc
2960, 2920, 2875	Aliphatic C–H stretch
2730, 2640	C–H stretch characteristic of an acid
1720	C=O stretch of aliphatic acid dimer (monomer ~1760 cm^{-1})
1510	Skeletal vibration of aromatic ring
1420	C=O bond and C–O stretch of carboxylic acid dimer
1235	C–O stretch

Ibuprofen

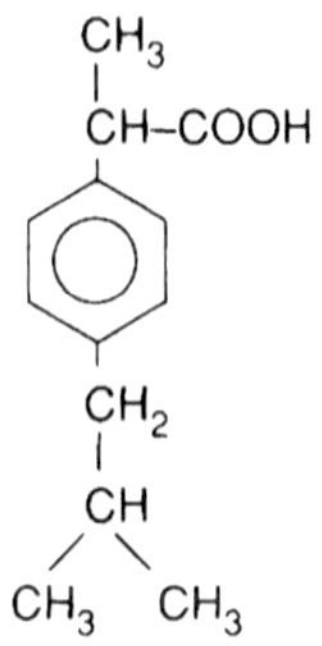

Figure 11 Infrared spectrum of ibuprofen

The high frequency end of this spectrum is dominated by the broad absorption peak of the hydrogen bonded OH group. This covers the 2500–3300 cm^{-1} range. The C–H vibrations in the 2800–3100 cm^{-1} region are still present, along with two absorptions not observed previously, characteristic of carboxylic acids at 2640 cm^{-1}and 2740 cm^{-1}respectively.

The absorption of the carbonyl group is not at the frequency expected for a carboxylic acid monomer (1760 cm^{-1}). Instead, it is observed at 1725 cm^{-1}. This is within the range expected for dimers of saturated aliphatic carboxylic acids, 1700–1725 cm^{-1}. In this case, any overtone will be lost in the signal from the edge of the OH vibration band.

THE ROYAL SOCIETY OF CHEMISTRY

Unilever

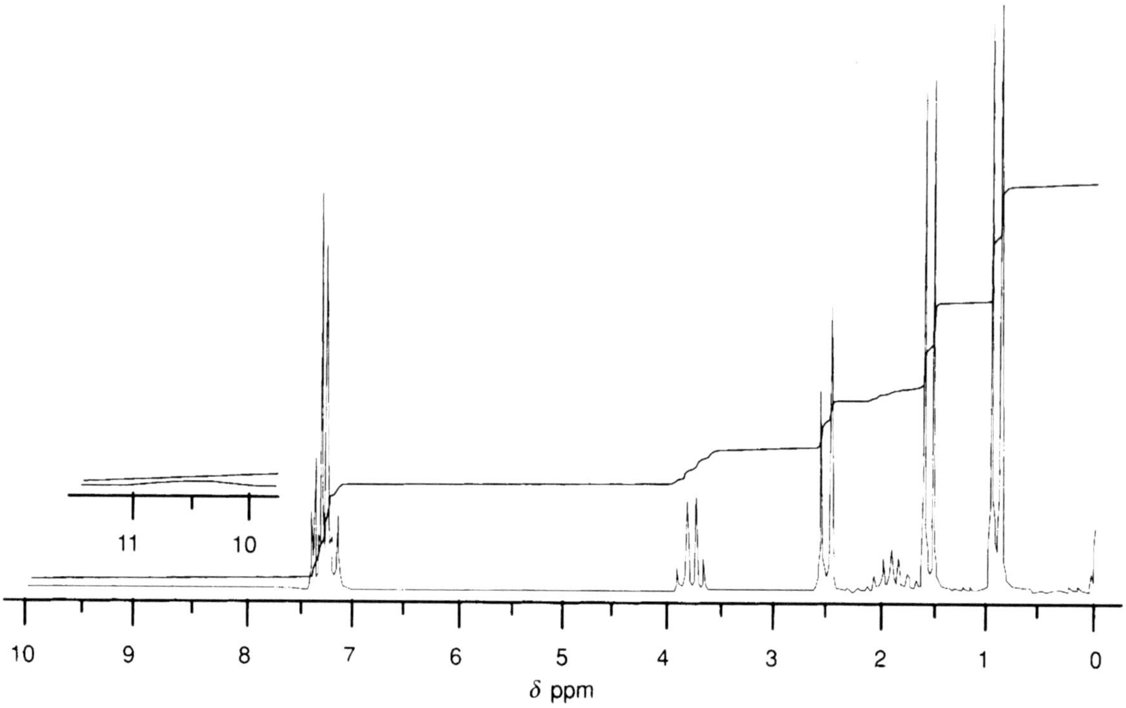

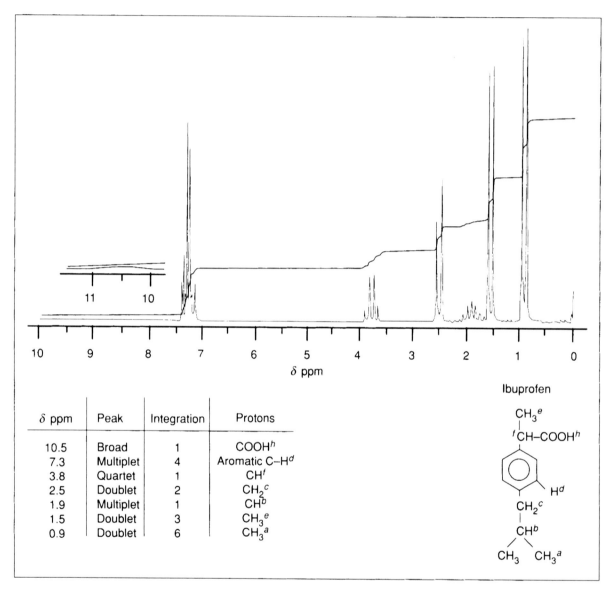

δ ppm	Peak	Integration	Protons
10.5	Broad	1	$COOH^h$
7.3	Multiplet	4	Aromatic $C–H^d$
3.8	Quartet	1	CH^f
2.5	Doublet	2	$CH_2{}^c$
1.9	Multiplet	1	CH^b
1.5	Doublet	3	$CH_3{}^e$
0.9	Doublet	6	$CH_3{}^a$

Ibuprofen

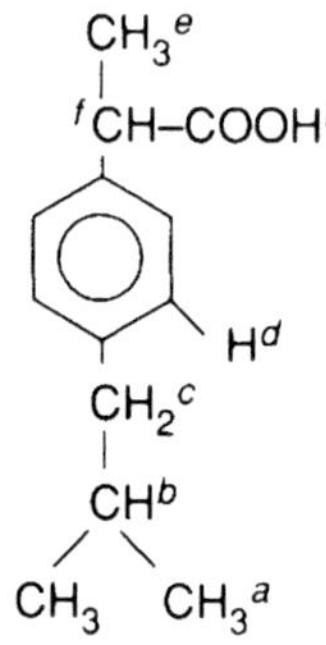

Figure 12 NMR spectrum of ibuprofen

Because the aldehyde has been oxidised, the methine proton (*f*) can couple with only the methyl protons (*e*), so its peak changes from a quartet of doublets to a simple quartet. Similarly the signal from the aldehydic proton at δ = 9.5 has been lost completely. Instead, a signal is detected further downfield at δ = 10.5. This is due to the acidic proton, and is not seen as a sharp peak, but a broad, weak band. This is because it exchanges rapidly with other acidic protons in the solution.

The ring protons previously labelled as d_1 and d_2 show some peak splitting because they are (again) in different electronic environments. The peak is complicated by the presence of an extra peak from the $CHCl_3$ impurity in the solvent.

Printed in the United Kingdom
by Lightning Source UK Ltd.
136413UK00001B/195/A

9 781870 343190